MONOGRAPHS ON
STATISTICS AND APPLIED PROBABILITY

General Editors

D. R. Cox, D. V. Hinkley, D. Rubin and B. W. Silverman

The Statistical Analysis of Compositional Data
J. Aitchison

Probability, Statistics and Time
M.S. Bartlett

The Statistical Analysis of Spatial Pattern
M.S. Bartlett

Stochastic Population Models in Ecology and Epidemiology
M.S. Bartlett

Risk Theory
R.E. Beard, T. Pentikäinen and E. Pesonen

Bandit Problems – Sequential Allocation of Experiments
D.A. Berry and B. Fristedt

Residuals and Influence in Regression
R.D. Cook and S. Weisberg

Point Processes
D.R. Cox and V. Isham

Analysis of Binary Data
D.R. Cox

The Statistical Analysis of Series of Events
D.R. Cox and P.A.W. Lewis

Analysis of Survival Data
D.R. Cox and D. Oakes

Queues
D.R. Cox and W.L. Smith

Stochastic Modelling and Control
M.H.A. Davis and R. Vinter

Stochastic Abundance Models
S. Engen

The Analysis of Contingency Tables
B.S. Everitt

An Introduction to Latent Variable Models
B.S. Everitt

Finite Mixture Distributions
B.S. Everitt and D.J. Hand

Population Genetics
W.J. Ewens

Classification
A.D. Gordon

Monte Carlo Methods
J.M. Hammersley and D.C. Handscomb

Identification of Outliers
D.M. Hawkins

Generalized Linear Models
P. McCullagh and J.A. Nelder

Distribution-free Statistical Methods
J.S. Maritz

Multivariate Analysis in Behavioural Research
A.E. Maxwell

Applications of Queueing Theory
G.F. Newell

Some Basic Theory for Statistical Inference
E.J.G. Pitman

Density Estimation for Statistics and Data Analysis
B.W. Silverman

Statistical Inference
S.D. Silvey

Models in Regression and Related Topics
P. Sprent

Regression Analysis with Applications
G.B. Wetherill

Sequential Methods in Statistics
G.B. Wetherill and K.D. Glazebrook

(Full details concerning this series are available from the publishers)

Residuals and Influence in Regression

R. Dennis Cook and Sanford Weisberg

School of Statistics
University of Minnesota

NEW YORK LONDON
CHAPMAN AND HALL

First published 1982 by
Chapman and Hall
29 West 35th Street, New York NY 10001
Published in Great Britain by
Chapman and Hall Ltd
11 New Fetter Lane, London EC4P 4EE

Printed in Great Britain by
J. W. Arrowsmith Ltd, Bristol
Reprinted 1986
ISBN 0 412 24280 X

Library of Congress Cataloging in Publication Data
Cook, R. Dennis.
Residuals and influence in regression.

(Monographs on statistics and applied
probability)
Bibliography: p.
Includes index.
1. Regression analysis. I. Weisberg,
Sanford, 1947– . II. Title. III. Series.
QA278.2.C665 519.5′36 82–4412
ISBN 0–412–24280–X AACR2

British Library Cataloguing in Publication Data
Cook, R. Dennis
Residuals and influence in regression.—(Monographs
on statistics and applied probability)
1. Regression analysis
I. Title II. Weisberg, Sanford III. Series
519.5′36 QA278.2

ISBN 0–412–24280–X

Contents

Preface

Residuals are used in many procedures designed to detect various types of disagreement between data and an assumed model. Many of the common methods of residual analysis are founded on work in the early 1960s by F. Anscombe, J. W. Tukey, G. E. P. Box, D. R. Cox, C. Daniel and K. S. Srikantan. The methodology grew steadily through the early 1970s and by 1975 residual analysis was widely regarded as an integral part of any regression problem, and many methods using residuals had been incorporated into generally distributed computer packages. An implicit presumption at that time seems to be that most deficiencies are correctable through proper choice of scales, weights, model and method of fitting, and that residual analysis was used only to produce stronger, compelling conclusions. During the late 1970s interest in residual analysis was renewed by the development and rapid acceptance of methods for assessing the influence of individual observations. These developments allow a more complete understanding of an analysis, and have stimulated an awareness that some deficiencies may not be removable and thus inherent weaknesses in conclusions may necessarily remain.

In the first part of this monograph, we present a detailed account of the residual based methods that we have found to be most useful, and brief summaries of other selected methods. Where possible, we present a unified treatment to allow standard options to be viewed in a larger context. Our emphasis is on graphical methods rather than on formal testing. In the remainder, we give a comprehensive account of a variety of methods for the study of influence.

In writing this book, we have assumed that the reader is familiar with, or concurrently studying, linear models and regression methods at the level of Seber (1977), or, with some supplementation, Draper and Smith (1981) or Weisberg (1980a). An early version of this monograph was used as the basis of a course in Winter 1981 at the

University of Minnesota, and many of the comments of the participants have resulted in substantial improvements. Norton Holschuh read the final version and corrected many errors that might otherwise remain. Typing and other organizational matters were ably handled by Carol Lieb and Linda D. Anderson-Courtney. Almost all of the figures in this work were computer drawn at the University of Minnesota.

St. Paul, Minnesota R. Dennis Cook
January 1982 Sanford Weisberg

CHAPTER 1

Introduction

'Complicated phenomena, in which several causes concurring, opposing, or quite independent of each other, operate at once, so as to produce a compound effect, may be simplified by subducting the effect of all the known causes, as well as the nature of the case permits, either by deductive reasoning or by appeal to experience, and thus leaving, as it were, a *residual phenomenon* to be explained. It is by this process, in fact, that science, in its present advanced state, is chiefly promoted.'

JOHN F. W. HERSCHEL (1830). *A Preliminary Discourse on the Study of Natural Philosophy*

The collection of statistical methods that has come to be associated with the term 'regression' is certainly valued and widely used. And yet, an annoying and often sizeable gap remains between the necessarily idealized theoretical basis for these methods and their routine application in practice. It is well known, for example, that inferences based on ordinary least squares regression can be strongly influenced by only a few cases in the data, and the fitted model may reflect unusual features of those cases rather than the overall relationship between the variables. Here, *case* refers to a particular observation on the response variable in combination with the associated values for the explanatory variables.

There appear to be two major ways in which the gap between theory and practice is being narrowed. One is by the continued development of robust or resistant methods of estimation and testing that require progressively fewer untenable assumptions. Robust regression methods, for example, are a step ahead of least squares regression in this regard. The other line of inquiry is through the development of diagnostic tools that identify aspects of a problem that do not conform to the hypothesized modeling process. For example, the scatterplot of residuals versus fitted values that accompanies a linear least square fit is a standard tool used to diagnose nonconstant variance, curvature, and outliers. Diagnostic tools such as this plot have two important uses.

First, they may result in the recognition of important phenomena that might otherwise have gone unnoticed. Outlier detection is an example of this, where an outlying case may indicate conditions under which a process works differently, possibly worse or better. It can happen that studies of the outlying cases have greater scientific importance than the study of the bulk of the data. Second, the diagnostic methods can be used to suggest appropriate remedial action to the analysis of the model.

These lines of development, robust methods and diagnostics, are not mutually exclusive. When robust regression is viewed as iteratively reweighted least squares, for example, the weights associated with the individual cases may be useful indicators of outliers (Hogg, 1979). While it seems true that these approaches are in some ways competitive, one is not likely to replace the other in the foreseeable future. As long as least squares methods are in widespread use, the need for corresponding diagnostics will exist. Indeed, the use of robust methods does not abrogate the usefulness of diagnostics in general, although it may render certain of them unnecessary.

This book is about diagnostics. The major emphasis is on diagnostic tools for data analyses based on linear models in combination with least squares methods of estimation. This material is given in Chapters 2–4. In Chapter 5 we discuss corresponding tools for other selected problems.

In the remainder of this chapter we introduce a data set that will be used for illustration throughout the rest of this book and suggest a basic paradigm for regression analysis. While many other data sets will be introduced in later chapters, a complete and detailed discussion of each of these is not possible. We hope that the following discussion can serve as a model for a useful, but perhaps not universally applicable, perspective on the use of diagnostics in data analyses.

1.1 Cloud seeding

Judging the success of cloud seeding experiments intended to increase rainfall is an important statistical problem (cf. Braham, 1979). Results from past experiments are mixed. It is generally recognized that, depending on various contributing environmental factors, seeding will produce an increase or decrease in rainfall, or have no effect. Moreover, the critical factors controlling the response are, for the most part, unknown. This fundamental treatment-unit nonadditivity makes judgments about the effects of seeding difficult.

In 1975 the Florida Area Cumulus Experiment (FACE) was conducted to determine the merits of using silver iodide to increase rainfall and to isolate some of the factors contributing to the treatment-unit nonadditivity (Woodley, Simpson, Biondini and Berkeley, 1977). The target consisted of an area of about 3000 square miles to the north and east of Coral Gables, Florida. In this experiment, 24 days in the summer of 1975 were judged suitable for seeding based on a daily suitability criterion of $S - Ne \geq 1.5$, where S (seedability) is the predicted difference between the maximum height of a cloud if seeded and the same cloud if not seeded, and Ne is a factor which increases with conditions leading to naturally rainy days. Generally, suitable days are those on which the seedability is large, and the natural rainfall early in the day is small. On each suitable day, the decision to seed was based on unrestricted randomization; as it happened, 12 days were seeded and 12 were unseeded.

The response variable Y is the amount of rain (in cubic meters $\times 10^7$) that fell in the target area for a 6 hour period on each suitable day. To provide for the possibilities of reducing the variability and discovering some factors that may be contributing to the nonadditivity, the following explanatory variables were recorded on each suitable day:

Echo coverage (C) = per cent cloud cover in the experimental area, measured using radar in Coral Gables, Florida,

Prewetness (P) = total rainfall in the target area 1 hour before seeding (in cubic meters $\times 10^7$),

Echo motion (E) = a classification indicating a moving radar echo (1) or a stationary radar echo (2),

Action (A) = a classification indicating seeding (1) or no seeding (0).

The data as presented by Woodley *et al.* (1977) are reproduced in Table 1.1.1.

In addition to selecting days based on suitability ($S - Ne$), the investigators attempted to use only days with $C \leq 13\%$. A *disturbed* day was defined as $C > 13\%$. From Table 1.1.1, the first two experimental days are disturbed with the second day being highly disturbed ($C = 37.9\%$).

As a first step in the analysis of the results of this experiment, we suppose that there exists a vector-valued function G such that the true or 'best' relationship between the response and the explanatory

Table 1.1.1 *Cloud seeding data. Source: Woodley et al.* (1977)

Case	*A*	*T*	*S − Ne*	*C*	*P*	*E*	*Y*
1	0	0	1.75	13.40	0.274	2	12.85
2	1	1	2.70	37.90	1.267	1	5.52
3	1	3	4.10	3.90	0.198	2	6.29
4	0	4	2.35	5.30	0.526	1	6.11
5	1	6	4.25	7.10	0.250	1	2.45
6	0	9	1.60	6.90	0.018	2	3.61
7	0	18	1.30	4.60	0.307	1	0.47
8	0	25	3.35	4.90	0.194	1	4.56
9	0	27	2.85	12.10	0.751	1	6.35
10	1	28	2.20	5.20	0.084	1	5.06
11	1	29	4.40	4.10	0.236	1	2.76
12	1	32	3.10	2.80	0.214	1	4.05
13	0	33	3.95	6.80	0.796	1	5.74
14	1	35	2.90	3.00	0.124	1	4.84
15	1	38	2.05	7.00	0.144	1	11.86
16	0	39	4.00	11.30	0.398	1	4.45
17	0	53	3.35	4.20	0.237	2	3.66
18	1	55	3.70	3.30	0.960	1	4.22
19	0	56	3.80	2.20	0.230	1	1.16
20	1	59	3.40	6.50	0.142	2	5.45
21	1	65	3.15	3.10	0.073	1	2.02
22	0	68	3.15	2.60	0.136	1	0.82
23	1	82	4.01	8.30	0.123	1	1.09
24	0	83	4.65	7.40	0.168	1	0.28

variables is of the form

$$\mathbf{Y} = G(A, C, E, P, S-Ne; \boldsymbol{\beta}; \boldsymbol{\varepsilon}) \tag{1.1.1}$$

where $\mathbf{Y}$ is the 24-vector of responses, $\boldsymbol{\beta}$ is the vector of unknown parameters whose dimension p' depends on G, $\boldsymbol{\varepsilon}$ is a 24-vector of unobservable random errors, and the remaining arguments indicate that G may depend on the values of the explanatory variables A, C, E, P and $S-Ne$. For further progress the form of G must be specified. Since theoretical considerations that might suggest a form are lacking, we proceed by imposing tentative assumptions that seem reasonable and are not contradicted by available information.

Initially, we suppose that G is of the form

$$G = \mathbf{X}\boldsymbol{\beta} + \boldsymbol{\varepsilon} \tag{1.1.2}$$

where $\mathbf{X}$ is an $24 \times p'$ full rank matrix whose columns correspond to explanatory variables, including but not limited to those given in (1.1.1). The choice of this form is based on convenience and the general notion that linear models with additive errors often serve as reasonable *local* approximations to more complex models; we have no firm information to support or deny this supposition.

We next choose the complete set of explanatory variables (that is, the columns of $\mathbf{X}$). First, since regression through the origin does not seem sensible here, we include a constant column of ones. Second, to allow for the possibility of nonadditivity, we include all cross-product terms between action A and the other explanatory variables listed in (1.1.1). Finally, we include the number of days T after the first day of the experiment (June 1, 1975 = 0) as an explanatory variable. This variable, which is also listed in Table 1.1.1, is potentially relevant because there may have been a trend in natural rainfall or modification in the experimental technique.

With the five explanatory variables given in (1.1.1), $\mathbf{X}$ now contains $p' = 11$ columns. In general, we set $p' = p + 1$ if $\mathbf{X}$ contains a column of ones and set $p' = p$ otherwise, so p is always the number of explanatory variables excluding the constant.

In scalar form, the model may be written as

$$\begin{aligned} Y = {} & \beta_0 + \beta_1 A + \beta_2 T + \beta_3 (S - Ne) + \beta_4 C + \beta_5 P + \beta_6 E \\ & + \beta_{13}(A \times (S - Ne)) + \beta_{14}(A \times C) + \beta_{15}(A \times P) \\ & + \beta_{16}(A \times E) + \varepsilon \end{aligned} \tag{1.1.3}$$

Now that the form of the model has been specified the goals of our analysis can be made more specific. The main goal is to describe the difference ΔY between the rainfall for seeded and unseeded days,

$$\begin{aligned} \Delta Y &= Y(A = 1) - Y(A = 0) \\ &= \beta_1 + \beta_{13}(S - Ne) + \beta_{14} C + \beta_{15} P + \beta_{16} E \end{aligned} \tag{1.1.4}$$

Thus, the additive effect and the four possible interaction terms are of primary interest. The prediction of rainfall by itself is of secondary interest.

Inferences concerning $\boldsymbol{\beta}$ will be conditional on $\mathbf{X}$ and our analysis will be based, at least initially, on least squares methods since these are by far the most convenient and straightforward. For this to be sensible, however, a number of additional assumptions are needed: for each $i = 1, 2, \ldots, 24$,

(1) $E\varepsilon_i = 0$ (appropriate model);
(2) var $(\varepsilon_i) = \sigma^2$ (constant variance);
(3) cov $(\varepsilon_i, \varepsilon_j) = 0$, $i \neq j$ (uncorrelated errors);
(4) any measurement errors in the elements of $\mathbf{X}$ are small relative to σ; and
(5) the errors ε_i are (approximately) normally distributed.

If all of the structure that we have imposed so far is appropriate then the usual normal theory inferences based on the fitted model given in Table 1.1.2 will be accurate. But much of this structure lacks substantive support and if we are to have faith in the conclusions we must be convinced that our assumptions are not seriously violated and that reasonable alternative structures will not produce severely different conclusions. Answers to the following questions will surely help:

(1) Case 2 is considered to be a disturbed day and thus the process under study may differ under the conditions of case 2. Is case 2 outside the local region of applicability of the assumed model? More generally, are there outliers in the data?
(2) Is there evidence to suggest that the variances are not constant or that the distribution of the errors deviates from normality in important ways?
(3) Is there evidence to suggest that the form of the model ($E\mathbf{Y} = \mathbf{X}\boldsymbol{\beta}$) is not appropriate?

Table 1.1.2 *Fitted model, cloud seeding data*

Variable	*Estimate*	*s.e.*	*t-value*
Intercept	−3.4991	4.0632	−0.86
A	16.2452	5.5216	2.94
T	−0.0450	0.0251	−1.80
$S-Ne$	0.4198	0.8445	0.50
C	0.3879	0.2179	1.78
P	4.1083	3.6010	1.14
E	3.1528	1.9325	1.63
$A \times (S-Ne)$	−3.1972	1.2671	−2.52
$A \times C$	−0.4863	0.2411	−2.02
$A \times P$	−2.5571	4.4809	−0.57
$A \times E$	−0.5622	2.6430	−0.21

df = 13; $\hat{\sigma}^2 = 4.8607$; $R^2 = 0.72$

(4) Will transformations of the response or explanatory variables result in a greatly improved fit? For example, the fact that the response variable is a measure of volume may suggest a log or cube root transformation.
(5) Will relatively minor variations in the model result in important changes in the conclusions? More specifically: Does the fitted model change drastically when the variance of one case (case 2, for example) is changed relative to that of the others?

These concerns are not intended to be exhaustive but are to illustrate ways of increasing our confidence and deciding on remedial action, if appropriate. Answers to these and related questions will be suggested in later chapters.

1.2 The basic paradigm

The introductory portion of the analysis of the cloud seeding data given in the previous section illustrates part of the basic paradigm that should be used in fitting regression models. The paradigm, adapted from discussions in Box (1979, 1980), is illustrated in Fig. 1.2.1. We begin in the left-hand square by formulating a problem of interest, deciding on an appropriate set of variables to study, choosing a functional form for a model, and making certain assumptions, such as 'errors are independent, identically distributed.' If Bayesian analysis is planned, we seek out prior information to aid in the process.

The upper arrow pointing to the right corresponds to fitting the model to observed data, obtaining estimates, posterior distributions, and so on. We will call this the estimation step. In principle, the method actually used in the estimation step should depend on the assumptions

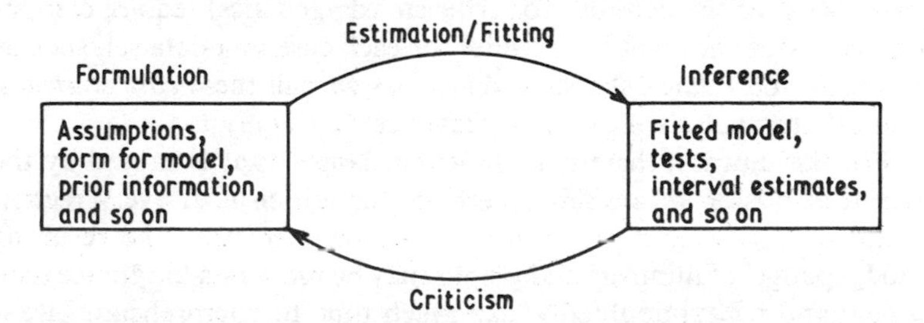

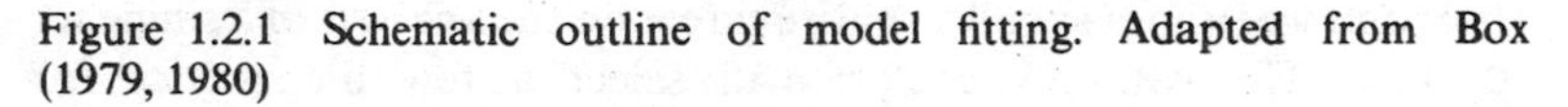

Figure 1.2.1 Schematic outline of model fitting. Adapted from Box (1979, 1980)

in the left-hand square. More frequently, the method of fitting is as much determined by available software as by assumptions. In the vast majority of regression analyses, least squares is used to carry out fitting. Whatever method of fitting is used, the right-hand square corresponds to the fitted model, including estimates, tests, and so forth. The treatment of the cloud seeding data in Section 1.1 is essentially an estimation step.

The bottom arrow in Fig. 1.2.1 is labeled *criticism*. It is meant to describe the act of critical assessment of the assumptions and the assumed model, given the fit in the right square and the actual data values. Criticism of a model may lead to modification of assumptions and thus further iteration through the system. The questions at the end of Section 1.1 may help guide this process.

Most of the work on model building, both for the statistician developing methods and for the scientist applying them, has concentrated on the upper estimation path. In precomputer days, the reason for this was clearly evident: fitting was laborious and time consuming. One of the earliest books on regression by Ezekiel (1930, 1941) rarely strays far from the computational problems of regression, and barely ventures beyond models with two predictors. Even more recent books (Ezekiel and Fox, 1958; Ostle and Mensing, 1975) still discuss time-consuming methods of inverting matrices via calculator. Since the fitting of models was inherently so difficult, it is not unreasonable that methods of criticism would be slow to develop and rarely used.

The availability of computers and the appearance of Draper and Smith (1966, 1981) changed this trend. The problems of the estimation step, at least via least squares, are now easily and quickly solved, and the analyst can consider inherently more complicated problems of criticism. Most of the methods for criticism (diagnostics) require computation of statistics that have values for each case in a data set, such as residuals and related statistics. As a class we call these *case statistics*, and call an analysis using these statistics *case analysis*.

For the unwary, there is an inherent danger that is caused by the recent explosion of available methods for criticism. If every recommended diagnostic is calculated for a single problem the resulting 'hodgepodge' of numbers and graphs may be more of a hindrance than a help and will undoubtedly take much time to comprehend. Life is short and we cannot spend an entire career on the analysis of a single set of data. The cautious analyst will select a few diagnostics for application in every problem and will make an additional parsimonious selection from the remaining diagnostics that correspond to the most probable or important potential failings in the problem at hand.

It is always possible, of course, that this procedure will overlook some problems that otherwise could be detected and that the urge to always apply 'just one more' diagnostic will be overwhelming. The truth is: If everything that can go wrong does go wrong, the situation is surely hopeless.

CHAPTER 2

Diagnostic methods using residuals

'Most of the phenomena which nature presents are very complicated; and when the effects of all known causes are estimated with exactness, and subducted, the residual facts are constantly appearing in the form of phenomena altogether new, and leading to the most important conclusions.'

HERSCHEL, *op. cit.*

The residuals carry important information concerning the appropriateness of assumptions. Analyses may include informal graphics to display general features of the residuals as well as formal tests to detect specific departures from underlying assumptions. Such formal and informal procedures are complementary, and both have a place in residual analysis.

Most residual based tests for specific alternatives for the errors are sensitive to several alternatives. These tests should be treated skeptically, or perhaps avoided entirely, until other alternatives that may account for an observed characteristic of the residuals have been eliminated. For example, outliers will affect all formal procedures that use residuals. Outlier detection procedures should usually be considered before any formal testing is done. On the other hand, informal graphical procedures can give a general impression of the acceptability of assumptions, even in the presence of outliers.

Anscombe (1961, Section 4.2) demonstrates that the whole of the data may contain relevant information about the errors beyond that available from the residuals alone. However, in the absence of specific alternative models or special design considerations, the residuals, or transformations thereof, provide the most useful single construction.

2.1 The ordinary residuals

The usual model for linear regression is summarized by

$$\begin{gathered} \mathbf{Y} = \mathbf{X}\boldsymbol{\beta} + \boldsymbol{\varepsilon} \\ E(\boldsymbol{\varepsilon}) = \mathbf{0}, \qquad \mathrm{Var}(\boldsymbol{\varepsilon}) = \sigma^2 \mathbf{I} \end{gathered} \tag{2.1.1}$$

where $\mathbf{X}$ is an $n \times p'$ full rank matrix of known constants, $\mathbf{Y}$ is an n-vector of observable responses, $\boldsymbol{\beta}$ is a p'-vector of unknown parameters, and $\boldsymbol{\varepsilon}$ is an n-vector of unobservable errors with the indicated distributional properties. To assess the appropriateness of this model for a given problem, it is necessary to determine if the assumptions about the errors are reasonable. Since the errors $\boldsymbol{\varepsilon}$ are unobservable, this must be done indirectly using residuals.

For linear least squares, the vector of ordinary residuals $\mathbf{e}$ is given by

$$\begin{aligned} \mathbf{e} &= (e_i) = \mathbf{Y} - \hat{\mathbf{Y}} \\ &= (\mathbf{I} - \mathbf{V})\mathbf{Y} \end{aligned} \tag{2.1.2}$$

where $\mathbf{V} = (v_{ij}) = \mathbf{X}(\mathbf{X}^{\mathrm{T}}\mathbf{X})^{-1}\mathbf{X}^{\mathrm{T}}$ and $\hat{\mathbf{Y}} = (\hat{y}_i)$ is the vector of fitted values. The relationship between $\mathbf{e}$ and $\boldsymbol{\varepsilon}$ is found by substituting $\mathbf{X}\boldsymbol{\beta} + \boldsymbol{\varepsilon}$ for $\mathbf{Y}$,

$$\begin{aligned} \mathbf{e} &= (\mathbf{I} - \mathbf{V})(\mathbf{X}\boldsymbol{\beta} + \boldsymbol{\varepsilon}) \\ &= (\mathbf{I} - \mathbf{V})\boldsymbol{\varepsilon} \end{aligned} \tag{2.1.3}$$

or, in scalar form, for $i = 1, 2, \ldots, n$,

$$e_i = \varepsilon_i - \sum_{j=1}^{n} v_{ij}\varepsilon_j \tag{2.1.4}$$

This identity demonstrates clearly that the relationship between $\mathbf{e}$ and $\boldsymbol{\varepsilon}$ depends only on $\mathbf{V}$. If the v_{ij}s are sufficiently small, $\mathbf{e}$ will serve as a reasonable substitute for $\boldsymbol{\varepsilon}$, otherwise the usefulness of $\mathbf{e}$ may be limited. For a sound understanding of the relationship between $\mathbf{e}$ and $\boldsymbol{\varepsilon}$, and most diagnostics in general, an understanding of the behavior of $\mathbf{V}$ is important.

2.1.1 THE HAT MATRIX

The matrix $\mathbf{V}$ is symmetric ($\mathbf{V}^{\mathrm{T}} = \mathbf{V}$) and idempotent ($\mathbf{V}^2 = \mathbf{V}$), and it is the linear transformation that orthogonally projects any n-vector onto the space spanned by the columns of $\mathbf{X}$. John W. Tukey has dubbed $\mathbf{V}$ the 'hat' matrix since it maps $\mathbf{Y}$ into $\hat{\mathbf{Y}}$, $\hat{\mathbf{Y}} = \mathbf{VY}$ (Hoaglin and Welsch, 1978). Since $\mathbf{V}$ is idempotent and symmetric it follows that

$$\operatorname{trace}(\mathbf{V}) = \operatorname{rank}(\mathbf{V}) = p', \qquad \sum_j v_{ij}^2 = v_{ii}$$

and that $\mathbf{V}$ is invariant under nonsingular linear reparameterizations. This latter property implies that, aside from computational concerns, collinearity between the columns of $\mathbf{X}$ is irrelevant to an understanding of how $\mathbf{V}$ behaves.

The projection onto the column space of $\mathbf{X}$ can be divided into the sum of two or more projections as follows: Partition $\mathbf{X} = (\mathbf{X}_1, \mathbf{X}_2)$, where $\mathbf{X}_1$ is $n \times q$ rank q, and let $\mathbf{U} = \mathbf{X}_1(\mathbf{X}_1^{\mathrm{T}}\mathbf{X}_1)^{-1}\mathbf{X}_1^{\mathrm{T}}$ be the projection matrix for the column space of $\mathbf{X}_1$. Next, let $\mathbf{X}_2^*$ be the component of $\mathbf{X}_2$ orthogonal to $\mathbf{X}_1$, $\mathbf{X}_2^* = (\mathbf{I} - \mathbf{U})\mathbf{X}_2$. Then,

$$\begin{aligned} \mathbf{T}^* &= \mathbf{X}_2^* (\mathbf{X}_2^{*\mathrm{T}}\mathbf{X}_2^*)^{-1}\mathbf{X}_2^{*\mathrm{T}} \\ &= (\mathbf{I} - \mathbf{U})\mathbf{X}_2 (\mathbf{X}_2^{\mathrm{T}}(\mathbf{I} - \mathbf{U})\mathbf{X}_2)^{-1}\mathbf{X}_2^{\mathrm{T}}(\mathbf{I} - \mathbf{U}) \end{aligned} \tag{2.1.5}$$

is the operator which projects onto the subspace of the column space of $\mathbf{X}$ orthogonal to the column space of $\mathbf{X}_1$, and

$$\mathbf{V} = \mathbf{U} + \mathbf{T}^* \tag{2.1.6}$$

This representation shows that the diagonal elements v_{ii} are nondecreasing in the number of explanatory variables p'. It can also be shown that, for fixed p', the v_{ii} are nonincreasing in n.

Let $\mathbf{X}_1 = \mathbf{1}$, an n-vector of ones. Then from (2.1.6) it follows immediately that

$$\mathbf{V} = \mathbf{1}\mathbf{1}^{\mathrm{T}}/n + \mathscr{X}(\mathscr{X}^{\mathrm{T}}\mathscr{X})^{-1}\mathscr{X}^{\mathrm{T}} \tag{2.1.7}$$

and

$$v_{ii} = \frac{1}{n} + \boldsymbol{x}_i^{\mathrm{T}}(\mathscr{X}^{\mathrm{T}}\mathscr{X})^{-1}\boldsymbol{x}_i \tag{2.1.8}$$

where $\mathscr{X}$ is the $n \times p$ matrix of centered explanatory variables and $\boldsymbol{x}_i^{\mathrm{T}}$ is the i-th row of $\mathscr{X}$. For simple regression, $y_i = \beta_0 + \beta_1 x_i + \varepsilon_i$, $v_{ii} = 1/n + (x_i - \bar{x})^2/\Sigma(x_j - \bar{x})^2$. For $p > 1$, contours of constant v_{ii} in p-dimensional space are ellipsoids, centered at the vector of sample averages.

The magnitudes of the diagonal elements of $\mathbf{V}$ play an important role in case analysis. From (2.1.8), $v_{ii} \geq 1/n$, $i = 1, 2, \ldots, n$, provided the model contains a constant. Upper bounds for v_{ii} depend on c, the number of times that the i-th row of $\mathbf{X}$, $\mathbf{x}_i^{\mathrm{T}}$, is replicated. If $\mathbf{x}_j = \mathbf{x}_i$, then $v_{ij} = v_{ii}$ and, using the symmetry and idempotency of $\mathbf{V}$,

$$v_{ii} = \sum_{j=1}^{n} v_{ij}v_{ji} = \sum_{j=1}^{n} v_{ij}^2 \geq c v_{ii}^2$$

which implies that $v_{ii} \leq 1/c$. Thus,

$$1/n \leq v_{ii} \leq 1/c \tag{2.1.9}$$

For models without a constant, the lower bound in (2.1.9) must be replaced by zero. The value of v_{ii} can attain its absolute maximum of 1 only if $\mathbf{x}_i$ occurs only once, and only if $v_{ij} = 0, j \neq i$. In this situation $\hat{y}_i = y_i$ and the i-th case will be fitted exactly. In effect, a single parameter is then devoted to a single case. This situation is pathological and will rarely occur in practice except when a variable is added to model an outlier as in Section 2.2.2. It can, however, occur with some frequency in multiple case generalizations.

The magnitude of v_{ii} depends on the relationship between $\boldsymbol{x}_i$ and the remaining rows of $\mathscr{X}$. Characteristics of $\boldsymbol{x}_i$ which cause v_{ii} to be relatively large or small can be seen as follows (Cook and Weisberg, 1980): Assuming that the intercept is included in the model, let $\mu_1 \geq \mu_2 \geq \ldots \geq \mu_p$ denote the eigenvalues of $\mathscr{X}^{\mathrm{T}}\mathscr{X}$, and let $\mathbf{p}_1, \ldots, \mathbf{p}_p$ denote the corresponding eigenvectors. Then, by the spectral decomposition of the corrected cross product matrix,

$$v_{ii} = \frac{1}{n} + \sum_{l=1}^{p} \left(\frac{\mathbf{p}_l^{\mathrm{T}} \boldsymbol{x}_i}{\sqrt{\mu_l}} \right)^2$$

Further, letting θ_{li} denote the angle between $\mathbf{p}_l$ and $\boldsymbol{x}_i$ we obtain

$$\cos(\theta_{li}) = \frac{\mathbf{p}_l^{\mathrm{T}} \boldsymbol{x}_i}{(\boldsymbol{x}_i^{\mathrm{T}} \boldsymbol{x}_i)^{1/2}}$$

and

$$v_{ii} = \frac{1}{n} + \boldsymbol{x}_i^{\mathrm{T}} \boldsymbol{x}_i \sum_{l=1}^{p} \frac{\cos^2(\theta_{li})}{\mu_l} \tag{2.1.10}$$

Thus, v_{ii} is large if: (1) $\boldsymbol{x}_i^{\mathrm{T}} \boldsymbol{x}_i$ is large, that is, $\boldsymbol{x}_i$ is well removed from the bulk of the cases; and (2) $\boldsymbol{x}_i$ is substantially in a direction of an eigenvector corresponding to a small eigenvalue of $\mathscr{X}^{\mathrm{T}}\mathscr{X}$. On the other hand, if $\boldsymbol{x}_i^{\mathrm{T}} \boldsymbol{x}_i$ is small, v_{ii} will be small regardless of its direction.

The elements of $\mathbf{V}$ are conveniently computed from any orthonormal basis for the column space of $\mathbf{X}$, such as that obtained from the singular value decomposition of $\mathbf{X}$, or the first p' columns of the matrix $\mathbf{Q}$ from the QR decomposition (see, for example, Stewart, 1973). If $\mathbf{q}_i^{\mathrm{T}}$ and $\mathbf{q}_j^{\mathrm{T}}$ are the i-th and j-th rows of the first p' columns of $\mathbf{Q}$, then $v_{ij} = \mathbf{q}_i^{\mathrm{T}} \mathbf{q}_j$.

Alternatively, the Choleski factor $\mathbf{R}$ (where $\mathbf{R}$ is upper triangular and $\mathbf{R}^{\mathrm{T}}\mathbf{R} = \mathbf{X}^{\mathrm{T}}\mathbf{X}$) can be used to compute the v_{ij} without inverting a matrix, since

$$\begin{aligned} v_{ij} &= \mathbf{x}_i^{\mathrm{T}} (\mathbf{X}^{\mathrm{T}}\mathbf{X})^{-1} \mathbf{x}_j \\ &= \mathbf{x}_i^{\mathrm{T}} (\mathbf{R}^{\mathrm{T}}\mathbf{R})^{-1} \mathbf{x}_j \\ &= \mathbf{a}_i^{\mathrm{T}} \mathbf{a}_j \end{aligned}$$

where $\mathbf{a}_i = \mathbf{R}^{-\mathrm{T}}\mathbf{x}_i$ is a p'-vector. Now $\mathbf{a}_i$ can be computed without inversion by the method of back substitution since $\mathbf{R}$ is upper triangular (see Dongarra, Bunch, Moler and Stewart, 1979, pp. 9.10).

2.1.2 THE ROLE OF V IN DATA ANALYSES

The distribution of $\mathbf{e}$, the vector of ordinary residuals, follows immediately from (2.1.3): If $\boldsymbol{\varepsilon} \sim N(\mathbf{0}, \sigma^2\mathbf{I})$ then $\mathbf{e}$ follows a singular normal distribution with $E(\mathbf{e}) = \mathbf{0}$ and $\mathrm{Var}(\mathbf{e}) = \sigma^2(\mathbf{I} - \mathbf{V})$, and the variation in $\mathbf{e}$ is controlled by $\mathbf{V}$.

The discussion of the previous section shows that cases remote in the factor space will have relatively large values of v_{ii}. Since $\mathrm{var}(\hat{y}_i) = v_{ii}\sigma^2$ and $\mathrm{var}(e_i) = (1 - v_{ii})\sigma^2$, fitted values at remote points will have relatively large variances and the corresponding residuals will have relatively small variances. Because of the analogy between $\mathrm{var}(\hat{y}_i)$ and the variance of the sample average based on a simple random sample (σ^2/n), Huber (1977) calls $1/v_{ii}$ the effective number of cases determining $\hat{y}_i$. Indeed, we have seen that when $v_{ii} = 1$, $\hat{y}_i = y_i$.

Many authors have hinted that the v_{ii} may play an important role in understanding an analysis based on (2.1.1). Behnken and Draper (1972) study the pattern of variation in the v_{ii} and note that wide variation reflects nonhomogeneous spacing of the rows of $\mathbf{X}$. Huber (1975) and Davies and Hutton (1975) point out that if $\max(v_{ii})$ is not considerably smaller than 1, it is probable that an outlier will go undetected when the residuals are examined. The average of the v_{ii} is p'/n and thus $\max(v_{ii}) \geq p'/n$. Accordingly, it may be difficult to identify outlying cases unless n is considerably larger than p'. Box and Draper (1975) suggest that for a designed experiment to be insensitive to outliers, the v_{ii} should be small, and approximately equal.

The importance of the v_{ii} is not limited to least squares analyses. Huber (1977) cautions that robust regression may not be effective, or work at all, if $\max(v_{ii})$ is close to 1. Huber's rationale is that it will be difficult for outliers to be identified and thus downweighted in robust regression if $\max(v_{ii})$ is large.

The $\max(v_{ii})$ is also important in determining the asymptotic character of least squares estimates: Let $\mathbf{z}$ denote a p'-vector with finite elements. Then a necessary and sufficient condition for all least squares estimates of the form $\mathbf{z}^{\mathrm{T}}\hat{\boldsymbol{\beta}}$ to be asymptotically normal is $\max(v_{ii}) \to 0$ as $n \to \infty$ (Huber, 1973). If $\max(v_{ii})$ is not small a normal approximation of the distribution of $\mathbf{z}^{\mathrm{T}}\hat{\boldsymbol{\beta}}$ may be suspect, at least for some $\mathbf{z}$ (see Wu, 1980, for further discussion).

Hoaglin and Welsch (1978) suggest a direct use of the v_{ii} as a diagnostic to identify 'high-leverage points'. The motivation behind this suggestion is based on the representation

$$\hat{y}_i = v_{ii} y_i + \sum_{j \neq i} v_{ij} y_j \tag{2.1.11}$$

The fitted value $\hat{y}_i$ will be dominated by $v_{ii} y_i$ if v_{ii} is large relative to the remaining terms. They interpret v_{ii} as the amount of leverage or influence exerted on $\hat{y}_i$ by y_i. It is clear, however, that for any $v_{ii} > 0$, $\hat{y}_i$ will be dominated by $v_{ii} y_i$ if y_i is sufficiently different from the other elements of $\mathbf{Y}$ (that is, an outlier).

2.1.3 USE OF THE ORDINARY RESIDUALS

When the fitted model is incorrect, the distribution of the unobservable errors $\boldsymbol{\varepsilon}$ and hence of the residual $\mathbf{e}$ will change. The goal in the study of the residuals is to infer any incorrect assumptions concerning $\boldsymbol{\varepsilon}$ from an examination of $\mathbf{e}$. Unfortunately, the correspondence between $\mathbf{e}$ and $\boldsymbol{\varepsilon}$ is less than perfect. In some problems, model failures will not be usefully transmitted to $\mathbf{e}$. In others, observed symptoms may be attributable to more than one incorrect assumption.

Consider as an alternative to (2.1.1) the model

$$\mathbf{Y} = \mathbf{X}\boldsymbol{\beta} + \mathbf{B} + \boldsymbol{\varepsilon} \tag{2.1.12}$$

where the n-vector $\mathbf{B} = (b_i)$ represents the bias in fitting (2.1.1) to a particular set of n cases. Often, the bias may be viewed, at least approximately, as $\mathbf{B} = \mathbf{Z}\boldsymbol{\phi}$, where $\boldsymbol{\phi}$ is an unobservable parameter vector. The columns of $\mathbf{Z}$ may represent important variables not included in $\mathbf{X}$, or nonlinear transformations of the columns of $\mathbf{X}$, perhaps polynomials or cross products. If (2.1.1) is fitted but (2.1.12) is the correct model, then

$$E(e_i) = (1 - v_{ii})b_i - \sum_{j \neq i} v_{ij} b_j \tag{2.1.13}$$

Bias would be diagnosed by a systematic feature in a plot of residuals against a column of $\mathbf{Z}$, if $\mathbf{Z}$ were known. However, the use of residuals for cases with large v_{ii} in this or other diagnostic procedures is likely to be limited, unless the bias at that case is extreme, since both terms on the right of (2.1.13) approach zero as $v_{ii} \to 1$. If v_{ii} is small, e_i may behave more like an average of the elements of $\mathbf{B}$ than like b_i. Most procedures that use $\mathbf{e}$ to detect model bias will tend to emphasize the fit of a model in the neighborhood of $\bar{\mathbf{x}}$ (v_{ii} small) while ignoring the relatively remote points (v_{ii} large).

EXAMPLE 2.1.1. ILLUSTRATION OF BIAS. Suppose that $\mathbf{X}$ is given by the first two columns of Table 2.1.1 and $\mathbf{Z}$ is given by the third column. If the correct model is $\mathbf{Y} = \mathbf{X}\boldsymbol{\beta} + \mathbf{Z}\boldsymbol{\phi} + \boldsymbol{\varepsilon}$, but the fitted model is $\mathbf{Y} = \mathbf{X}\boldsymbol{\beta} + \boldsymbol{\varepsilon}$, the v_{ii}, b_i, and $E(e_i)$ are as given in the next three columns of the table. Even in this small example, the differences between b_i and $E(e_i)$ are clear. Cases with small v_{ii} will have b_i accurately reflected (on the average) by the e_i, but cases with large v_{ii} do not share this property. If the bias in the model was largest at extreme cases (cases with larger values of v_{ii}), we would not expect the residuals to diagnose this problem accurately. □

Table 2.1.1 *Data set illustrating bias when some v_{ii} are large*

X		Z	v_{ii}	b_i	$E(e_i)$
1	−1	−4	0.217	-4ϕ	-3.2ϕ
1	−1	−3	0.217	-3ϕ	-2.2ϕ
1	−1	−1	0.217	-1ϕ	-0.2ϕ
1	−1	0	0.217	0	0.8ϕ
1	0	6	0.167	6ϕ	6.0ϕ
1	4	2	0.967	2ϕ	-1.2ϕ

Now suppose that (2.1.1) is correct except that $\text{Var}(\boldsymbol{\varepsilon}) = \sigma^2\mathbf{W}^{-1}$, for some unknown positive definite symmetric matrix $\mathbf{W}$. If (2.1.1) is fitted assuming that $\text{Var}(\boldsymbol{\varepsilon}) = \sigma^2\mathbf{I}$, then $E(\mathbf{e}) = \mathbf{0}$, but $\text{Var}(\mathbf{e}) = \sigma^2 (\mathbf{I}-\mathbf{V})\mathbf{W}^{-1}(\mathbf{I}-\mathbf{V})$. Depending on $\mathbf{W}^{-1}$, the actual variances of the residuals may be quite different from $\sigma^2(1-v_{ii})$, and from the variances for the residuals that would be obtained if the correct weighted least squares model were used (see Appendix A.1). For example, suppose that

$$\mathbf{W} = \begin{pmatrix} w_1 & \mathbf{0} \\ \mathbf{0} & \mathbf{I} \end{pmatrix}$$

so only case 1 has variance potentially different from σ^2. Let $\rho_{ij}^2 = v_{ij}^2/[(1-v_{ii})(1-v_{jj})]$, the squared correlation between the i-th and j-th residuals (Appendix A.3). Then an easy calculation shows that

$$\operatorname{var}(e_1) = \sigma^2(1 - v_{11})\left(1 + (1 - v_{11})\left(\frac{1 - w_1}{w_1}\right)\right)$$

and

$$\operatorname{var}(e_j) = \sigma^2(1 - v_{jj})\left(1 + \rho_{1j}^2(1 - v_{11})\left(\frac{1 - w_1}{w_1}\right)\right), \quad j \neq 1 \quad (2.1.14)$$

The effect of $w_1 \neq 1$ depends on the values of w_1, v_{11}, and ρ_{1j}^2. If ρ_{1j}^2 is small, var (e_1) will be the only term seriously affected by the noncons-tant variance. However, if ρ_{1j}^2 is large, then the change in var (e_j) will be comparable to the change in var (e_1). If w_1 is large, so ε_1 is less variable than the other errors, the true variances of the residuals will be smaller than their nominal values (in addition, the residual mean square will underestimate σ^2). If w_1 is small, then all the variances can become large. Analogous results for general $\mathbf{W}$ are more complicated, but it is clear that the residuals need not reflect nonconstant variances in the ε_j if some of the ρ_{ij}^2 are large.

When both bias and nonconstant variance are present, the residuals will have both nonzero means and variances other than those given by the usual formulae. However, examination of the residuals will not generally allow the analyst to distinguish between these two problems, since both can lead to the same symptoms in the residuals.

2.2 Other types of residuals

For use in diagnostic procedures, several transformations of the ordinary residuals have been suggested to overcome partially some of their shortcomings. We first consider in Section 2.2.1 the Studentization of residuals to obtain a set of residuals that have null distributions that are independent of the scale parameters. These residuals are shown to be closely related to a mean shift model for outliers (Section 2.2.2) and to the residuals obtained when each case in turn is left out of the data (Section 2.2.3). Alternatively, the residuals can be transformed to have a selected covariance structure: The usual suggestion is to obtain a vector of length $n - p'$ of residuals with uncorrelated elements. The methodology and usefulness of these residuals is briefly outlined in Section 2.2.4.

2.2.1 STUDENTIZED RESIDUALS

The ordinary residuals have a distribution that is scale dependent since the variance of each e_i is a function of both σ^2 and v_{ii}. For many diagnostic procedures, it is useful to define a Studentized version of the residuals that does not depend on either of these quantities. Following Margolin (1977), we use the term *Studentization* to describe the division of a scale dependent statistic, say U, by a scale estimate T so that the resulting ratio $S = U/T$ has a distribution that is free of the nuisance scale parameters. David (1981) makes a further distinction between *internal* Studentization, in which U and T are generally derived from the same data and are dependent, and *external* Studentization, where U and T are independent.

Internal Studentization

In least squares regression, the internally Studentized residuals are defined by

$$r_i = \frac{e_i}{\hat{\sigma}\sqrt{(1-v_{ii})}}, \qquad i = 1, 2, \ldots, n \tag{2.2.1}$$

where $\hat{\sigma}^2 = \Sigma e_i^2/(n-p')$ is the residual mean square. We reserve the term Studentized residual to refer to (2.2.1). This transformation of residuals has been studied by Srikantan (1961), Anscombe and Tukey (1963), Ellenberg (1973, 1976), Beckman and Trussell (1974), Prescott (1975), and many others. Many of these studies were motivated by a concern about outliers.

Ellenberg (1973) provides the joint distribution of a subset of $m < n - p'$ Studentized residuals, assuming that (2.1.1) holds and $\boldsymbol{\varepsilon} \sim \mathrm{N}(\mathbf{0}, \sigma^2\mathbf{I})$. The derivation of the joint distribution uses some interesting properties of the residuals and proceeds as follows. Suppose that an m-vector $\mathrm{I} = (i_1, i_2, \ldots, i_m)^{\mathrm{T}}$ indexes the m Studentized residuals of interest, and define $\mathbf{R}_{\mathrm{I}}$ and $\mathbf{e}_{\mathrm{I}}$ to be m-vectors whose j-th elements are r_{i_j} and e_{i_j}, respectively. Also, define $\mathbf{V}_{\mathrm{I}}$ to be the $m \times m$ minor of $\mathbf{V}$ given by the intersection of the rows and columns indexed by I. The rank of $\mathbf{V}_{\mathrm{I}}$ is no greater than p', and its eigenvalues are bounded between 0 and 1. The $m \times m$ matrix $\mathbf{I} - \mathbf{V}_{\mathrm{I}}$ is positive definite whenever the maximum eigenvalue of $\mathbf{V}_{\mathrm{I}}$ is less than 1.

The random vector $\mathbf{e}_{\mathrm{I}}$ follows a $N(\mathbf{0}, \sigma^2(\mathbf{I} - \mathbf{V}_{\mathrm{I}}))$ distribution. If we can find a quadratic form in $\mathbf{e}$ that is independent of $\mathbf{e}_{\mathrm{I}}$, then the joint distribution of $\mathbf{e}_{\mathrm{I}}$ and that quadratic form can be easily written. The joint distribution of $\mathbf{R}_\text{I}$ is then found by a change of variables and integration.

Provided that the inverse exists, the required quadratic form is given by

$$S^2_{(\text{I})} = \mathbf{e}^\text{T}\mathbf{e} - \mathbf{e}_\text{I}^\text{T}(\mathbf{I} - \mathbf{V}_\text{I})^{-1}\mathbf{e}_\text{I} \tag{2.2.2}$$

where $S^2_{(\text{I})} \sim \sigma^2\chi^2(n - p' - m)$ and $S^2_{(\text{I})}$ and $\mathbf{e}_\text{I}$ are independent. These facts can be proved using Appendix A.2 to show that $S^2_{(\text{I})}$ is the residual sum of squares for (2.1.1) with the cases indexed by I removed from the data.

The joint density of $\mathbf{e}_\text{I}$ and $S^2_{(\text{I})}$ is then

$$f(\mathbf{e}_\text{I}, S^2_{(\text{I})}) = \frac{|\mathbf{I} - \mathbf{V}_\text{I}|^{-\frac{1}{2}}(2\sigma^2)^{-(\nu + m/2)}}{\pi^{m/2}\Gamma(\nu)}(S^2_{(\text{I})})^{\nu - 1}$$

$$\exp\left\{-\frac{1}{2\sigma^2}[\mathbf{e}_\text{I}^\text{T}(\mathbf{I} - \mathbf{V}_\text{I})^{-1}\mathbf{e}_\text{I} + S^2_{(\text{I})}]\right\} \tag{2.2.3}$$

where $\nu = (n - p' - m)/2$. Next, let $\mathbf{D} = \text{diag}(1 - v_{i_1 i_1}, \ldots, 1 - v_{i_m i_m})$, and make the transformations

$$\mathbf{R}_\text{I} = \hat{\sigma}^{-1}\mathbf{D}^{-1/2}\mathbf{e}_\text{I} \tag{2.2.4}$$

and

$$\hat{\sigma} = [(S^2_{(\text{I})} + \mathbf{e}_\text{I}^\text{T}(\mathbf{I} - \mathbf{V}_\text{I})^{-1}\mathbf{e}_\text{I})/(n - p')]^{1/2}$$

Computing the Jacobian, substituting (2.2.4) into (2.2.3), and integrating over $\hat{\sigma}$ will give the density of $\mathbf{R}_\text{I}$. If $\mathbf{C}_\text{I} = \mathbf{D}^{-1/2}(\mathbf{I} - \mathbf{V}_\text{I})\mathbf{D}^{-1/2}$, the correlation matrix of the residulas indexed by I, then the density of $\mathbf{R}_\text{I}$ is

$$f(\mathbf{r}) = \frac{\Gamma(\nu + m/2)}{\Gamma(\nu)} \frac{|\mathbf{I} - \mathbf{V}_\text{I}|^{-1/2}}{((n - p')\pi)^{m/2}} \Pi(1 - v_{ii})^{1/2}$$

$$[1 - (n - p')^{-1}\mathbf{r}^\text{T}\mathbf{C}_\text{I}^{-1}\mathbf{r}]^{\nu - 1} \tag{2.2.5}$$

over the region $\mathbf{r}^\text{T}\mathbf{C}_\text{I}^{-1}\mathbf{r} \le n - p'$, and zero elsewhere. Form (2.2.5) can be recognized as an inverted Student function. Contours of constant density of (2.2.5) are ellipsoids of the form $\mathbf{r}^\text{T}\mathbf{C}_\text{I}^{-1}\mathbf{r} = c$. For the special case of $m = 1$, (2.2.5) reduces to

$$f(r) = \frac{\Gamma(\nu + \frac{1}{2})}{\Gamma(\nu)\Gamma(\frac{1}{2})(n - p')^{1/2}}\left(1 - \frac{r^2}{n - p'}\right)^{\nu - 1}; \quad |r| \le (n - p')^{1/2} \tag{2.2.6}$$

Hence, $r_i^2/(n - p')$ follows a Beta distribution, with parameters 1/2 and $(n - p' - 1)/2$, and it follows that $E(r_i) = 0$, $\text{var}(r_i) = 1$ and, from (2.2.5), $\text{cov}(r_i, r_j) = -v_{ij}/[(1 - v_{ii})(1 - v_{jj})]^{1/2}$, $i \ne j$.

The Studentized residuals are used as replacements for the ordinary residuals in graphical procedures, such as the plot against fitted values (Andrews and Pregibon, 1978; Anscombe and Tukey, 1963). They are also basic building blocks for most of the case statistics to be discussed in this and later chapters.

External Studentization

For externally Studentized residuals, an estimator of σ^2 that is independent of e_i is required. Under normality of the errors, Equation (2.2.2) provides such an estimate. Defining $\hat{\sigma}^2_{(i)}$ to be the residual mean square computed without the i-th case, it follows from Equation (2.2.2) that

$$\hat{\sigma}^2_{(i)} = \frac{(n-p')\hat{\sigma}^2 - e_i^2/(1-v_{ii})}{n-p'-1} \tag{2.2.7}$$

$$= \hat{\sigma}^2\left(\frac{n-p'-r_i^2}{n-p'-1}\right) \tag{2.2.8}$$

Under normality, $\hat{\sigma}^2_{(i)}$ and e_i are independent, and the externally Studentized residuals are defined by

$$t_i = \frac{e_i}{\hat{\sigma}_{(i)}(1-v_{ii})^{1/2}} \tag{2.2.9}$$

The distribution of t_i is Student's t with $n-p'-1$ degrees of freedom. The relationship between t_i and r_i is found by substituting (2.2.8) into (2.2.9),

$$t_i = r_i\left(\frac{n-p'-1}{n-p'-r_i^2}\right)^{1/2} \tag{2.2.10}$$

which shows that t_i^2 is a monotonic transformation of r_i^2.

2.2.2 MEAN SHIFT OUTLIER MODEL

Suppose that the i-th case is suspected as being an outlier. A useful framework used to study outliers is the mean shift outlier model,

$$\mathbf{Y} = \mathbf{X}\boldsymbol{\beta} + \mathbf{d}_i\phi + \boldsymbol{\varepsilon}$$
$$E(\boldsymbol{\varepsilon}) = \mathbf{0}, \qquad \mathrm{Var}(\boldsymbol{\varepsilon}) = \sigma^2\mathbf{I} \tag{2.2.11}$$

where $\mathbf{d}_i$ is an n-vector with i-th element equal to one, and all other elements equal to zero. Nonzero values of ϕ imply the i-th case is an outlier.

Under this model an outlier may occur in y_i, $\mathbf{x}_i$, or both. Suppose, for example, that y_i is not an outlier while the i-th row of $\mathbf{X}$ is in error by an unknown amount $\boldsymbol{\delta}_i$; that is, observed $(\mathbf{x}_i) = \mathbf{x}_i - \boldsymbol{\delta}_i$. Then,

$$\mathbf{Y} = \left[\mathbf{X} + \begin{pmatrix} \mathbf{0} \\ \boldsymbol{\delta}_i^{\mathrm{T}} \\ \mathbf{0} \end{pmatrix}\right] \boldsymbol{\beta} + \boldsymbol{\varepsilon} \tag{2.2.12}$$

which is in the form of (2.2.11) with $\phi = \boldsymbol{\delta}_i^{\mathrm{T}} \boldsymbol{\beta}$.

It is instructive to rewrite (2.2.11) by making the added variable orthogonal to the columns of $\mathbf{X}$ (as described near (2.1.5)),

$$\mathbf{Y} = \mathbf{X}\boldsymbol{\beta}^* + (\mathbf{I} - \mathbf{V})\mathbf{d}_i \phi + \boldsymbol{\varepsilon} \tag{2.2.13}$$

where $\boldsymbol{\beta}^*$ is not the same as $\boldsymbol{\beta}$ in (2.2.11), but ϕ is the same in both formulations. Because of the orthogonality, (2.2.13) can be fitted in two steps. First, fit the usual regression of $\mathbf{Y}$ on $\mathbf{X}$, ignoring the additional variable. Next, estimate ϕ from the regression of the residuals $\mathbf{e} = (\mathbf{I} - \mathbf{V})\mathbf{Y}$ computed in the first step on the added variable $(\mathbf{I} - \mathbf{V})\mathbf{d}_i$

$$\hat{\phi} = \frac{\mathbf{d}_i^{\mathrm{T}}(\mathbf{I} - \mathbf{V})(\mathbf{I} - \mathbf{V})\mathbf{Y}}{\mathbf{d}_i^{\mathrm{T}}(\mathbf{I} - \mathbf{V})(\mathbf{I} - \mathbf{V})\mathbf{d}_i} = \frac{e_i}{1 - v_{ii}} \tag{2.2.14}$$

The sum of squares for regression on $\mathbf{X}$ is $\mathbf{Y}^{\mathrm{T}}\mathbf{V}\mathbf{Y}$, while the additional sum of squares for regression on $(\mathbf{I} - \mathbf{V})\mathbf{d}_i$ is $\hat{\phi}^2(\mathbf{d}_i^{\mathrm{T}}(\mathbf{I} - \mathbf{V})^2\mathbf{d}_i)$ $= e_i^2/(1 - v_{ii})$. Hence, the residual sum squares for (2.2.13) is $\mathbf{Y}^{\mathrm{T}}(\mathbf{I} - \mathbf{V})\mathbf{Y} - e_i^2/(1 - v_{ii})$. Assuming normality, the t-statistic for a test of $\phi = 0$ is

$$t_i = \frac{e_i/(1 - v_{ii})^{1/2}}{\{[\mathbf{Y}^{\mathrm{T}}(\mathbf{I} - \mathbf{V})\mathbf{Y} - e_i^2/(1 - v_{ii})]/[n - p' - 1]\}^{1/2}} \tag{2.2.15}$$

which follows a $t(n - p' - 1)$ distribution under the null hypothesis. However, comparison of (2.2.15) with (2.2.7) and (2.2.9) shows that this test statistic for the shift model is identical to the externally Studentized residual.

Under the mean shift outlier model, the nonnull distribution of t_i^2 when $\phi \neq 0$ is noncentral F with noncentrality parameter $\phi^2(1 - v_{ii})/\sigma^2$. Since the noncentrality parameter is relatively small for v_{ii} near 1, finding outliers at remote points will be more difficult than finding outliers at cases with v_{ii} small. Yet it is precisely the former cases where interest in outliers is greatest. Also, since v_{ii} is increasing in p', outliers become more difficult to detect as the model is enlarged.

When the candidate case for an outlier is unknown, the test is usually based on the maximum of the t_i^2 over all i. A multiple testing procedure, such as one based on the first Bonferroni inequality (Miller, 1966), must be used to find significance levels. A nominal level α, two-tailed test for a single outlier will reject if $\max_i |t_i| > t(\alpha/n; n-p'-1)$. Cook and Weisberg (1980) suggest the alternative rule $\max_j |t_j| = |t_i| > t(v_{ii}\alpha/p'; n-p'-1)$. This rule maintains the overall significance level but provides an increase in power at cases with large v_{ii}. Special tables for the outlier test are provided by Lund (1975), Bailey (1977) and Weisberg (1980a). Moses (1979) provides useful charts. Tietjen, Moore and Beckman (1973) give critical values for simple linear regression.

EXAMPLE 2.2.1. ADAPTIVE SCORE DATA NO. 1. The simple regression data shown in Table 2.2.1 are from a study carried out at the University of Calfornia at Los Angeles on cyanotic heart disease in children. Here, x is the age of a child in months at first word and y is the Gesell adaptive score for each of $n = 21$ children. The data are given by Mickey, Dunn and Clark (1967) and have since been analyzed extensively in the statistical literature.

Table 2.2.1 *Gesell adaptive score* (y) *and age at first word* (x), *in months, for* 21 *children. Source: Mickey et al.* (1967)

Case	*x*	*y*	*Case*	*x*	*y*
1	15	95	11	7	113
2	26	71	12	9	96
3	10	83	13	10	83
4	9	91	14	11	84
5	15	102	15	11	102
6	20	87	16	10	100
7	18	93	17	12	105
8	11	100	18	42	57
9	8	104	19	17	121
10	20	94	20	11	86
			21	10	100

The lower triangular part of the symmetric matrix $\mathbf{V}$ is given in Table 2.2.2. Since even for simple regression $\mathbf{V}$ is $n \times n$, it is rarely computed in full, but we present it here for completeness. Examination of this matrix indicates that most of the v_{ii} are small (19 of the 21 are in

Table 2.2.2 *Projection matrix*, **V**, for adaptive score data

	1	2	3	4	5	6	7	8	9	10	11	12	13	14	15	16	17	18	19	20	21
1	0.05																				
2	0.05	0.15																			
3	0.05	0.01	0.06																		
4	0.04	−0.00	0.07	0.07																	
5	0.05	0.05	0.05	0.04	0.05																
6	0.05	0.10	0.03	0.02	0.05	0.07															
7	0.05	0.03	0.04	0.03	0.05	0.06	0.06														
8	0.05	0.02	0.06	0.06	0.05	0.03	0.04	0.06													
9	0.04	−0.01	0.07	0.07	0.04	0.02	0.03	0.06	0.08												
10	0.05	0.10	0.03	0.02	0.05	0.07	0.06	0.03	0.02	0.07											
11	0.04	−0.02	0.07	0.08	0.04	0.01	0.03	0.07	0.08	0.01	0.09										
12	0.04	−0.00	0.07	0.07	0.04	0.02	0.03	0.06	0.07	0.02	0.08	0.07									
13	0.05	0.01	0.06	0.07	0.05	0.03	0.04	0.06	0.07	0.03	0.07	0.07	0.06								
14	0.05	0.02	0.06	0.06	0.05	0.03	0.04	0.06	0.06	0.03	0.07	0.06	0.06	0.06							
15	0.05	0.02	0.06	0.06	0.05	0.03	0.04	0.06	0.06	0.03	0.07	0.06	0.06	0.06	0.06						
16	0.05	0.01	0.06	0.07	0.05	0.03	0.04	0.06	0.07	0.03	0.07	0.07	0.06	0.06	0.06	0.06					
17	0.05	0.03	0.06	0.06	0.05	0.04	0.04	0.05	0.06	0.04	0.06	0.06	0.06	0.05	0.05	0.06	0.05				
18	0.06	0.30	−0.05	−0.07	0.06	0.17	0.13	−0.03	−0.09	0.17	−0.11	−0.07	−0.05	−0.03	−0.03	−0.05	−0.00	0.65			
19	0.05	0.07	0.04	0.04	0.05	0.06	0.06	0.04	0.03	0.06	0.03	0.04	0.04	0.04	0.04	0.04	0.04	0.10	0.05		
20	0.05	0.02	0.06	0.06	0.05	0.03	0.04	0.06	0.06	0.03	0.07	0.06	0.06	0.06	0.06	0.06	0.05	−0.03	0.04	0.06	
21	0.05	0.01	0.06	0.07	0.05	0.03	0.04	0.06	0.07	0.03	0.07	0.07	0.06	0.06	0.06	0.06	0.06	−0.05	0.04	0.06	0.06

the range 0.05–0.09), the only exceptions being $v_{2,2} = 0.15$ and $v_{18,18} = 0.65$. Since $\Sigma v_{ii} = 2$, the diagonal for case 18 is relatively large. (The role of case 18 in this data set will be discussed at some length in succeeding sections.) Also, the v_{ij}, $i \neq j$, are generally small and positive, the exceptions again being associated with cases 2 and 18.

Next consider the linear regression model, $y_i = \beta_0 + \beta_1 x_i + \varepsilon_i$ for the data in Table 2.2.1. A scatter plot of the data is given in Fig. 2.2.1; the numbers on the graph give the case number of the closest points. From this graph, a straight line model appears plausible although cases 19, 18, and possibly 2 appear to dominate our perception of this plot. If the points for these three cases were removed, the perceived linearity would be less pronounced. Cases 18 and 2 fall near the perceived (and the fitted) regression line, while case 19 is quite distant.

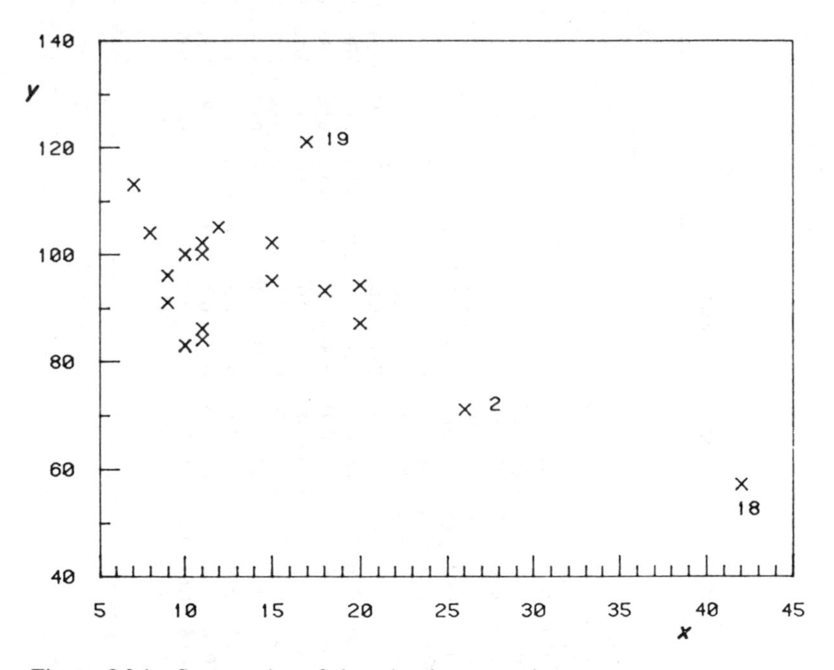

Figure 2.2.1 Scatter plot of the adaptive score data

Figure 2.2.2, an index plot (plot against case number) of the e_i, reflects the comments of the last paragraph, with the residual for case 19 clearly larger than the others. Figure 2.2.3 provides a plot of r_i versus $\hat{y}_i$, a

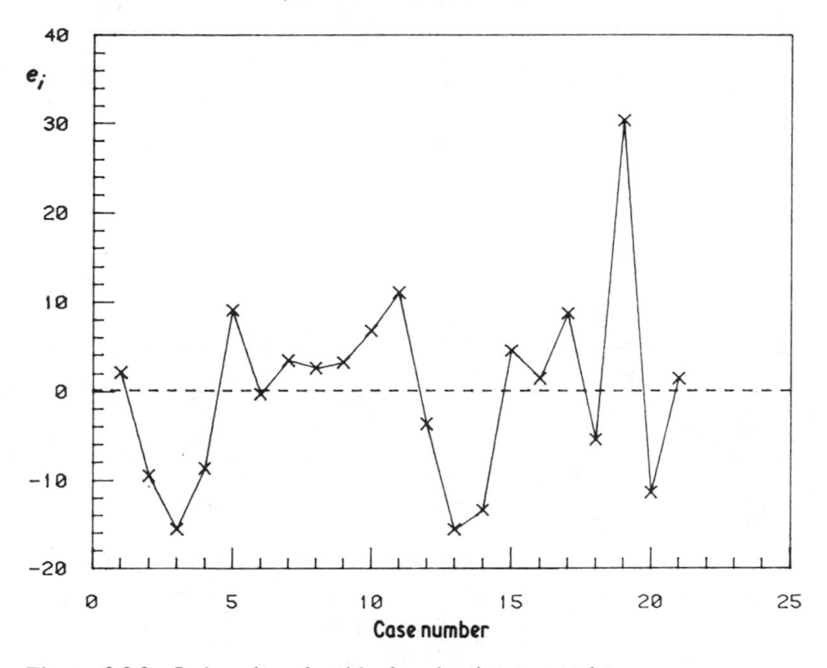

Figure 2.2.2 Index plot of residuals, adaptive scores data

standard plot used to find various problems that might be a function of the fitted values. Cases 18 and 19 stand apart in this data since $v_{18,18} = 0.65$ and $r_{19} = 2.823$. The statistic $t_{19} = 3.607$ computed from r_i can be used to test case 19 as an outlier; the Bonferroni upper bound for the p-value for this test is 0.0425.

The importance of case 19 in fitting the model can best be judged by deleting it and refitting the line, as summarized in Table 2.2.3. Deletion of the case has little effect on the estimated slope and intercept but it does clearly reduce the estimated variance. The role or influence of this case, as contrasted with cases 2 and 18, will be pursued in Chapter 3. □

EXAMPLE 2.2.2. CLOUD SEEDING NO. 2. As pointed out in Chapter 1, case 2 is an extremely disturbed day, and we may have prior interest in testing case 2 as an outlier. Because of the prior interest, the outlier statistic for case 2, $t_2 = 1.60$, can be compared to $t(n - p' - 1)$ to obtain significance levels. However, since $v_{2,2} = 0.9766$, the power of the test for this case is relatively small, and we cannot expect to detect anything but extreme deviations from model (1.1.3). □

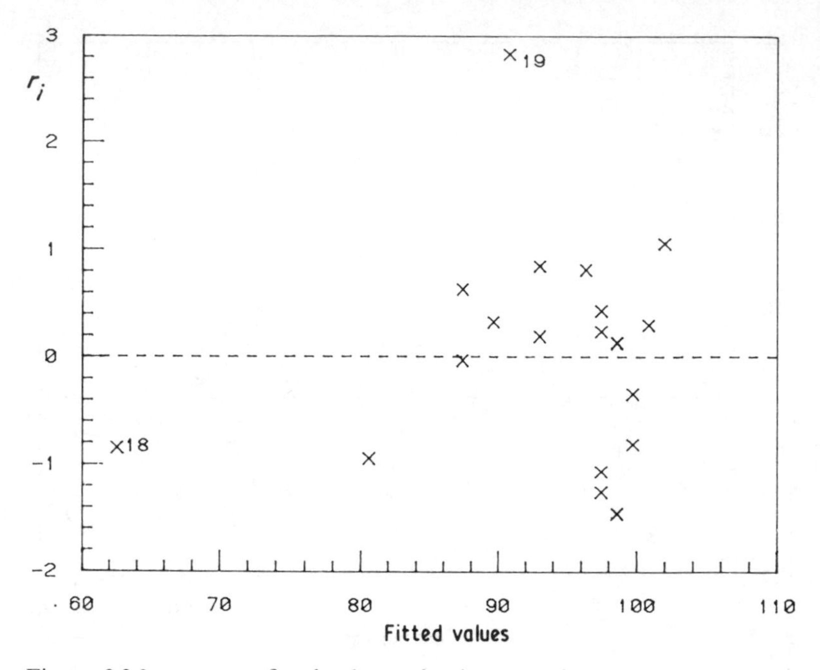

Figure 2.2.3 r_i versus fitted values, adaptive score data

Table 2.2.3 *Regression summaries with and without case 19, adaptive score data*

	Full data		*Case 19 deleted*	
	Estimate	*s.e.*	*Estimate*	*s.e.*
Intercept	109.87	5.06	109.30	3.97
Slope	−1.13	0.31	−1.19	0.24
	df = 19; $\hat{\sigma}^2 = 121.50$; $R^2 = 0.41$		df = 18; $\hat{\sigma}^2 = 74.45$; $R^2 = 0.57$	

Accuracy of the Bonferroni bound for the outlier test

Under the outlier test that uses the rejection rule $\max(t_i^2) > F(\alpha/n; 1, n - p' - 1)$, the first Bonferroni upper bound for the true p-value is p-value $\leq n \Pr(F > t_m^2)$ where F follows an $F(1, n - p' - 1)$ distribution and t_m^2 is the observed value of $\max(t_i^2)$. Cook and Prescott (1981) provide a relatively simple method for assessing the accuracy of this upper bound. The advantage of this method is that numerical integration is not required.

Let ρ_{ij} denote the correlation between e_i and e_j $(i \neq j)$,

$$\rho_{ij} = -v_{ij}/[(1-v_{ii})(1-v_{jj})]^{1/2}, \qquad (2.2.16)$$

let r_m denote the observed value of the Studentized residual corresponding to $\max(t_i^2)$ and define

$$c(\pm) = \{(i,j) | i < j,\ r_m^2 < \tfrac{1}{2}(n-p')(1 \pm \rho_{ij})\}$$

Then

$$\alpha - \beta^+ - \beta^- \leq p\text{-value} \leq \alpha \qquad (2.2.17)$$

where

$$\alpha = n \Pr[F \geq t_m^2]$$

$$\beta^+ = \sum_{c(+)} \Pr[F > r_m^2(n-p'-1)/(\tfrac{1}{2}(n-p')(1+\rho_{ij}) - r_m^2)]$$

and

$$\beta^- = \sum_{c(-)} \Pr[F > r_m^2(n-p'-1)/(\tfrac{1}{2}(n-p')(1-\rho_{ij}) - r_m^2)]$$

It follows immediately from (2.2.17) that the upper bound is exact when $c(+)$ and $c(-)$ are empty, or equivalently if

$$1 + \max_{i<j} |\rho_{ij}| < 2r_m^2/(n-p')$$

This is equivalent to the sufficient conditions given by Prescott (1977), Stefansky (1972a, b), and Srikantan (1961). Note also that since $r_m^2/(n-p') < 1$, the upper bound can never be exact if $\rho_{ij} = \pm 1$ for some $i \neq j$.

Calculation of the lower bound in (2.2.17) requires knowledge of the ρ_{ij}s. In many designed experiments, these will have a simple structure so that the lower bound can be calculated without difficulty. For example, in a two-way table with one observation per cell there are only three distinct residual correlations. Residual correlations for selected models of 2^k designs are given by Cook and Prescott (1981). In other cases, the lower bound may be approximated further by replacing ρ_{ij} in β^+ and β^- by $\max_{c(+)}(\rho_{ij})$ and $\min_{c(-)}(\rho_{ij})$, respectively. Our experience suggests that this will often be adequate.

EXAMPLE 2.2.3. ADAPTIVE SCORE DATA NO. 2. We have seen previously that the upper bound on the p-value for the outlier test for case 19 is 0.0425. While refining this value may be unnecessary from a hypothesis testing point of view, it may be desirable to judge the accuracy of the upper bound when p-values are used to assess the weight of evidence against the null hypothesis. This can be done using the lower bound (2.2.17).

Direct application of this bound would require evaluation of about 420 probability statements. While it would be straightforward to write a code to perform the required calculations, it will usually be sufficient to employ the further approximation, so that the number of probability statements that must be evaluated is reduced. A small number of evaluations can be handled easily on many hand-held calculators.

Inspection of the residual correlations in Table 2.2.4 shows that all correlations are in the interval $[-0.556, 0.202]$. A first lower bound on $\alpha-\beta^{+}-\beta^{-}$ can be obtained by replacing each ρ_{ij} in the expressions for β^{-} and β^{+} by the lower and upper bounds, respectively. However, this results in negative values for the lower bound on $\alpha-\beta^{+}-\beta^{-}$ at $r_{19}=2.923$, so a closer approximation is required.

A second inspection of the residual correlations reveals that one pair has a correlation of -0.556, two other pairs have correlations of -0.300, and of the remaining pairs 17 correlations lie in the interval $[0.002, 0.202]$ and 190 lie in $[-0.221, -0.016]$. A second lower bound on $\alpha-\beta^{+}-\beta^{-}$ can be obtained by using the four values $\{-0.556, -0.300, -0.016, 0.202\}$ in combination with their respective frequencies $\{1, 2, 190, 17\}$ to evaluate β^{+} and the four values $\{-0.556, -0.300, -0.221, 0.002\}$ in combination with the same respective frequencies to evaluate β^{-}. This procedure, which requires the evaluation of only eight probability statements, produces $\beta^{+}+\beta^{-} < 0.0016$. In short, the true p-value corresponding to $r_{19}=2.823$ is between 0.0409 and 0.0425.

As further illustration, the lower bounds obtained by using this procedure for $\alpha = 0.01$, 0.05, and 0.1 are 0.00997, 0.0476, and 0.086, respectively. Clearly, this procedure produces useful bounds in each case. □

Multiple cases

As before, let I be an m-vector of case subscripts and let $\mathbf{e}_{\text{I}}$, $\mathbf{V}_{\text{I}}$ be as defined previously. Multiple outlying cases can be modeled under a shift model by

$$\mathbf{Y} = \mathbf{X}\boldsymbol{\beta} + \mathbf{D}\boldsymbol{\phi} + \boldsymbol{\varepsilon}; \qquad E(\boldsymbol{\varepsilon}) = \mathbf{0}, \qquad \text{Var}(\boldsymbol{\varepsilon}) = \sigma^2\mathbf{I} \quad (2.2.18)$$

where $\mathbf{D}$ is $n \times m$ with k-th column $\mathbf{d}_{i_k}$ and $\boldsymbol{\phi}$ is an m-vector of unknown parameters. The normal theory statistic t_{I}^2 for testing $\boldsymbol{\phi} = 0$ is obtained

Table 2.2.4 *Residual correlations for the adaptive score data*

	1	2	3	4	5	6	7	8	9	10	11	12	13	14	15	16	17	18	19	20	21
1	1.000																				
2	−0.059	1.000																			
3	−0.048	−0.008	1.000																		
4	−0.048	0.002	−0.071	1.000																	
5	−0.050	−0.059	−0.048	−0.048	1.000																
6	−0.054	−0.112	−0.030	−0.026	−0.054	1.000															
7	−0.052	−0.091	−0.037	−0.034	−0.052	−0.068	1.000														
8	−0.048	−0.018	−0.063	−0.066	−0.048	−0.035	−0.040	1.000													
9	−0.048	0.013	−0.075	−0.081	−0.048	−0.021	−0.032	−0.069	1.000												
10	−0.054	−0.112	−0.030	−0.026	−0.054	−0.078	−0.068	−0.035	−0.021	1.000											
11	−0.047	0.023	−0.079	−0.086	−0.047	−0.016	−0.029	−0.073	−0.093	−0.016	1.000										
12	−0.048	0.002	−0.071	−0.076	−0.048	−0.026	−0.034	−0.066	−0.081	−0.026	−0.086	1.000									
13	−0.048	−0.008	−0.067	−0.071	−0.048	−0.030	−0.037	−0.063	−0.075	−0.030	−0.079	−0.071	1.000								
14	−0.048	−0.018	−0.063	−0.066	−0.048	−0.035	−0.040	−0.060	−0.069	−0.035	−0.073	−0.066	−0.063	1.000							
15	−0.048	−0.018	−0.063	−0.066	−0.048	−0.035	−0.040	−0.060	−0.069	−0.035	−0.073	−0.066	−0.063	−0.060	1.000						
16	−0.048	−0.008	−0.067	−0.071	−0.048	−0.030	−0.037	−0.063	−0.075	−0.030	−0.079	−0.071	−0.067	−0.063	−0.063	1.000					
17	−0.049	−0.029	−0.059	−0.062	−0.049	−0.039	−0.043	−0.057	−0.064	−0.039	−0.066	−0.062	−0.059	−0.057	−0.057	−0.059	1.000				
18	−0.106	−0.556	0.084	0.123	−0.106	−0.300	−0.221	0.046	0.162	−0.300	0.202	0.123	0.084	0.046	0.046	0.084	0.008	1.000			
19	−0.052	−0.080	−0.041	−0.039	−0.052	−0.063	−0.058	−0.043	−0.037	−0.063	−0.035	−0.039	−0.041	−0.043	−0.043	−0.041	−0.045	−0.183	1.000		
20	−0.048	−0.018	−0.063	−0.066	−0.048	−0.035	−0.040	−0.060	−0.069	−0.035	−0.073	−0.066	−0.063	−0.060	−0.060	−0.063	−0.057	0.046	−0.043	1.000	
21	−0.048	−0.008	−0.067	−0.071	−0.048	−0.030	−0.037	−0.063	−0.075	−0.030	−0.079	−0.071	−0.067	−0.063	−0.063	−0.067	−0.059	0.084	−0.041	−0.063	1.000

in analogy to the development leading to (2.2.15) (Gentleman and Wilk, 1975b; Cook and Weisberg, 1980),

$$t_{\mathrm{I}}^2 = \frac{(\mathbf{e}_{\mathrm{I}}^{\mathrm{T}}(\mathbf{I}-\mathbf{V}_{\mathrm{I}})^{-1}\mathbf{e}_{\mathrm{I}})(n-p'-m)}{((n-p')\hat{\sigma}^2 - \mathbf{e}_{\mathrm{I}}^{\mathrm{T}}(\mathbf{I}-\mathbf{V}_{\mathrm{I}})^{-1}\mathbf{e}_{\mathrm{I}})(m)}$$

The null distribution of this statistic under normality is $F(m, n-p'-m)$. Critical values for the multiple case outlier test can be based on the Bonferroni inequality, but these critical values are likely to be very conservative.

The multiple case analogue of the internally Studentized residual, since $\mathrm{Var}(\mathbf{e}_{\mathrm{I}}) = \sigma^2(\mathbf{I}-\mathbf{V}_{\mathrm{I}})$, is

$$r_{\mathrm{I}}^2 = \frac{\mathbf{e}_{\mathrm{I}}^{\mathrm{T}}(\mathbf{I}-\mathbf{V}_{\mathrm{I}})^{-1}\mathbf{e}_{\mathrm{I}}}{\hat{\sigma}^2} \tag{2.2.19}$$

The relationship between t_{I}^2 and r_{I}^2 is given by

$$t_{\mathrm{I}}^2 = \frac{r_{\mathrm{I}}^2(n-p'-m)}{m(n-p'-r_{\mathrm{I}}^2)} \tag{2.2.20}$$

Computations. Computing r_{I}^2 can be simplified by the use of appropriate matrix factorization. Gentleman (1980), for example, has used a Choleski factorization of $\mathbf{I}-\mathbf{V}_{\mathrm{I}}$: There is an $m \times m$ upper triangular matrix $\mathbf{R}$ such that $\mathbf{R}^{\mathrm{T}}\mathbf{R} = \mathbf{I}-\mathbf{V}_{\mathrm{I}}$ (for necessary software, see Dongarra *et al.*, 1979). Given this factorization, the Studentized residual can be computed in two steps. First, solve for $\mathbf{a}$ in the triangular system $\mathbf{R}^{\mathrm{T}}\mathbf{a} = \mathbf{e}_{\mathrm{I}}$. Then, compute $r_{\mathrm{I}}^2 = \mathbf{a}^{\mathrm{T}}\mathbf{a}/\hat{\sigma}^2$. This method has the advantage that if I^* is the subset consisting of the first $m^* < m$ cases included in I, then $r_{\mathrm{I}^*}^2 = \mathbf{a}_*^{\mathrm{T}}\mathbf{a}_*/\hat{\sigma}^2$ where $\mathbf{a}_*$ is the first m^* elements of $\mathbf{a}$.

Alternatively, let $\mathbf{V}_{\mathrm{I}} = \mathbf{\Gamma}\mathbf{\Lambda}\mathbf{\Gamma}^{\mathrm{T}}$ be the spectral decomposition of $\mathbf{V}_{\mathrm{I}}$, with the columns of $\mathbf{\Gamma}$ (eigenvectors) denoted by $\boldsymbol{\gamma}_1, \boldsymbol{\gamma}_2, \ldots, \boldsymbol{\gamma}_m$ and the diagonals of $\mathbf{\Lambda}$ denoted by $\lambda_1 \leq \ldots \leq \lambda_m$. Following Cook and Weisberg (1980),

$$r_{\mathrm{I}}^2 = \frac{1}{\hat{\sigma}^2}\sum_{l=1}^{m}\frac{(\boldsymbol{\gamma}_l^{\mathrm{T}}\mathbf{e}_{\mathrm{I}})^2}{1-\lambda_l} \tag{2.2.21}$$

provided that $\lambda_m < 1$. If $\lambda_m = 1$, deletion of the cases in I results in a rank deficit model and a test to see if I is an outlying set is not possible using the mean shift outlier model.

Finding the set I of m cases most likely to be an outlying set requires finding I to maximize r_I^2 over all $\binom{n}{m}$ possible subsets of size m. Even for modest n, if m is bigger than 2 or 3, this can be very expensive. This problem can be approached in at least two ways. First, some linear models have a special structure for $\mathbf{V}$ (and thus also for the $\mathbf{V}_I$) and this structure can be exploited. Gentleman (1980) has used this idea to obtain an algorithm for outliers in an $r \times c$ table with one observation per cell. She finds, for example, if $m = 2$, $\mathbf{V}_I$ can only be one of three possible 2×2 matrices, while for $m = 5$, $\mathbf{V}_I$ will be one of 354 possible 5×5 matrices. The factorization of $\mathbf{V}_I$ or $\mathbf{I} - \mathbf{V}_I$ need only be computed once, and then r_I^2 can be calculated for all I with a common value for $\mathbf{V}_I$. Alternatively, sequential outlier detection methods can be used. These methods have the disadvantage of failing to account for the signs of the residuals and their relative position in the observation space; the m cases with the largest residuals need not be the best candidates for an m-case outlier. Furthermore, residuals of opposite sign or on opposite sides of the observation space can mask each other so none appear as outliers if considered one at a time. Sequential outlier methods have a long history, dating at least to Pearson and Sekar (1936); see also Grubbs (1950) and Dixon (1950). Mickey *et al.* (1967) provide a modification of the sequential methods based on fitting models using stepwise regression methods that add dummy variables to delete outliers. The Furnival and Wilson (1974) algorithm can be used to perform the same function.

EXAMPLE 2.2.4. ADAPTIVE SCORE DATA NO. 3 The eight pairs of cases with largest r_I^2 or t_I^2 are listed in Table 2.2.5. All these pairs

Table 2.2.5 *Eight largest r_I^2 for the adaptive score data*

I	r_I^2	t_I^2
(3, 19)	9.788	9.032
(13, 19)	9.788	9.032
(11, 19)	9.287	8.127
(14, 19)	9:268	8.094
(5, 19)	8.944	7.561
(20, 19)	8.877	7.454
(17, 19)	8.844	7.402
(10, 19)	8.624	7.066

include case 19; in fact, the subsets with the 20 largest r_I^2 all include case 19. However, since the critical value for the $m = 2$ outlier test at level 0.05 based on the Bonferroni bound is 14.18, none of the pairs would be declared as an outlying pair by this conservative test. Clearly little is gained here by considering cases in pairs. Case 19 is found to be an outlier by itself, and we should not be surprised to find that it remains an outlier when considered with other cases. In problems where the cases have a natural relationship, perhaps in space or in time, pairs of cases that include individual outliers may well be of interest. This is not so in this example. □

Other outlier models

The mean shift outlier model is not the only plausible model for outliers. As might be expected, alternative formulations can lead to different procedures. For example, Cook, Holschuh, and Weisberg (1982) consider a variance-shift model in which the homoscedastic model (2.1.1) holds for all but one unknown case with variance $w\sigma^2$, $w > 1$. Assuming normality and maximum likelihood estimation for $(w, \boldsymbol{\beta}, \sigma^2)$, the case selected as the most likely outlier need not be the case with largest e_i or r_i, and thus at least the maximum likelihood procedure based on this model is not equivalent to the mean-shift model. However, if the case with the largest Studentized residual r_i also has the largest ordinary residual e_i, then that case will be identified as the most likely outlier under both the mean and variance shift models.

Another outlier model assumes that data is sampled from a mixture $g(\mathbf{x}) = \pi f_1(\mathbf{x}) + (1-\pi) f_2(\mathbf{x})$, with mixing parameter π. This formulation can include both location and scale shift models by appropriate choice of f_1 and f_2. Aitkin and Wilson (1980) consider maximum likelihood estimation of π and the parameters of f_1 and f_2 assuming that the densities are normal. Marks and Rao (1979) present a similar example with π assumed known. Since for this problem the likelihood function is often multimodal, the solution obtained, necessarily by an iterative method, will be sensitive to choice of starting values. Such mixture models have also been considered in a Bayesian framework (Box and Tiao, 1968; Box, 1980).

All of these methods differ from the outlier procedure based on the maximum Studentized residual in the philosophy of handling outliers since they are designed to accommodate outliers in the process of making inferences about the other parameters. Our approach is to identify outliers for further study. The action to be taken as a result of finding an outlier, such as case deletion or downweighting, will depend on the context of the problem at hand. This approach is more consistent with the overall goal of identifying interesting cases.

The outlier problem has recently received more detailed treatment by Barnett and Lewis (1978) and Hawkins (1980).

2.2.3 PREDICTED RESIDUALS

The ordinary and Studentized residuals are based on a fit to all the data. In contrast, the i-th predicted residual $e_{(i)}$ is based on a fit to the data with the i-th case excluded. Let $\hat{\boldsymbol{\beta}}$ denote the least squares estimate of $\boldsymbol{\beta}$ based on the full data and let $\hat{\boldsymbol{\beta}}_{(i)}$ be the corresponding estimate with the i-th case excluded. Then, the i-th predicted residual is

$$e_{(i)} = y_i - \mathbf{x}_i^{\mathrm{T}} \hat{\boldsymbol{\beta}}_{(i)}, \qquad i = 1, 2, \ldots, n \tag{2.2.22}$$

Each $e_{(i)}$ has several interpretations. First, one may think of it as a prediction error, since the data on the i-th case is not used in obtaining its fitted value. Anderson, Allen, and Cady (1972) and Allen (1974) use $PRESS = \Sigma e_{(i)}^2$ (the *pre*dicted *re*sidual *s*um of *s*quares) as a criterion for model selection, better models corresponding to relatively small values of *PRESS*. Much the same motivation, except from a Bayesian-predictivist point of view, is provided by Lee and Geisser (1972, 1975). Stone (1974) and Mosteller and Tukey (1977) discuss the related ideas of cross validation in which the data are split into two or more subsets, and parameters estimated on one subset are used to obtain fitted values for the other subsets to validate the model. A limit of this process, which gives rise to $e_{(i)}$, is obtained by dividing the data into n subsets, each consisting of a single case.

A relationship between $e_{(i)}$ and e_i is easily obtained using the formulae in Appendix A.2,

$$\begin{aligned} e_{(i)} &= y_i - \mathbf{x}_i^{\mathrm{T}} \hat{\boldsymbol{\beta}}_{(i)} \\ &= y_i - \mathbf{x}_i^{\mathrm{T}} \left(\hat{\boldsymbol{\beta}} - \frac{e_i (\mathbf{X}^{\mathrm{T}}\mathbf{X})^{-1} \mathbf{x}_i}{1 - v_{ii}} \right) \\ &= \frac{e_i}{1 - v_{ii}} \end{aligned} \tag{2.2.23}$$

which is identical to the estimate of ϕ (2.2.14) under the mean shift model. Deleting case i and predicting at $\mathbf{x}_i$ is therefore equivalent to adding a dummy variable $\mathbf{d}_i$ to the model and estimating a coefficient.

Moreover, the i-th predicted residual divided by the least squares estimate of the standard error of prediction based on the reduced data is equal to the i-th externally Studentized residual,

$$t_i = \frac{e_{(i)}}{\hat{\sigma}_{(i)}[1 + \mathbf{x}_i^{\mathrm{T}}(\mathbf{X}_{(i)}^{\mathrm{T}}\mathbf{X}_{(i)})^{-1}\mathbf{x}_i]^{1/2}} \tag{2.2.24}$$

where $\mathbf{X}_{(i)}$ is obtained from $\mathbf{X}$ by deleting the i-th row $\mathbf{x}_i^{\mathrm{T}}$.

It is clear that the $e_{(i)}$ are normally distributed (if the ε_i are normally distributed) with mean zero and variance $\sigma^2/(1 - v_{ii})$, and have the same correlation structure as the e_i. Use of the predicted residuals in place of the ordinary residuals in case analysis will tend to emphasize cases with large v_{ii} while use of e_i tends to emphasize cases with small v_{ii}. Using *PRESS* as a criterion for model selection will result in preference for models that fit relatively well at remote rows of $\mathbf{X}$. To correct for this, Studentized versions of the predicted residuals and of *PRESS* can be suggested. Not unexpectedly, these will get us back to r_i and t_i:

$$\frac{e_{(i)}}{\hat{\sigma}/(1 - v_{ii})^{1/2}} = r_i \tag{2.2.25}$$

$$\frac{e_{(i)}}{\hat{\sigma}_{(i)}/(1 - v_{ii})^{1/2}} = t_i \tag{2.2.26}$$

Alternative versions of *PRESS* may be defined as Σr_i^2 or Σt_i^2 in the same spirit as the weighted jackknife suggested by Hinkley (1977). See Geisser and Eddy (1979) and Picard (1981) for other uses of these residuals in model selection.

2.2.4 UNCORRELATED RESIDUALS

While the Studentized residuals do correct the residuals for equal variance, the correlation structure of the residuals is not changed. Clearly, $\mathbf{e}$ can be transformed to have a different correlation structure. Since the distribution of $\mathbf{e}$ is singular, the obvious goal of transforming so that the elements of the resulting vector are uncorrelated can be met only if we are satisfied with a lower-dimensional vector. This, in turn, has the serious drawback that the identification of residuals with cases becomes blurred, and interpretation of these transformed residuals as case statistics is generally not possible. However, for some special purposes, such as formal tests for normality, change points, or nonconstant variance, transformation to uncorrelated residuals has a certain intuitive appeal.

Suppose that an $n \times (n-p')$ matrix $\mathbf{C}$ defines a linear transformation $\tilde{\mathbf{e}} = \mathbf{C}^{\mathrm{T}}\mathbf{Y}$. We will call $\tilde{\mathbf{e}}$ a vector of linear unbiased scalar (or LUS) residuals if

$$E(\tilde{\mathbf{e}}) = \mathbf{0} \quad \text{(unbiased condition)} \tag{2.2.27}$$

$$\text{Var}(\tilde{\mathbf{e}}) = \sigma^2\mathbf{I} \quad \text{(scalar covariance matrix condition)} \tag{2.2.28}$$

These conditions require only that $\mathbf{C}^{\mathrm{T}}\mathbf{X} = \mathbf{0}$ and $\mathbf{C}^{\mathrm{T}}\mathbf{C} = \mathbf{I}$. The two common methods of choosing $\mathbf{C}$ both require that p' cases be nominated to have zero residuals. The choice of the nominated cases may be arbitrary, so that the definition of the uncorrelated residuals is not unique.

Suppose we partition $\mathbf{e}^{\mathrm{T}} = (\mathbf{e}_1^{\mathrm{T}}, \mathbf{e}_2^{\mathrm{T}})$, $\mathbf{X}^{\mathrm{T}} = (\mathbf{X}_1^{\mathrm{T}}, \mathbf{X}_2^{\mathrm{T}})$, $\mathbf{C}^{\mathrm{T}} = (\mathbf{C}_1^{\mathrm{T}}, \mathbf{C}_2^{\mathrm{T}})$ such that the subscript 1 corresponds to the p' cases nominated to have zero residuals, and subscript 2 corresponds to the remaining $n-p'$ cases. We assume $\mathbf{X}_1$ to be nonsingular. It follows from (2.2.27) and Appendix A.2 that $\mathbf{C}_2$ must satisfy

$$\mathbf{I} = \mathbf{C}_2^{\mathrm{T}}[\mathbf{I} - \mathbf{X}_2(\mathbf{X}^{\mathrm{T}}\mathbf{X})^{-1}\mathbf{X}_2^{\mathrm{T}}]^{-1}\mathbf{C}_2 \tag{2.2.29}$$

$\mathbf{C}_2$ can be chosen to be any factorization of the matrix in square brackets in (2.2.29). $\mathbf{C}_1$ is then determined uniquely from $\mathbf{C}_1^{\mathrm{T}} = -\mathbf{C}_2^{\mathrm{T}}\mathbf{X}_2\mathbf{X}_1^{-1}$.

BLUS residuals

Theil (1965) added the requirement of best, to get BLUS residuals, by requiring $\tilde{\mathbf{e}}$ to minimize $E[(\tilde{\mathbf{e}} - \mathbf{e}_2)^{\mathrm{T}}(\tilde{\mathbf{e}} - \mathbf{e}_2)]$. Theil (1968) showed that this is equivalent to using a spectral decomposition to find $\mathbf{C}_2$. Computational methods are given by Theil (1968) and Farebrother (1976a).

Using the BLUS residuals, Theil proposed a competitor to the Durbin–Watson (1950, 1951) test for serial correlation; critical values of Theil's test are given by Koerts and Abrahamse (1968, 1969). These two tests have been compared in several studies (Abrahamse and Koerts, 1969; Smith, 1976; Durbin and Watson, 1971) and are generally comparable in power, although the Durbin–Watson statistic has superior theoretical properties. Variants of Theil's method are given by Durbin (1970), Abrahamse and Louter (1971), Abrahamse and Koerts (1971), and Sims (1975).

Huang and Bloch (1974) used $\tilde{\mathbf{e}}$ in place of $\mathbf{e}$ in testing for normality. They point out that the independence of the BLUS residuals holds if and only if the errors are normally distributed, and thus, under a non-normal distribution for $\boldsymbol{\varepsilon}$, the apparent advantages of using the BLUS residuals disappears. Furthermore, they point out that the independence of the BLUS residuals is lost if heteroscedasticity is present. Thus, it should be no surprise that $\mathbf{e}$ appears to be more useful in normality tests than $\tilde{\mathbf{e}}$. Hedayat, Raktoe, and Telwar (1977) use the BLUS residuals in a test for nonconstant variance.

Recursive residuals

To construct the recursive residuals (Brown, Durbin, and Evans, 1975) it is necessary to first order the cases, typically by time. With the cases ordered, the k-th recursive residual $\tilde{\tilde{\mathbf{e}}}_k$ is defined as

$$\tilde{\tilde{\mathbf{e}}}_k = 0, \qquad k = 1, 2, \ldots, p'$$

$$\tilde{\tilde{\mathbf{e}}}_k = \frac{y_k - \mathbf{x}_k^{\mathrm{T}} \hat{\boldsymbol{\beta}}_{k-1}}{\sqrt{[1 + \mathbf{x}_k^{\mathrm{T}} (\mathbf{X}_{k-1}^{\mathrm{T}} \mathbf{X}_{k-1})^{-1} \mathbf{x}_k]}}, \qquad k = p' + 1, \ldots, n \tag{2.2.30}$$

where $\hat{\boldsymbol{\beta}}_{k-1}$ and $\mathbf{X}_{k-1}$ are computed using the first $k-1$ cases only. The term recursive is applied because $\hat{\boldsymbol{\beta}}_k$ can be computed from $\hat{\boldsymbol{\beta}}_{k-1}$ by use of an updating formula. Under (2.1.1) and normality, it is straightforward to show that the $\tilde{\tilde{e}}_k$ for $k > p'$ are independent and $N(0, \sigma^2)$. Equivalent versions of (2.2.30) have been proposed as early as Pizetti (1891). Algorithms for their construction are given by Brown *et al.* and Farebrother (1976b).

The recursive residuals, which correspond to using a Choleski factorization to choose $\mathbf{C}_2$ (Fisk, 1975), are appropriate for examining assumptions that depend on the order of the cases. Brown *et al.* (1975) consider two tests for a change point in the parameter vector $\boldsymbol{\beta}$ as a function of k via cumulative sums of recursive residuals. Phillips and Harvey (1974) use the recursive residuals in developing a test for serial correlation. Tests for heteroscedasticity using recursive residuals are discussed by Hedayat and Robson (1970) and Harvey and Phillips (1974).

2.3 Plotting methods

Residuals can be used in a variety of graphical and nongraphical summaries to identify inappropriate assumptions. Generally, a number of different plots will be required to extract the available information.

2.3.1 STANDARD RESIDUAL PLOTS

Standard residual plots are those in which the r_i or e_i are plotted against fitted values $\hat{y}_i$ or other functions of $\mathbf{X}$ that are approximately orthogonal to r_i (exactly orthogonal to the e_i). Anscombe (1973) gives an interesting discussion of the motivation for these graphical procedures. The plots are commonly used to diagnose nonlinearity and nonconstant error variance. Patterns, such as those in Fig. 2.3.1(b)–(d), are indicative of these problems, since under a correctly specified model the plots will appear haphazard, as in Fig. 2.3.1(a).

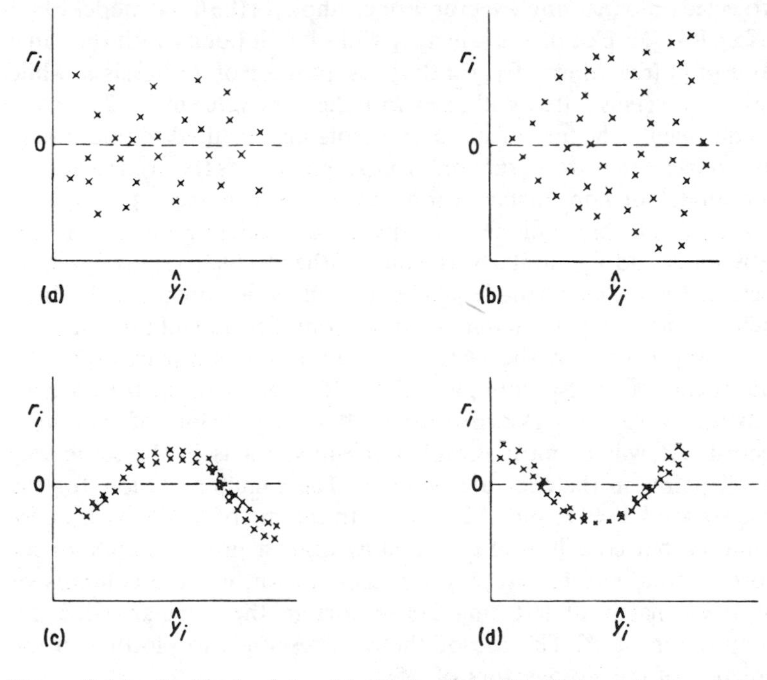

Figure 2.3.1 Residual plots. (a) Null plot. (b) Nonconstant variance. (c) Nonlinearity. (d) Nonlinearity. Source: Weisberg (1980a)

Historically, the ordinary residuals have been used most frequently in standard residual plots. Recently, however, a number of authors, including Andrews and Pregibon (1978), have indicated a preference for the Studentized residuals. The patterns in plots using the r_i will not be complicated by the nonconstant residual variances and will generally be more revealing than those using e_i. It is possible, for example, for a plot using the e_i to show a pattern similar to that in Fig. 2.3.1(b) simply because the v_{ii} are not constant.

In simple regression, the plot of residuals against $\hat{y}_i$ provides the relevant information about the fit of the model that is available without use of extra information such as time or spatial ordering of the cases. In multiple regression, the proper choice of horizontal axis for this plot is more problematic, as the two-dimensional plot is used to represent a model in a p'-dimensional space. In essence, a vector in p'-dimensional space is chosen and the data points are projected onto that single vector. For example, in the fitted model $\hat{y} = 3 + 2x_1 + 4x_2$, a plot of r_i against x_1 will plot all points with the same value of x_1 (regardless of x_2) at the same position of the abscissa, while a plot of r_i versus $\hat{y}_i$ treats all cases with the same value of $3 + 2x_1 + 4x_2$ as equivalent. The first of these two plots may be used to find model inadequacies that are a function of x_1 alone, such as the need to add x_1^2 to a model, or nonconstant variances of the form $\text{var}(\varepsilon_i) = x_{1i}\sigma^2$, $i = 1, 2, \ldots, n$, but will be inadequate for detecting an interaction between x_1 and x_2. Similarly, the plot of the residuals versus $\hat{y}_i$ will be useful in finding model inadequacies in the direction of the fitted values, such as a variance pattern that is a monotonic function of the response.

For any n-vector $\mathbf{Z}$, the vector $\mathbf{VZ}$ is in the column space of $\mathbf{X}$. The equivalence class of points plotted at the same place on the abscissa consists of a $(p-1)$-dimensional flat. The plot of residuals against $\mathbf{VZ}$ will be most useful if the model acts in the same way on all points in the equivalence class. The common choices for the abscissa are $\mathbf{VY} = \hat{\mathbf{Y}}$, and, if $\mathbf{X}_j$ is the j-th column of $\mathbf{X}$, $\mathbf{VX}_j = \mathbf{X}_j$. Less common, but equally useful, are plots against principal component score vectors, which, except for a scale factor, are the columns of the $n \times p$ matrix of left singular vectors in the singular value decomposition of $\mathscr{X}$. The use of these corresponds to plotting in the direction of the eigenvectors of $\mathscr{X}^{\mathrm{T}} \mathscr{X}$.

EXAMPLE 2.3.1. CLOUD SEEDING NO. 3. Figure 2.3.2 is a plot of r_i versus $\hat{y}_i$ for the cloud seeding data. This plot is clearly indicative of some problem, since cases 1 and 15 are well separated from the others, predicted rainfall is negative for two cases, and the general pattern of the residuals appears to decrease as $\hat{y}_i$ increases. It may show the need to transform Y to correct possible nonlinearity and perhaps to eliminate negative predicted rainfalls, or it may suggest other remedies such as

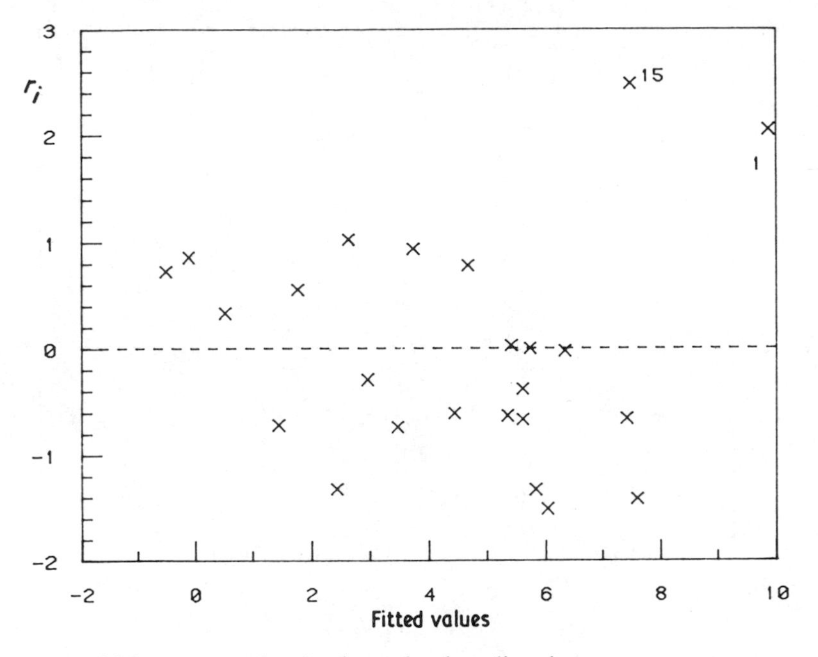

Figure 2.3.2 r_i versus fitted values, cloud seeding data

transforming predictors, or giving special attention to the cases that are separated from the rest of the data.

Figure 2.3.3, a plot of r_i versus $S - Ne$, suggests that the variance is a decreasing function of $S - Ne$ since most of the large residuals correspond to small values of $S - Ne$. In combination, the two plots clearly suggest that the original model is inadequate, but the appropriate remedial action is not clear. □

Plots of residuals against **VZ** are often difficult to interpret because informative patterns can be masked by the general scatter of points. As an aid to using these plots for relatively large data sets, Cleveland and Kleiner (1975) suggest superimposing robust reference lines. Let the values plotted on the abscissa be denoted by a_k, $k = 1, 2, \ldots, n$, with the a_k ordered from the smallest to largest, and let $a_{k1}, \ldots, a_{kl}$ be the l values of a with the smallest absolute deviation from a_k. Let b_{kj}, $j = 1, \ldots, l$, be the corresponding values of the ordinate (usually the residuals or Studentized residuals). For each k, robust estimates of the 0.25, 0.50, and 0.75 quantiles of $b_{kj}, j = 1, \ldots, l$ are plotted against a

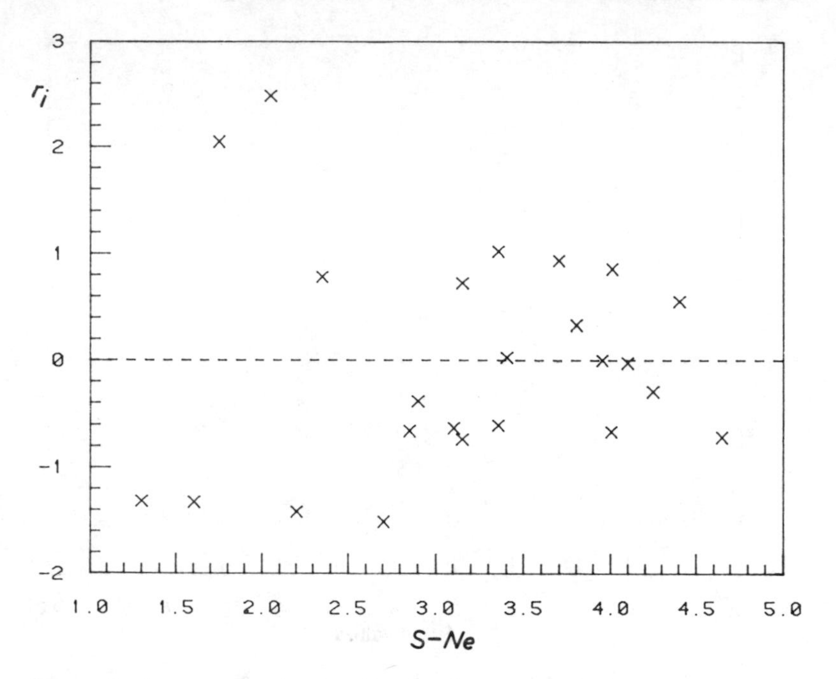

Figure 2.3.3 r_i versus $S - Ne$, cloud seeding data

robust estimate of the median of the a_{kj} (computer code is given by Gentleman, 1978; see also Cleveland, 1981). The window length l must be chosen to balance resolution and stability, and is often chosen by trial and error.

EXAMPLE 2.3.2. OLD FAITHFUL GEYSER. A *geyser* is a hot spring that occasionally becomes unstable and erupts hot water and steam into the air. One particular geyser, Old Faithful in Yellowstone National Park, is particularly well known and is one of the major tourist attractions in the United States. It erupts at an interval of about 40–100 min, with eruptions lasting from 1–6 min, to heights of near 35 m. National Park personnel predict eruption times based on the length of the last eruption. Their predictions are based on the empirical linear equation (minutes to the next eruption) = 30 + 10 × (duration of current eruption in minutes). Because the physical mechanisms that govern eruptions of the geyser are unknown, the prediction problem is one of statistical modeling based on observed values of intervals and durations only.

Figure 2.3.4 contains a scatter plot of $y =$ interval versus $x =$ duration for 272 eruptions of Old Faithful in October, 1980. These data were provided by Roderick A. Hutchinson, the Yellowstone Park geologist. Following standard park procedure, intervals are measured from the beginning of one eruption to the beginning of the next. The figure indicates that a simple regression model is at least plausible for this prediction problem, although the clustering of points into two groups is clearly evident. Figure 2.3.5 gives a plot of r_i versus x_i. While the clustering is clear, there is no obvious problem. However, if the robust reference lines are superimposed, as in Fig. 2.3.6 (with $l = 30$), slight curvature in the plot becomes apparent: extreme durations lead, on the average, to predictions that are too long. With the reference lines superimposed we recognize the possible need for a transformation of this data. □

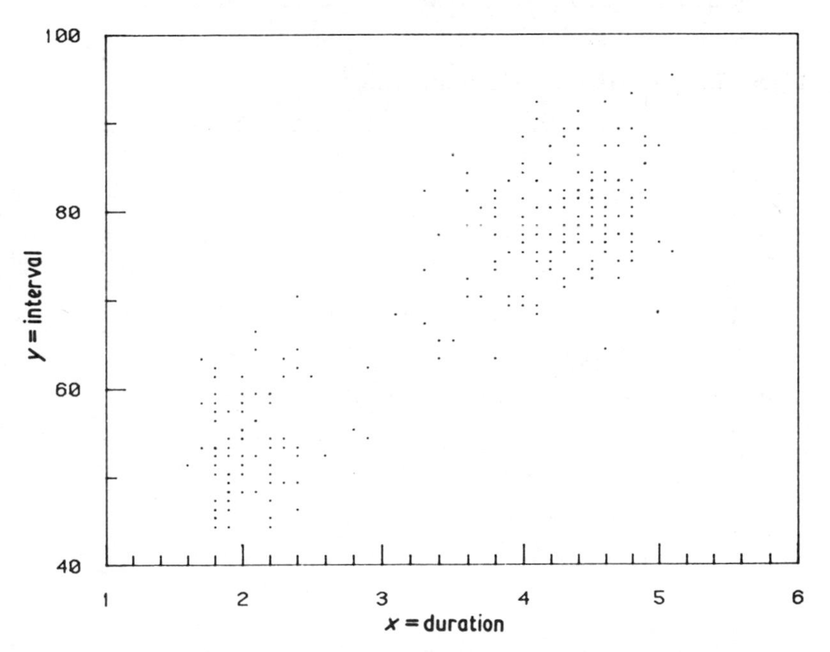

Figure 2.3.4 y (interval to next eruption in minutes) versus x (duration of current eruption to the nearest 0.1 min) for 272 eruptions of Old Faithful Geyser, October, 1980. Source: Roderick A. Hutchinson, Yellowstone National Park

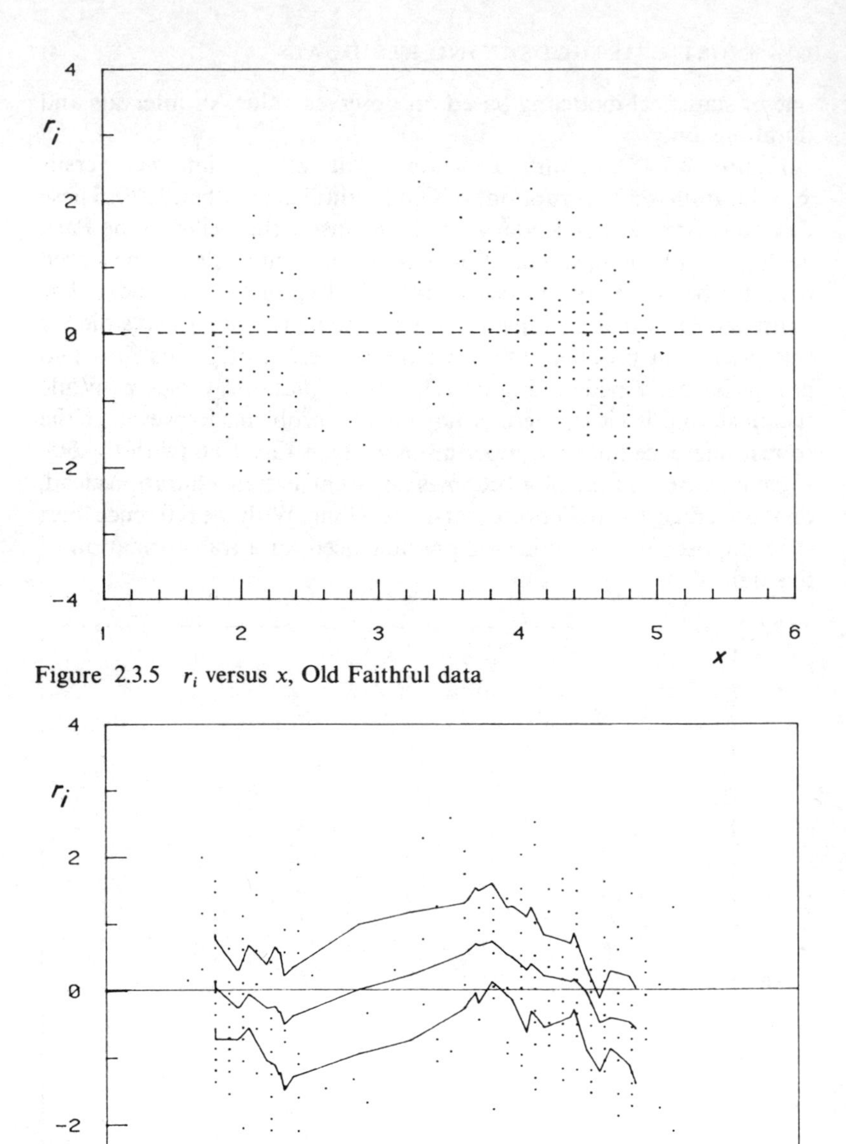

Figure 2.3.5 r_i versus x, Old Faithful data

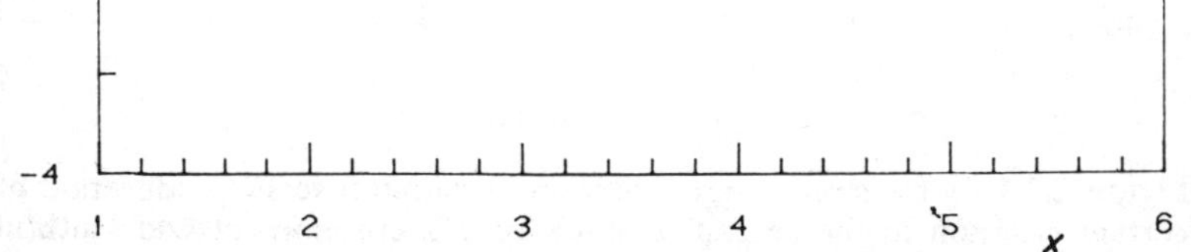

Figure 2.3.6 Enhanced residual plot, Old Faithful data, window width = 30

EXAMPLE 2.3.3. RESIDUALS IN PLACE. When the cases in a data set have identifiable physical locations, useful information about a model may be obtained by a semigraphical display obtained by plotting the residuals in their physical locations. An example of this is given by Daniel (1976) who discusses a classic 2^5 experiment on beans reported by Yates (1937). The experiment was carried out in blocks of 8, with two 3-factor and one 4-factor interaction confounded with blocks. Fitting a model including one block effect, four main effects and one 2-factor interaction, Daniel obtained residuals, and plotted them in their locations in the field (for a balanced design, all the v_{ii} are equal, so a plot of the residuals is equivalent to a plot of the Studentized residuals), as reproduced in Fig. 2.3.7. This plot indicates a region of apparent high fertility that extends into all four blocks, and is therefore not removed by the blocking effects. Daniel reanalyzed the data, using only blocks I and III and found that the estimated residual variation is reduced by a factor of 3. □

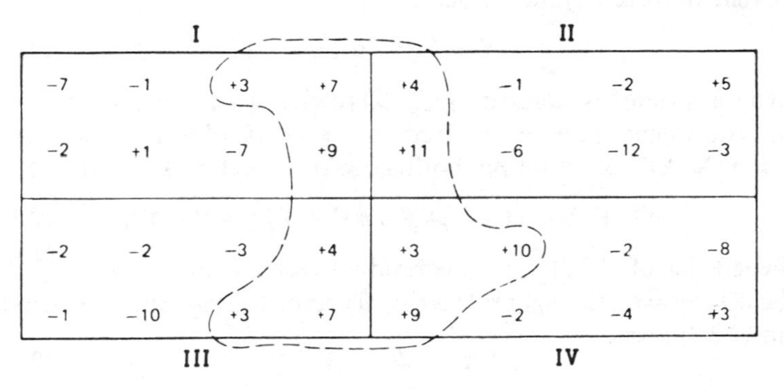

Figure 2.3.7 Residuals in place. Source: Daniel (1976), reprinted with permission

When a constant term is not included in a model, plots of residuals versus $\mathbf{VZ}$ are complicated by the fact that the simple regression of $\mathbf{e}$ on $\mathbf{VZ}$ is nonzero. If $\bar{e}$, $\bar{y}$, and $\bar{\mathbf{x}}$ are, respectively, the average of the residuals, the ys, and the vector of averages of the xs, then $\bar{e} = \bar{y} - \bar{\mathbf{x}}^{\mathrm{T}}\hat{\boldsymbol{\beta}}$ must be zero only if the constant is in the model. The slope of the regression of $\mathbf{e}$ on $\mathbf{VZ}$ is

$$-\bar{e}\frac{\mathbf{1}^{\mathrm{T}}\mathbf{VZ}}{\mathbf{Z}^{\mathrm{T}}\mathbf{VZ} - (\mathbf{1}^{\mathrm{T}}\mathbf{VZ})^2/n} \tag{2.3.1}$$

Thus, even a null plot will exhibit systematic features, especially if $\bar{e}$ is far from zero.

2.3.2 ADDED VARIABLE PLOTS

For models with a constant, the standard plot of $\mathbf{e}$ versus $\mathbf{VZ}$ exploits the orthogonality (or near orthogonality if the r_i are used) between plotted variables. Systematic nonlinear features of such plots suggest model inadequacies, and may be useful when specific alternative models are not available. However, they do suffer from the visual difficulty that is often apparent in attempting to detect systematic features of a swarm of points. This difficulty can be overcome by using plots in which a systematic *linear* feature indicates an incorrect model. To obtain these plots, we must choose a specific alternative for the fitted model. From the alternative, a plot, and usually a test, can be derived that compares the two models. These plots are often easy to interpret and can be very useful.

Consider first an alternative model that differs from (2.1.1) by the inclusion of a new explanatory variable $\mathbf{Z}$. We hypothesize as an alternative to (2.1.1) the model

$$\mathbf{Y} = \mathbf{X}\boldsymbol{\beta} + \phi\mathbf{Z} + \boldsymbol{\varepsilon} \tag{2.3.2}$$

An appropriate test comparing (2.3.2) to (2.1.1) is the F-test for $\phi = 0$. An equivalent plot is derived as follows. Defining as usual $\mathbf{V} = \mathbf{X}(\mathbf{X}^{\mathrm{T}}\mathbf{X})^{-1}\mathbf{X}^{\mathrm{T}}$, multiply both sides of (2.3.2) by $\mathbf{I}-\mathbf{V}$, to get

$$(\mathbf{I}-\mathbf{V})\mathbf{Y} = (\mathbf{I}-\mathbf{V})\mathbf{X}\boldsymbol{\beta} + \phi(\mathbf{I}-\mathbf{V})\mathbf{Z} + (\mathbf{I}-\mathbf{V})\boldsymbol{\varepsilon} \tag{2.3.3}$$

The left side of (2.3.3) is just the residual vector $\mathbf{e}$ for the model (2.1.1). The first term on the right side is exactly zero. Taking expectations over $\boldsymbol{\varepsilon}$ in (2.3.3) gives

$$E(\mathbf{e}) = \phi(\mathbf{I}-\mathbf{V})\mathbf{Z} \tag{2.3.4}$$

which suggests that a plot of $\mathbf{e}$ versus $(\mathbf{I}-\mathbf{V})\mathbf{Z}$ will be linear, through the origin. We call the plot of $\mathbf{e}$ versus $(\mathbf{I}-\mathbf{V})\mathbf{Z}$ an *added variable plot*, since it is designed to measure the effect of adding a variable to a model. These plots have been discussed or illustrated by Draper and Smith (1966, 1981), Anscombe (1967), Mosteller and Tukey (1977), Belsley, Kuh and Welsch (1980), and Weisberg (1980a).

In the regression of $\mathbf{e}$ on $(\mathbf{I}-\mathbf{V})\mathbf{Z}$, the estimated slope is

$$\hat{\phi} = \frac{\mathbf{Z}^{\mathrm{T}}(\mathbf{I}-\mathbf{V})\mathbf{Y}}{\mathbf{Z}^{\mathrm{T}}(\mathbf{I}-\mathbf{V})\mathbf{Z}} \tag{2.3.5}$$

and the intercept is 0 if there is a constant in the model.

By the conditions given in Kruskal (1968), the correct generalized least squares estimate obtained using the covariance matrix implied in (2.3.3) is identical to the ordinary least squares estimate given by (2.3.5). Using the results near (2.1.5), it can be shown that $\hat{\phi}$ is identical to the least squares estimate of ϕ obtained from the regression of $\mathbf{Y}$ on both $\mathbf{X}$ and $\mathbf{Z}$. From this it follows immediately that the residuals in the added variable plot are the same as the residuals for the regression of $\mathbf{Y}$ on both $\mathbf{X}$ and $\mathbf{Z}$.

Added variable plots are very useful for studying the role of a variable $\mathbf{Z}$ if it enters linearly into a model. The general scatter of the points gives an overall impression of the strength of the relationship. Individual points that are well separated from the rest of the data give heuristic information about the effects of outlying points on individual coefficients, and may suggest cases for special study.

The added variable $\mathbf{Z}$ can represent either a constructed variable that is defined by a specific alternative model, as will be discussed later in this chapter, or one of the variables in the model. If $\mathbf{U}_k$ is the projection matrix on all the columns of $\mathbf{X}$ except $\mathbf{X}_k$, then the k added variable plots of $(\mathbf{I}-\mathbf{U}_k)\mathbf{Y}$ versus $(\mathbf{I}-\mathbf{U}_k)\mathbf{X}_k$ have been advocated by Belsley *et al.* (1980), who call them *partial leverage regression plots.*

Non-null behavior

When the appropriate model for the relationship between $\mathbf{Y}$ and $(\mathbf{X}, \mathbf{Z})$ is more complicated than model (2.3.2), the usefulness of the added variable plot depends on $\mathbf{V}$. To see this, consider the model

$$\mathbf{Y} = \mathbf{X}\boldsymbol{\beta} + \phi\mathbf{Z}^{(\lambda)} + \boldsymbol{\varepsilon} \tag{2.3.6}$$

where $\mathbf{Z}^{(\lambda)}$ has i-th element

$$z_i^{(\lambda)} = \begin{cases} z_i^{\lambda}, & \lambda \neq 0 \\ \log(z_i), & \lambda = 0 \end{cases} \tag{2.3.7}$$

Power transformations are used in several places in this chapter, and provide a rich and interesting class of nonlinear functions. Using a linear Taylor series expansion about $\lambda = 1$, $z_i^{(\lambda)} \simeq z_i + (\lambda - 1)z_i \log(z_i)$, so the model (2.3.6) is approximately

$$\mathbf{Y} = \mathbf{X}\boldsymbol{\beta} + \phi\mathbf{Z} + \phi(\lambda - 1)\mathbf{L} + \boldsymbol{\varepsilon} \tag{2.3.8}$$

where $\mathbf{L}$ is an n-vector with i-th element $z_i \log(z_i)$. Multiplying by $(\mathbf{I}-\mathbf{V})$ and taking expectations,

$$E(\mathbf{e}) = \phi(\mathbf{I}-\mathbf{V})\mathbf{Z} + \phi(\lambda - 1)(\mathbf{I}-\mathbf{V})\mathbf{L} \tag{2.3.9}$$

Thus the regression of $\mathbf{e}$ on $(\mathbf{I}-\mathbf{V})\mathbf{Z}$ may have any shape if $\mathbf{V}$ is chosen appropriately and $\phi \neq 0$, $\lambda \neq 1$. Similar results are obtained if $\mathbf{Y}$ is a nonlinear function of $\mathbf{X}$.

Computations. The added variable plots for each $\mathbf{X}_k$ are potentially expensive to compute since for each plot two sets of residuals must be computed. However, Mosteller and Tukey (1977) and Velleman and Welsch (1981) outline a method to obtain these plots in a relatively simple way. For the variable $\mathbf{X}_k$ Equation (2.3.5) implies that

$$\hat{\beta}_k = \frac{\mathbf{X}_k^{\mathrm{T}}(\mathbf{I}-\mathbf{U}_k)\mathbf{Y}}{\mathbf{X}_k^{\mathrm{T}}(\mathbf{I}-\mathbf{U}_k)\mathbf{X}_k} = \mathbf{A}_k^{\mathrm{T}}\mathbf{Y}, \qquad k = 1, 2, \ldots, p' \tag{2.3.10}$$

where $\mathbf{A}_k = (\mathbf{I}-\mathbf{U}_k)\mathbf{X}_k/\mathbf{X}_k^{\mathrm{T}}(\mathbf{I}-\mathbf{U}_k)\mathbf{X}_k$ is the k-th column of an $n \times p'$ matrix $\mathbf{A}$. In matrix form, (2.3.10) is simply $\hat{\boldsymbol{\beta}} = \mathbf{A}^{\mathrm{T}}\mathbf{Y}$. But, since $\hat{\boldsymbol{\beta}} = (\mathbf{X}^{\mathrm{T}}\mathbf{X})^{-1}\mathbf{X}^{\mathrm{T}}\mathbf{Y}$, it follows that $\mathbf{A}^{\mathrm{T}} = (\mathbf{X}^{\mathrm{T}}\mathbf{X})^{-1}\mathbf{X}^{\mathrm{T}}$, the Moore–Penrose generalized inverse of $\mathbf{X}$. Except for a scale factor, $(\mathbf{I}-\mathbf{U}_k)\mathbf{X}_k$ is the k-th column of $\mathbf{A}$: If a_{ij} is the (i, j)-th element of $\mathbf{A}$, the i-th element of $(\mathbf{I}-\mathbf{U}_k)\mathbf{X}_k$ is $a_{ik}/\Sigma_l a_{lk}^2$. Given $(\mathbf{I}-\mathbf{U}_k)\mathbf{X}_k$ and $\mathbf{e} = (\mathbf{I}-\mathbf{V})\mathbf{Y}$, the vector $(\mathbf{I}-\mathbf{U}_k)\mathbf{Y}$ is computed from the identity

$$(\mathbf{I}-\mathbf{U}_k)\mathbf{Y} = \mathbf{e} + (\mathbf{I}-\mathbf{U}_k)\mathbf{X}_k\hat{\beta}_k \tag{2.3.11}$$

which is proved by writing $\mathbf{V}$ as a sum of projections, $\mathbf{V} = \mathbf{U}_k + \mathbf{T}_k$, where $\mathbf{T}_k$ is the orthogonal projection on $(\mathbf{I}-\mathbf{U}_k)\mathbf{X}_k$. Then $(\mathbf{I}-\mathbf{U}_k)\mathbf{Y} = (\mathbf{I}-\mathbf{V})\mathbf{Y} + \mathbf{T}_k\mathbf{Y}$, which upon simplifying gives (2.3.11).

As long as sufficient computer storage is available, the Moore–Penrose inverse can be computed to obtain added variable plots. However, if $\mathbf{X}$ is illconditioned, the Moore–Penrose inverse can be numerically unstable. G. W. Stewart (personal communication, 1981) suggests that a stable algorithm can be based on the QR decomposition (Stewart, 1973). If $\mathbf{X}_{p'}$ is the last column of $\mathbf{X}$, $\mathbf{Q}_{p'}$ is the corresponding column of $\mathbf{Q}$, and $r_{p'p'}$ is the indicated element of $\mathbf{R}$, then $(\mathbf{I}-\mathbf{U}_{p'})\mathbf{X}_{p'} = r_{p'p'}\mathbf{Q}_{p'}$. $(\mathbf{I}-\mathbf{U}_k)\mathbf{X}_k$ can be computed for other columns by using routines SQRDC, SQRSL, and SCHEX in LINPACK (Dongarra *et al.*, 1979).

EXAMPLE 2.3.4. JET FIGHTERS NO. 1. Stanley and Miller, in a 1979 RAND Corporation technical report, have attempted to build a descriptive model of the role of various design and performance factors in modeling technological innovation in jet fighter aircraft. Using data on American jet fighters built since 1940, they used the date of the first flight as a stand-in for a measure of technology; presumably, the level of technology is increasing with time. In some of their work, they considered the following variables:

FFD = first flight date, in months after January 1940;
SPR = specific power, proportional to power per unit weight;
RGF = flight range factor;
PLF = payload as a fraction of gross weight of aircraft;
SLF = sustained load factor;
CAR = 1 if aircraft can land on a carrier; 0 otherwise.

Exact definitions of all these quantities can be found in Stanley and Miller (1979). Between 1940 and 1979, 39 American jet fighters were flown. Of these, 14 aircraft were modifications of earlier aircraft, and for three others, the F-14A, F-15A, and F-16A, data are not available. Data on the 22 remaining planes are given in Table 2.3.1. Following Stanley and Miller we will fit models with FFD (or transformations of it) as a linear function of the other variables.

Table 2.3.1 *Jet fighter data. Source: Stanley and Miller (1979)*

Case	*ID*	*FFD*	*SPR*	*RGF*	*PLF*	*SLF*	*CAR*
1	FH-1	82	1.468	3.300	0.166	0.10	0
2	FJ-1	89	1.605	3.640	0.154	0.10	0
3	F-86A	101	2.168	4.870	0.177	2.90	1
4	F9F-2	107	2.054	4.720	0.275	1.10	0
5	F-94A	115	2.467	4.110	0.298	1.00	1
6	F3D-1	122	1.294	3.750	0.150	0.90	0
7	F-89A	127	2.183	3.970	0.000	2.40	1
8	XF10F-1	137	2.426	4.650	0.117	1.80	0
9	F9F-6	147	2.607	3.840	0.155	2.30	0
10	F-100A	166	4.567	4.920	0.138	3.20	1
11	F4D-1	174	4.588	3.820	0.249	3.50	0
12	F11F-1	175	3.618	4.320	0.143	2.80	0
13	F-101A	177	5.855	4.530	0.172	2.50	1
14	F3H-2	184	2.898	4.480	0.178	3.00	0
15	F-102A	187	3.880	5.390	0.101	3.00	1
16	F-8A	189	0.455	4.990	0.008	2.64	0
17	F-104A	194	8.088	4.500	0.251	2.70	1
18	F-105B	197	6.502	5.200	0.366	2.90	1
19	YF-107A	201	6.081	5.650	0.106	2.90	1
20	F-106A	204	7.105	5.400	0.089	3.20	1
21	F-4B	255	8.548	4.200	0.222	2.90	0
22	F-111A	328	6.321	6.450	0.187	2.00	1

An issue in building a model for these data is the choice of appropriate scaling for the response and for the explanatory variables. The use of *FFD* as the response suggests the unlikely assumption that technological innovation is constant over time. It is perhaps more reasonable to transform *FFD* so that the rate of change decreases with time, since we are measuring innovation in one general technology. A possible alternative scaling is the logarithm of *FFD* as a response, but the value of log (*FFD*) will depend on the choice of origin for *FFD*. If *FFD* is measured in months after January 1, 1900, then log (*FFD*) for the range of first flight dates in the data would represent rates of change that are nearly constant, while using January 1, 1940 as an origin will allow greater variation. Following Miller and Stanley, we tentatively adopt this as an origin both to allow for greater variation in the rate of change and because 1940 represents a reasonable origin for the jet age. In this example, we will define $LFFD = \log(FFD)$, and use natural logarithms. We return to the problem of scaling *FFD* later.

The regression of *LFFD* on the five predictors is summarized in Table 2.3.2. Three of the five variables are associated with large t-values, and the coefficients for *CAR* and *PLF* are negative, indicating that the ability to land on carriers, and the payload size adjusted for the other variables, are negatively related to *LFFD*. The aircraft with the largest v_{ii} is the F-111A with $v_{22,22} = 0.496$, although several of the v_{ii} are of comparable magnitude. The F-111A also has the largest Studentized residual, $r_{22} = 2.337$, with corresponding $t_{22} = 2.77$.

The added variable plots for *SPR*, *RGF*, and *SLF* are given as Figs. 2.3.8–2.3.10. The apparent linear trends in the first two of these

Table 2.3.2 *Fitted models for jet fighter data*

	Full data		*Case 22 (F-111A) deleted*	
Variable	*Estimate*	*t-value*	*Estimate*	*t-value*
Intercept	3.72	13.96	4.12	15.45
SPR	0.085	3.94	0.075	4.01
RGF	0.22	3.55	0.10	1.56
PLF	−0.48	−1.02	−0.45	−1.15
SLF	0.084	1.82	0.14	3.19
CAR	−0.23	−2.66	−0.20	−2.66
df = 16; $\hat{\sigma}^2 = 0.026$; $R^2 = 0.83$			df = 15; $\hat{\sigma}^2 = 0.019$; $R^2 = 0.85$	

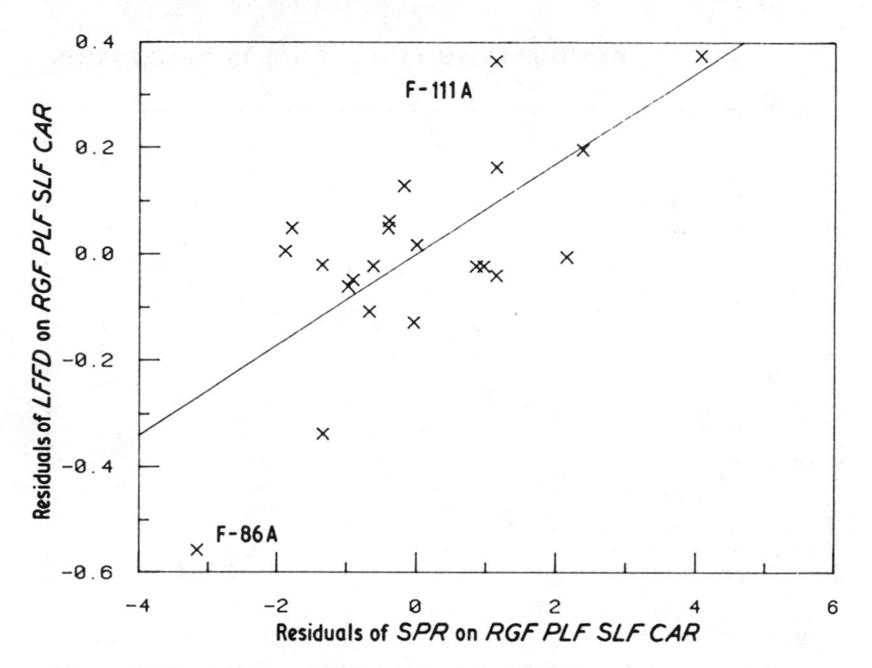

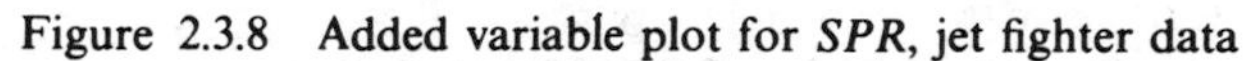
Figure 2.3.8 Added variable plot for *SPR*, jet fighter data

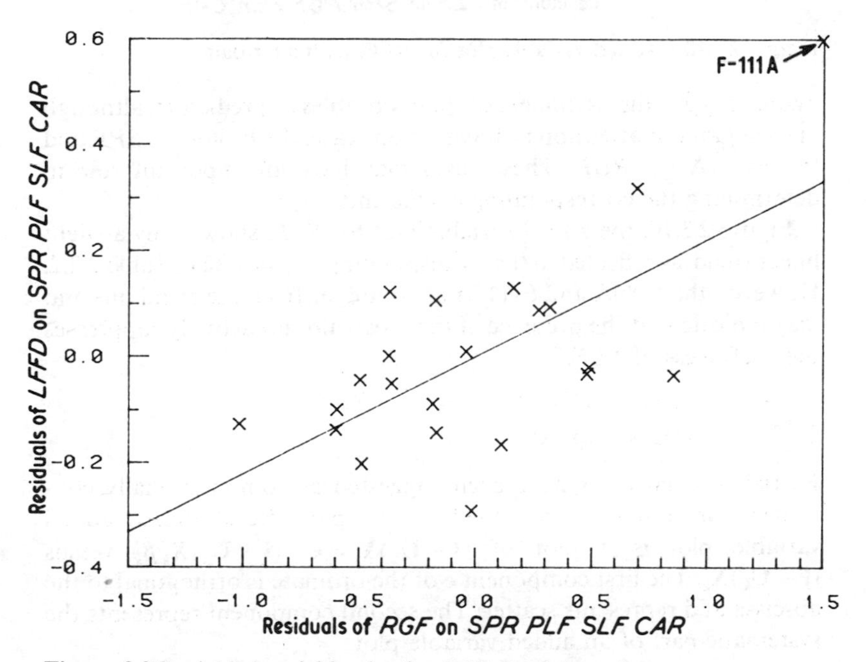

Figure 2.3.9 Added variable plot for *RGF*, jet fighter data

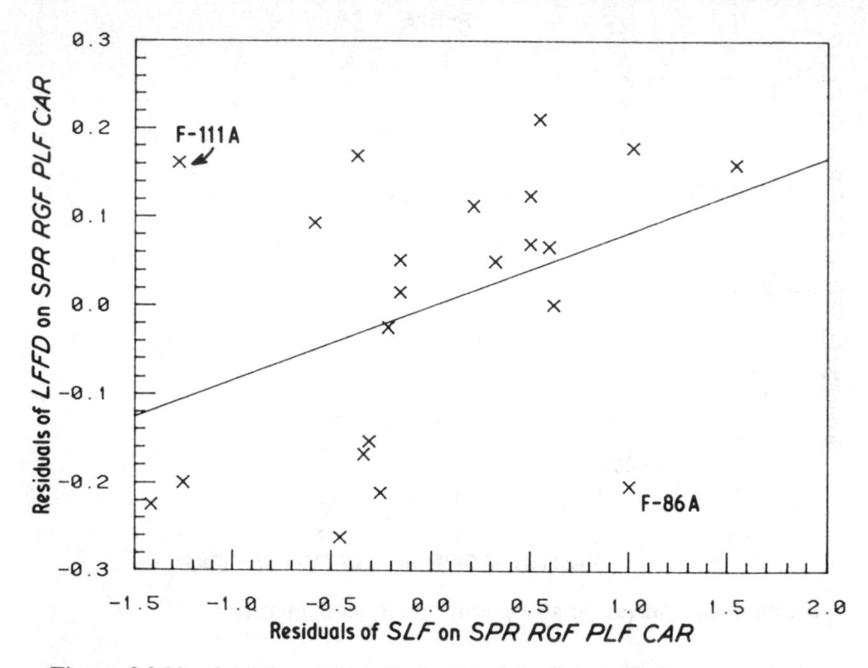

Figure 2.3.10 Added variable plot for *SLF*, jet fighter data

figures suggest the usefulness of these variables as predictors, although in each plot our attention is drawn to one case, the F-86A for *SPR* and the F-111A for *RGF*. These cases may have an important role in determining the corresponding coefficients.

Figure 2.3.10, the added variable plot for *SLF*, shows only a slight linear trend, as reflected in the corresponding $t_{SLF} = 1.82$ in Table 2.3.2. However, the F-86A and F-111A are quite far from the trend line and may indicate that the presence of these two aircraft actually suppresses the usefulness of *SLF*.□

2.3.3 PARTIAL RESIDUAL PLOTS

Partial residual plots have been suggested as computationally convenient substitutes for the added variable plots. Recall that an added variable plot is a plot of $(\mathbf{I}-\mathbf{U}_k)\mathbf{Y} = \mathbf{e} + (\mathbf{I}-\mathbf{U}_k)\mathbf{X}_k\hat{\beta}_k$ versus $(\mathbf{I}-\mathbf{U}_k)\mathbf{X}_k$. The first component $\mathbf{e}$ of the ordinate is orthogonal to the abscissa and represents scatter. The second component represents the systematic part of an added variable plot.

Computationally, the most difficult part of an added variable plot is obtaining $\mathbf{U}_k$. If this matrix is replaced by zero the result is a *partial residual* plot of $\mathbf{e}+\mathbf{X}_k\hat{\beta}_k$ versus $\mathbf{X}_k$. Ezekiel (1924) used such a plot to diagnose the need to transform an explanatory variable. As with the added variable plot, the two terms that make up the ordinate are orthogonal, the first term representing scatter and the second giving the systemmatic component. Again, the slope of the regression in this plot is $\hat{\beta}_k$, and the residuals from the regression line are given by the elements of $\mathbf{e}$. This plot was called a partial residual plot by Larsen and McCleary (1972), and a residual plus component plot by Wood (1973).

Although both the added variable plot and the partial residual plot have the same slope and the same residuals, their appearance can be markedly different. In the added variable plot, for example, the estimated variance of the slope is

$$\left(\frac{n-p'}{n-2}\right)\hat{\sigma}^2\frac{1}{\sum_i (x_{ik}-\bar{x}_k)^2(1-R_k^2)} \tag{2.3.12}$$

where R_k^2 is the square of the multiple correlation between $\mathbf{X}_k$ and $\mathbf{X}_1$, the matrix containing the other Xs. Apart from the multiplier $(n-p')/(n-2)$, the apparent estimated variance of $\hat{\beta}_k$ in the added variable plot is the same as the estimated variance of $\hat{\beta}_k$ from the full regression. In the partial residual plot the apparent variance of $\hat{\beta}_k$ is

$$\left(\frac{n-p'}{n-2}\right)\hat{\sigma}^2\frac{1}{\sum_i (x_{ik}-\bar{x}_k)^2} \tag{2.3.13}$$

which ignores any effect due to fitting the other variables. If R_k^2 is large, then (2.3.13) can be much smaller than (2.3.12), and the partial residual plot will present an incorrect image of the strength of the relationship between $\mathbf{Y}$ and $\mathbf{X}_k$ (conditional on the other Xs). In fact, it can be seen that the partial residual plot is a hybrid, reflecting the systematic trend of $\mathbf{X}_k$ adjusted for $\mathbf{X}_1$, but the scatter of $\mathbf{X}_k$ ignoring $\mathbf{X}_1$.

EXAMPLE 2.3.5. JET FIGHTERS NO. 2. The partial residual plots corresponding to the added variable plots for *SPR*, *RGF*, and

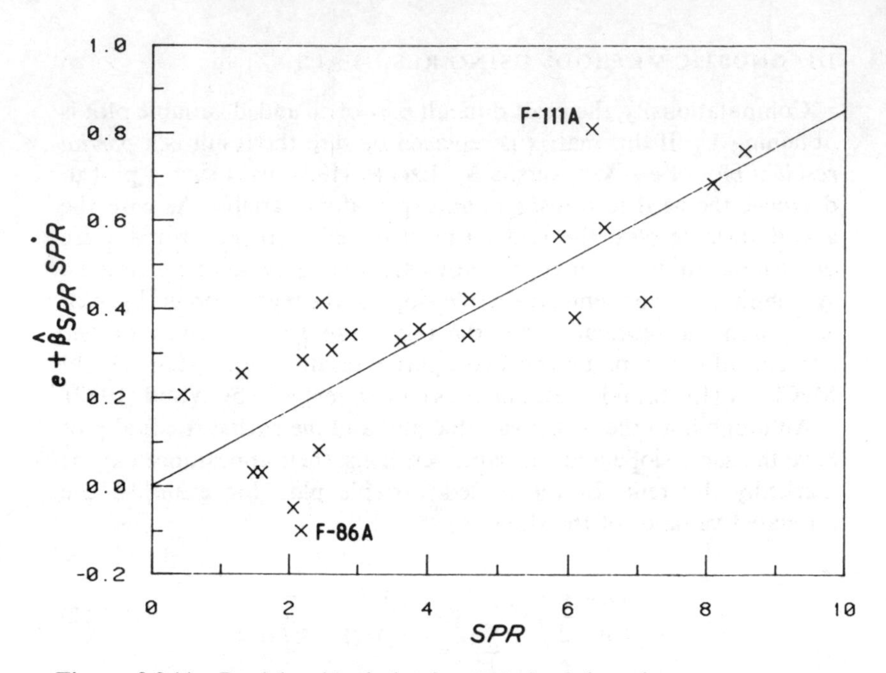

Figure 2.3.11 Partial residual plot for *SPR*, jet fighter data

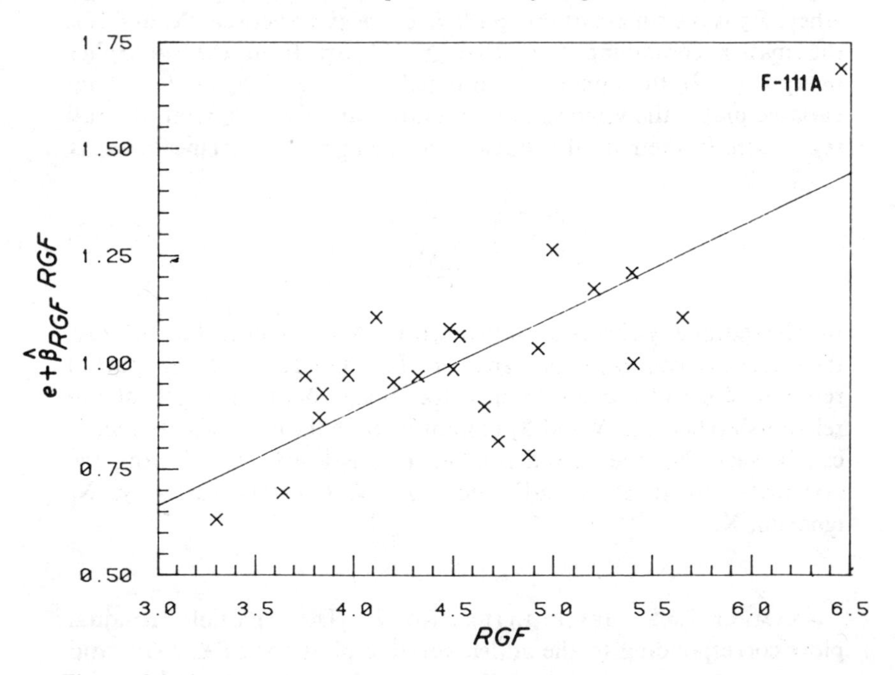

Figure 2.3.12 Partial residual plot for *RGF*, jet fighter data

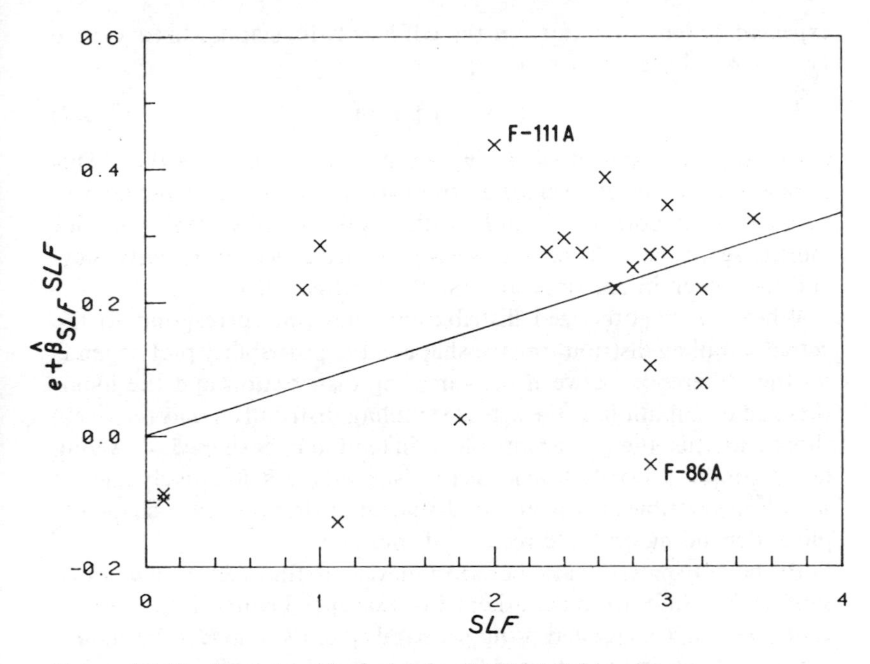

Figure 2.3.13 Partial residual plot for *SLF*, jet fighter data

SLF given previously are shown as Figs. 2.3.11–2.3.13. The two plots for *SPR* (Figs. 2.3.8 and 2.3.11) are not too different, although the overall impression of the partial residual plot is of a stronger relationship than is shown in Fig. 2.3.8, and the F-86A is no longer an extreme point. The two plots for *RGF* are very similar, and would lead to the same conclusions. The two plots for *SLF*, however, are quite different. In particular neither the F-111A nor the F-86A stand apart from the rest of the data in Fig. 2.3.13, and the general swarm of points is shifted right. □

2.3.4 PROBABILITY PLOTS

Let $y_1, y_2, \ldots, y_n$ denote n independent, univariate observations and let F denote a cdf from a location/scale invariant family with mean μ and variance σ^2. Under the hypothesis that the y_is are an identically distributed sample from F, the regression of the vector of observed order statistics $\mathbf{u}^{\mathrm{T}} = (u_1, u_2, \ldots, u_n)$, $u_n = \max(y_i)$, on the vector of expected order statistics from the cdf F_0 of the standardized variate $(y-\mu)/\sigma$ is linear (Lloyd, 1952),

$$E(\mathbf{u}) = \mu\mathbf{1} + \sigma\mathbf{a} \tag{2.3.14}$$

where the i-th element of the n-vector $\mathbf{a}$ is $a_i = E((u_i - \mu)/\sigma)$. This implies that a plot of $\mathbf{u}$ versus $\mathbf{a}$ can be used to check the appropriateness of the hypothesized cdf F, with a substantially nonlinear plot indicating an incorrect choice. Such plots are called probability plots and have been in use since at least 1934 (Bliss, 1934).

When the hypothesized distribution does not correspond to the actual sampling distribution, the shape of the probability plot depends on the 'difference' between the sampling distribution and the hypothesized distribution. If the actual sampling distribution has relatively short tails, then the probability plot will tend to be S-shaped. A long tailed sampling distribution leads to shaped plots. Relatively skewed sampling distributions result in J-shaped or inverted J-shaped plots, depending on the direction of the skew.

Probability plots can also be used as devices to find a few elements of a sample that differ from the others. For example, Daniel (1959; see also Zahn, 1975a, b) suggested using probability plots to assess the significance of effects in unreplicated factorial designs with all factors at two levels. If the absolute values of the usual contrasts are plotted against half-normal order statistics then the large or significant contrasts will be plotted near the upper right corner of the plot, while the smaller or nonsignificant contrasts will more or less fall on a line; see Zahn (1975a) for details. The identical method can be used to detect outliers in general: outlying elements of a sample will tend to fall toward the extremes of the plot, while most of the points will fall on a line that does not point toward the apparent outliers.

In general judging the adequacy of a probability plot requires experience. For the normal distribution, Daniel and Wood (1980) and Daniel (1976) provide many pages of training plots that may help the analyst gain the necessary experience.

The construction of probability plots may be hindered by the unavailability of exact values for expected order statistics. However, adequate approximations can usually be constructed from F^{-1}. For the standard normal cdf Φ, for example, $a_i = \Phi^{-1}((i-3/8)/(n+1/4))$ provides an excellent approximation for $n \geq 5$ (Blom, 1958). Approximations for the half-normal distribution are given by Sparks (1970), and for the gamma distribution by Roy, Gnanadesikan and Srivastava (1971, pp. 286–98). Wilk and Gnanadesikan (1968) coined the term Q–Q plot (for quantile versus quantile) for these probability plots to reflect the practical manner in which they are constructed.

In some problems, it may be useful to have a summary statistic for a probability plot. An intuitively reasonable summary for symmetric families is the squared correlation between the plotted quantities,

$$W' = \frac{(\mathbf{u}^{\mathrm{T}}\mathbf{a})^2}{(\mathbf{a}^{\mathrm{T}}\mathbf{a})\sum(u_i-\bar{u})^2} \tag{2.3.15}$$

Small values of W' would give evidence against the assumed distribution.

The statistic W' was suggested as a test for normality by Shapiro and Francia (1972, see also Weisberg and Bingham, 1975); a similar statistic was suggested by Filliben (1975). W' was originally suggested as an approximation to the Shapiro and Wilk (1965) W statistic,

$$W = \frac{(\mathbf{a}^{\mathrm{T}}\mathbf{\Omega}^{-1}\mathbf{u})^2/(\mathbf{a}^{\mathrm{T}}\mathbf{\Omega}^{-2}\mathbf{a})}{\sum(u_i-\bar{u})^2} \tag{2.3.16}$$

where Ω is the variance–covariance matrix of the order statistics from the standard distribution ($\mathbf{\Omega}$ is given for $n \leq 50$ for the normal by Tietjen, Kahaner and Beckman, 1977). Weisberg (1974) pointed out that for the normal distribution W and W' are essentially identical. Both statistics have reasonable power against a wide class of alternatives. Critical values for the normal distribution are given by Shapiro and Wilk (1965) and Shapiro and Francia (1972), and have been widely reprinted elsewhere. Prescott (1976) has studied the behavior of W in the presence of one or two outliers. Shapiro, Wilk and Chen (1968) and Pearson, D'Agostino and Bowman (1977) compare various tests for normality.

In regression, the probability plot and W (or W') are usually applied to $\mathbf{e}$ or to the r_i since $\boldsymbol{\varepsilon}$ is unobservable. For example, normal plots of residuals or Studentized residuals are a standard feature of most regression packages. Unfortunately, normal plots and the corresponding tests may not be effective when applied to residuals. Recall from Equation (2.1.4) that $e_i = \varepsilon_i - \Sigma_j v_{ij}\varepsilon_j$. As long as the ε_js have finite variance, $\Sigma_j v_{ij}\varepsilon_j$ will tend toward normality and in some cases may dominate ε_i. Thus, the e_i or the r_i may exhibit a *supernormality* property (Gentleman and Wilk, 1975a; Gnanadesikan, 1977) and behave more like a normal sample than would the ε_i. In small samples, the usefulness of normal plots is unclear, and depends on n, p', and on $\mathbf{V}$ (Weisberg, 1980b). In larger samples, however, if $\max(v_{ii}) \to 0$ (as required for asymptotic normality) W and W' applied to residuals is the same as applying them to the unobserved errors (Pierce and Kopecky, 1979). In such cases, a normal plot of residuals may be interpreted in a way equivalent to a normal plot of a univariate sample.

Atkinson (1981), following Gentleman and Wilk (1975a, b), suggests a method of interpreting probability plots of residuals, even in small samples. The technique presented here is a straightforward extension of Atkinson's basic idea.

For a problem with $(\mathbf{I}-\mathbf{V})$ fixed, m pseudo-random n-vectors $\boldsymbol{\varepsilon}_1, \ldots, \boldsymbol{\varepsilon}_m$ are generated from F (usually, F will be taken as standard normal). The pseudo residuals $\mathbf{e}_k = (\mathbf{I}-\mathbf{V})\boldsymbol{\varepsilon}_k$, $k = 1, 2, \ldots, m$, are then computed. Let the ordered elements of $\mathbf{e}_k$ be denoted by $e_{(i)k}$ and, for each i, let $e_{(i)}^{\eta}$, $0 < \eta < 1$, denote the $\eta \times 100$ percentile of the empirical distribution of $\{e_{(i)k}, k = 1, 2, \ldots, m\}$. Simultaneous probability plots of the two n-vectors with elements $(e_{(i)}^{\alpha/n})$ and $(e_{(i)}^{1-\alpha/n})$ describe an envelope, roughly like a $(1-2\alpha) \times 100\%$ simultaneous confidence region. The probability plot of the data is then plotted along with the corresponding envelope. If the observed residuals fall beyond or near the boundary of the envelope, the assumption that $\boldsymbol{\varepsilon}$ is sampled from F is called into doubt. If it is desired to use the envelope as an exact test, further simulation may be necessary to determine the size. Atkinson (1981) uses a transformation of residuals in this plot, and chooses to use F = half-normal distribution, but the ideas are the same regardless of the transformation and choice of F.

EXAMPLE 2.3.6 CLOUD SEEDING NO. 4. To illustrate probability plots, we again use the cloud seeding data. Figures 2.3.14 and 2.3.15 are normal probability plots for the variables $S - Ne$ and P, respectively. These plots are included for illustration only, since the sampling plan outlined in Chapter 1 would not lead us to expect the predictors to behave as a normally distributed sample. However, the plot for $S - Ne$ is approximately linear, as one would obtain from a normal sample. The value of $W' = 0.972$ is well above the 10% point of its distribution given normality. The plot for P is clearly not straight, indicating

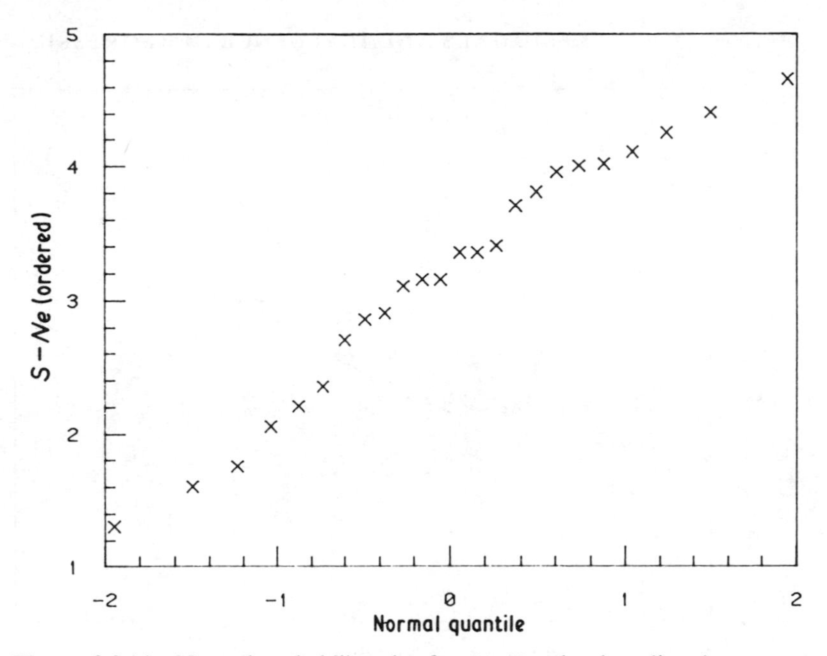

Figure 2.3.14 Normal probability plot for $S-Ne$, cloud seeding data

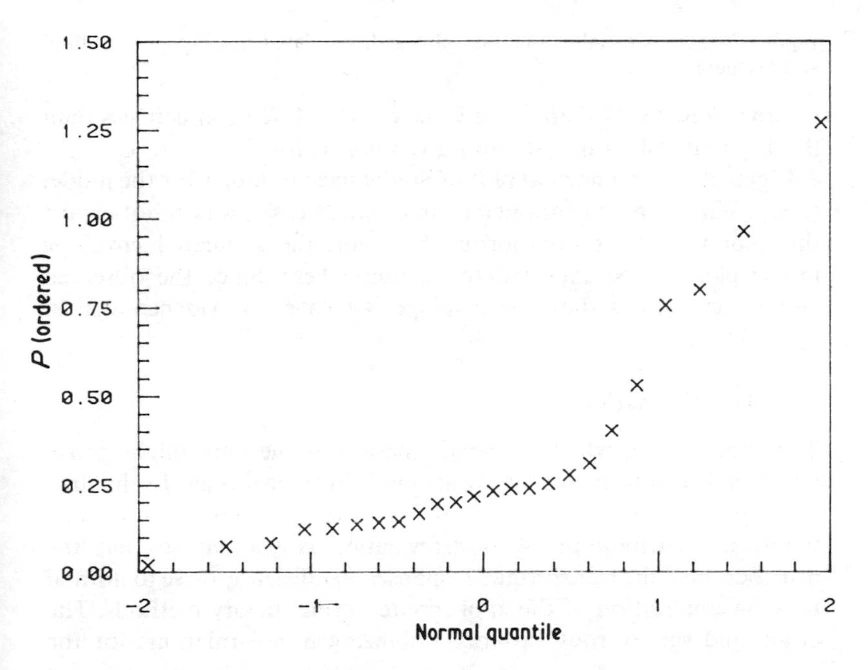

Figure 2.3.15 Normal probability plot for P, cloud seeding data

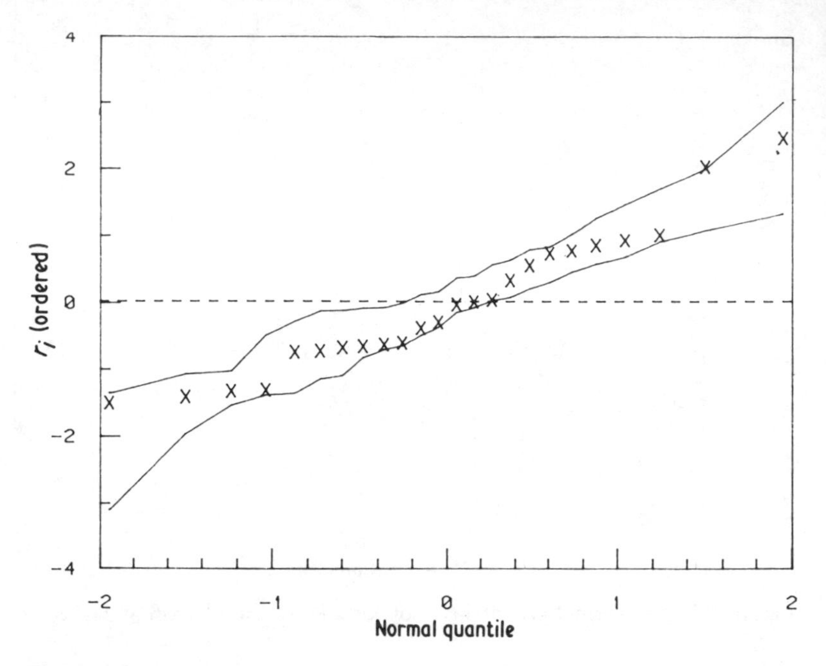

Figure 2.3.16 Normal probability plot with simulated envelope for r_i, cloud seeding data

positive skew by its shape. The value of $W' = 0.757$ is much less than the 1% point of its distribution given normality.

Figure 2.3.16 is a normal plot of Studentized residuals for the model (1.1.3). With so many parameters and only 24 cases, we cannot expect this plot to exhibit non-normal behavior; the simulated envelope in the plot can be expected to be useful here. Since the observed plot is generally within the envelope, we have no evidence against normality.□

2.4 Transformations

The situations in which a transformation of the data might prove worthwhile can be conveniently arranged in three classes. In the first, the responses y_i are independent and come from a known non-normal family of distributions. A transformation is selected so that the distribution of the transformed responses is sufficiently close to normal to allow application of the appropriate normal theory methods. The arcsin and square root variance stabilizing transformations for the binomial and Poisson distributions are typical examples. The important point here is that the selection of the transformation is based on the known distribution of the response variables.

In the second class, the expected responses Ey_i are related to the explanatory variables $x_1, \ldots, x_p$ by a known nonlinear function of the parameters. A transformation is selected to linearize the response function. If the distribution of the errors is sufficiently well behaved, the transformed data can be analyzed using standard linear least squares. For example, if theory suggests the relationship $Ey = \beta_0 \exp(\beta_1 x)$, then it is reasonable to expect an approximately linear relationship between $\log(y)$ and (x), $\log(y) = \log(\beta_0) + \beta_1 x$. It will, of course, be important to perform various diagnostic checks on the transformed model since there is no guarantee that the standard least squares methods will be appropriate. If, for example, the errors ε in the original model have mean zero, constant variance, and are such that $y = \beta_0 \exp(\beta_1 x)(1+\varepsilon)$ then the centered errors in the transformed model will also have mean zero and constant variance, $y_i = [\log(\beta_0) + E\log(1+\varepsilon)] + \beta_1 x + [\log(1+\varepsilon) - E\log(1+\varepsilon)]$. On the other hand, if the errors in the original model are additive, $y = \beta_0 \exp(\beta_1 x) + \varepsilon$, then the error variances in the transformed model will depend on Ey.

In the final class, neither the distribution of the errors nor the functional form of the relationship between Ey and the explanatory variables is known precisely. This situation is perhaps the most difficult to handle since a specific single rationale for choosing a transformation is lacking. Generally, we would like a transformation to result in a model with constant error variance, approximately normal errors, and an easily interpreted and scientifically meaningful structure. One method of proceeding in this situation is to specify a family $y^{(\lambda)}$ of transformations indexed by a possibly vector-valued parameter λ and then use the data to select a specific transformation that may result in a model that has all the desirable properties.

Methods of selecting a transformation in situations falling in the first or second class are well known and good discussions can be found in many standard references. For example, Scheffé (1959, Section 10.7) discusses a general method of choosing variance stabilizing transformations. Daniel and Wood (1980, Chapter 3) give plots of a variety of nonlinear forms that can be transformed to linear forms. In this section, we concentrate on the third class of situations. We first present a number of families of transformations and sketch a method of analysis based on likelihood considerations. Several related graphical and approximate methods are discussed later.

2.4.1 FAMILIES OF TRANSFORMATIONS

For a positive response variable $y > 0$, Box and Cox (1964) studied a slight generalization of the family of monotonic *power transformations* used earlier by Tukey (1957),

$$y^{(\lambda)} = \begin{cases} \dfrac{y^{\lambda} - 1}{\lambda}, & \lambda \neq 0 \\ \log(y), & \lambda = 0 \end{cases} \tag{2.4.1}$$

This family contains the usual log, square root, and inverse transformations as special cases and is scaled to be continuous at $\lambda = 0$. $y^{(\lambda)}$ is convex in y for $\lambda \geq 1$ and concave in y for $\lambda \leq 1$, and is increasing in both y and λ. It will be useful for inducing approximate symmetry when the response is skewed. One effect of the log transformation, for example, is to lighten one tail of the distribution. Generally, (2.4.1) will be sensible in situations where the origin occurs naturally and the response is skewed and positive. Since most robust methods of estimation are dependent on symmetry, (2.4.1) might be used prior to the application of such methods.

If the origin is artificial or negative responses occur, added flexibility is provided by the *extended power family*,

$$y^{(\lambda)} = \begin{cases} \dfrac{(y + \lambda_2)^{\lambda_1} - 1}{\lambda_1}, & \lambda_1 \neq 0 \\ \log(y + \lambda_2), & \lambda_1 = 0 \end{cases} \tag{2.4.2}$$

Here, $y + \lambda_2 > 0$. In some situations, it may be sufficient to substitute a convenient value for λ_2 and then proceed using (2.4.1) in combination with the shifted response $y + \lambda_2$.

John and Draper (1980) propose the family of *modulus transformations*

$$y^{(\lambda)} = \begin{cases} \operatorname{sign}(y)\left[\dfrac{(|y| + 1)^{\lambda} - 1}{\lambda}\right], & \lambda \neq 0 \\ \operatorname{sign}(y)[\log(|y| + 1)], & \lambda = 0 \end{cases} \tag{2.4.3}$$

for obtaining approximate normality from symmetric long-tailed distributions. This family is monotonic, continuous at $\lambda = 0$, and applicable in the presence of negative responses. When the responses are all positive the modulus family reduces to a special case of the extended power family (2.4.2). Basically, (2.4.3) applies the same power transformation to both tails of a distribution symmetric about zero. If desirable, an arbitrary point of symmetry can be included by adding a parameter λ_2 as in (2.4.2). If $\lambda < 0$, then $y^{(\lambda)}$ is restricted to the interval $[\lambda^{-1}, -\lambda^{-1}]$.

The family of power transformations can be applied in any problem with positive responses. As mentioned before, however, this family will be most useful for removing skewness and, thus, may not work well when the response is bounded above as well as below. For responses constrained to the interval $[0, b]$ some improvement might be realized by using the family of *folded-power transformations* (Mosteller and Tukey, 1977, p. 92; Atkinson, 1982)

$$y^{(\lambda)} = \begin{cases} \dfrac{y^{\lambda} - (b-y)^{\lambda}}{\lambda}, & \lambda \neq 0 \\ \log[y/(b-y)], & \lambda = 0 \end{cases} \tag{2.4.4}$$

which contains the usual logit transformation ($\lambda = 0$) as a special case. If the responses are concentrated near 0 or b, this family will behave like the power family.

2.4.2 SELECTING A TRANSFORMATION

In their original paper, Box and Cox (1964) discuss both likelihood and Bayesian methods for selecting a particular transformation from the chosen family. Following this account, the development of the specific methods for any of the transformations families discussed above is straightforward. Here we consider only likelihood based methods.

It is assumed that for each λ, $y^{(\lambda)}$ is a monotonic function of y and that for some unknown λ the vector of transformed responses $\mathbf{Y}^{(\lambda)} = (y_i^{(\lambda)})$ can be written as

$$\mathbf{Y}^{(\lambda)} = \mathbf{X}\boldsymbol{\beta} + \boldsymbol{\varepsilon} \tag{2.4.5}$$

where the quantitites on the right are consistent with previous notation and, in addition, the elements of $\boldsymbol{\varepsilon}$ are independent and (approximately) normally distributed with mean zero and constant variance σ^2. The probability density of the untransformed observations is

$$J\,(2\pi\sigma^2)^{-n/2}\exp\left\{-\frac{1}{2\sigma^2}(\mathbf{Y}^{(\lambda)}-\mathbf{X}\boldsymbol{\beta})^{\mathrm{T}}(\mathbf{Y}^{(\lambda)}-\mathbf{X}\boldsymbol{\beta})\right\} \qquad (2.4.6)$$

where J is the Jacobian of the transformation

$$J=\prod_{i=1}^{n}\left|\frac{\mathrm{d}y_i^{(\lambda)}}{\mathrm{d}y_i}\right|$$

For fixed λ, (2.4.6) is the standard normal likelihood and thus the log likelihood maximized over $\boldsymbol{\beta}$ and σ^2, apart from an unimportant constant, is

$$L_{\max}(\lambda)=-\tfrac{1}{2}n\log[RSS(\lambda;\mathbf{Y})/n]+\log(J) \qquad (2.4.7)$$

where RSS denotes the residual sum of squares from a fit using the transformed responses,

$$RSS(\lambda;\mathbf{Y})=\mathbf{Y}^{(\lambda)\mathrm{T}}[\mathbf{I}-\mathbf{V}]\mathbf{Y}^{(\lambda)}.$$

Equivalently, the maximized log likelihood can be written as

$$L_{\max}(\lambda)=-\tfrac{1}{2}n\log[RSS(\lambda,\mathbf{Z})/n] \qquad (2.4.8)$$

where the n-vector $\mathbf{Z}$ has elements

$$z_i^{(\lambda)}=y_i^{(\lambda)}/J^{1/n}$$

In this latter form, the correction for change of scale is apparent. If more than one model is to be considered, the analyses are conveniently studied using the normalized transformation $z_i^{(\lambda)}$, so the residual sum of squares for each λ are on the same scale and can thus be compared. The normalized transformation should also provide better computational accuracy, particularly for large λ.

For an arbitrary collection of n positive scalars $a_1, a_2, \ldots, a_n$, let $g(a)$ denote the geometric mean function

$$g(a)=\left(\prod_{i=1}^{n}a_i\right)^{1/n}$$

The normalized transformation for the extended power family is then

$$z_i^{(\lambda)}=\begin{cases}\dfrac{(y_i+\lambda_2)^{\lambda_1}-1}{\lambda_1 g^{\lambda_1-1}(y+\lambda_2)}, & \lambda_1\neq 0\\ g(y+\lambda_2)\log(y_i+\lambda_2), & \lambda_1=0\end{cases} \qquad (2.4.9)$$

which gives the corresponding transformation for the power family by setting $\lambda_2 = 0$. The normalized transformations for the modulus and folded-power families are

$$z_i^{(\lambda)} = \begin{cases} \operatorname{sign}(y_i)\left[\dfrac{(|y_i|+1)^\lambda - 1}{\lambda g^{\lambda-1}(|y|+1)}\right], & \lambda \neq 0 \\ \operatorname{sign}(y_i) g(|y|+1)\log(|y_i|+1), & \lambda = 0 \end{cases} \tag{2.4.10}$$

and

$$z_i^{(\lambda)} = \begin{cases} \dfrac{y_i^\lambda - (b-y_i)^\lambda}{\lambda g(y^{\lambda-1} + (b-y)^{\lambda-1})}, & \lambda \neq 0 \\ b^{-1} g[y(b-y)]\log[y_i/(b-y_i)], & \lambda = 0 \end{cases} \tag{2.4.11}$$

respectively.

The maximum likelihood estimate of λ can be obtained by maximizing (2.4.7) or (2.4.8), or by finding the solution to $\mathrm{d}L_{\max}(\lambda)/\mathrm{d}\lambda = 0$. Alternatively, when λ is a scalar, Box and Cox suggest reading $\hat{\lambda}$ from a plot of $L_{\max}(\lambda)$ against λ for a few selected values of λ. Unless special software is available, such plots will require one regression for each value of λ chosen. The accuracy of the estimate of λ obtained in this way will usually be acceptable since in practice it is desirable to round $\hat{\lambda}$ to a convenient or theoretically justifiable value.

An approximate $(1-\alpha) \times 100\%$ confidence region for λ is given by the set of all λ^* satisfying

$$2[L_{\max}(\hat{\lambda}) - L_{\max}(\lambda^*)] \leq \chi^2(\alpha, \nu) \tag{2.4.12}$$

where $\chi^2(\alpha, \nu)$ is the $(1-\alpha) \times 100$ percentile of a chi-squared distribution with degrees of freedom ν equal to the number of components in λ. When λ is a scalar such confidence regions are easily constructed from the plot of $L_{\max}(\lambda)$ against λ.

Invariance

Before turning to examples, a few general comments may remove some of the concerns about this procedure that are likely to arise in practice. We first comment on invariance under rescaling the responses and then briefly discuss normality, the choice of a model, and methods of inference.

From (2.4.11) it is easily seen that for the family of folded-power transformations the estimate $\hat{\lambda}$ will be unchanged under rescaling of the responses, $y_i \overset{c}{\to} cy_i, c > 0$. Thus, without loss of generality, the responses may be scaled so that $b = 1$. If $\mathbf{X}$ contains a column of 1s, the extended power family will be invariant under rescaling in the sense that $(\hat{\lambda}_1, \hat{\lambda}_2) \xrightarrow{c} (\hat{\lambda}_1, c\hat{\lambda}_2)$. If $\mathbf{X}$ does not contain a column of 1s, this family is not invariant under rescaling when $\lambda_1 \neq 0$. Schlesselman (1971) discusses this problem and suggests modifications of the power family that yield scale invariant estimates when regression is through the origin. The transformations obtained from the modulus family are not invariant under rescaling, a characteristic that is likely to be annoying in practice. However, Shih (1981) suggests a generalized two-parameter modulus transformation that is scale invariant.

Normality

The Box–Cox procedure for choosing a transformation is based on the assumption that $\mathbf{Y}^{(\lambda)}$ is normally distributed. It is clear, however, that in general this assumption cannot be true, although it may hold in certain special cases ($\lambda = 0$ in the power family). Hernandez and Johnson (1980) investigate the consequences of this inconsistency for the power family. Their results suggest that asympotically $\hat{\lambda}$ and the least squares estimates of $\boldsymbol{\beta}$ and σ^2 based on the transformed data are chosen to make the distribution of the transformed data as close as possible to a normal distribution, as measured by Kullback–Leibler divergence. They emphasize that appropriate diagnostic checks should always be applied to the transformed data since an adequate approximation to normality is not guaranteed by this procedure.

Choice of model and scaling the predictors

The role of $\mathbf{X}$ in selecting a transformation for $\mathbf{Y}$ can be crucial since the likelihood procedure tries to achieve $E\mathbf{Y}^{(\lambda)} = \mathbf{X}\boldsymbol{\beta}$ in addition to normality and constant variance. The indicated transformation for one $\mathbf{X}$-structure may not be the same as that for another and the selection of $\mathbf{X}$ may well be the most important step. Generally, $\mathbf{X}$ should be selected so that the resulting model can be interpreted without great difficulty, is flexible enough to describe important possibilities, and is scientifically meaningful.

Box and Cox suggest the following technique as an aid to understanding the importance of selected columns of $\mathbf{X}$ in determining a transformation. Partition $\boldsymbol{\beta}^{\mathrm{T}} = (\boldsymbol{\beta}_1^{\mathrm{T}}, \boldsymbol{\beta}_2^{\mathrm{T}})$ where $\boldsymbol{\beta}_2$ is $q \times 1$ and for fixed λ let $L_{\max}(\lambda \,|\, \boldsymbol{\beta}_2 = 0)$ denote the maximized log likelihood for the model with $\boldsymbol{\beta}_2 = 0$. Then

$$L_{\max}(\lambda \,|\, \boldsymbol{\beta}_2 = 0) = L_{\max}(\lambda) - \tfrac{1}{2} n \log\left[1 + \frac{q}{n - p'} F(\lambda; \mathbf{Z})\right] \qquad (2.4.13)$$

where $F(\lambda; \mathbf{Z})$ is the usual F-ratio for H: $\boldsymbol{\beta}_2 = 0$ in terms of the normalized response $z_i^{(\lambda)}$. If both $L_{max}(\lambda \mid \boldsymbol{\beta}_2 = 0)$ and $L_{max}(\lambda)$ are plotted against λ on the same graph then the difference between the heights of the two curves at a selected value of λ is a monotonic function of $F(\lambda; \mathbf{Z})$. Large differences indicate that inclusion of $\boldsymbol{\beta}_2$ may yield an improved fit. If the maxima of the likelihoods occur at substantially different values of λ then the transformation under $\boldsymbol{\beta}_2 = 0$ may be attempting to compensate for inadequacies in the reduced model.

Inference

Once an appropriate transformation has been selected, the analyst must choose between conditional and unconditional methods of inference for the transformed data. In the conditional approach the fact that the data are used to select a transformation is ignored and the analysis proceeds as if the appropriate scale were known *a priori*. In contrast, unconditional methods include λ as an unknown parameter and allow for the appropriate modification of confidence statements.

Historically, conditional methods of inference seem to dominate the literature on transformations. Bickel and Doksum (1981) provide a comprehensive account of the unconditional approach and demonstrate that the unconditional variances of parameter estimates can be much larger than those from the conditional approach. If, for example, the power family is used to select a transformation of a simple random sample and $\lambda = 0$, then (Hinkley, 1975)

$$\text{var}\,(\bar{y}^{(\hat{\lambda})}) = \frac{\sigma^2}{n} + \frac{1}{6n}\left[1 + \frac{(Ey^{(\lambda)})^2}{\sigma^2}\right]^2$$

The second term on the right is the amount that the variance is inflated due to estimation of λ. Hinkley and Runger (1980) provide a number of compelling arguments in favor of the conditional approach. They comment that unconditional confidence statements must logically take a rather useless form. For example, an unconditional confidence statement based on the average of a transformed simple random sample might read: 'On some unknown scale λ, which is probably around $\hat{\lambda}$, a 95% confidence interval for $Ey^{(\lambda)}$ is $\bar{y}^{(\hat{\lambda})} \pm a$.' Such statements relate to unknown parameters in unknown scales and cannot be very helpful.

Carroll and Ruppert (1981) investigate the variance inflation due to estimating λ when prediction of future observations is the primary goal and the data are back-transformed so that the predictions are always made in the original scale. They conclude that, while there is some inflation in this problem, it is generally not severe or important. For further discussion, see Box and Cox (1982).

In this monograph we adopt the conditional approach.

EXAMPLE 2.4.1. TREE DATA NO. 1. To provide a first illustration of the use of the Box–Cox procedure, we use the power family in combination with the tree data from the *Minitab Student Handbook* (Ryan, Joiner and Ryan, 1976, p. 278). The data, given in Table 2.4.1, consist of measurements on the volume *Vol*, height *H*, and diameter *D*

Table 2.4.1 *Tree data. Source: Ryan et al.* (1976)

D = *Diameter*	H = *Height*	*Vol* = *Volume*
8.3	70	10.3
8.6	65	10.3
8.8	63	10.2
10.5	72	16.4
10.7	81	18.8
10.8	83	19.7
11.0	66	15.6
11.0	75	18.2
11.1	80	22.6
11.2	75	19.9
11.3	79	24.2
11.4	76	21.0
11.4	76	21.4
11.7	69	21.3
12.0	75	19.1
12.9	74	22.2
12.9	85	33.8
13.3	86	27.4
13.7	71	25.7
13.8	64	24.9
14.0	78	34.5
14.2	80	31.7
14.5	74	36.3
16.0	72	38.3
16.3	77	42.6
17.3	81	55.4
17.5	82	55.7
17.9	80	58.3
18.0	80	51.5
18.0	80	51.0
20.6	87	77.0

at 4.5 ft above ground level for a sample of 31 black cherry trees in the Allegheny National Forest, Pennsylvania. The data were collected to provide a basis for determining an easy way of estimating the volume of a tree (and eventually the amount of timber in a specified area of the forest) using its height and diameter. Since the volume of a cone or cylinder is not a linear function of diameter, a transformation of *Vol* is likely to result in a fit superior to that provided by the untransformed data.

Generally, a straightforward method of proceeding is to consider the simple additive model for the transformed response, here $(Vol)^{(\lambda)}$ on D and H. For illustration, we consider also a second model $(Vol)^{(\lambda)}$ on H and D^2, since it is not unreasonable to suppose that the area of a cross section of the tree rather than its diameter was reported. As a common reference for these two models, we include the third and final model $(Vol)^{(\lambda)}$ on H, D, and D^2 which was investigated by Ryan *et al.* (1976, p. 279). We refer to these as Models 1, 2, and 3, respectively.

With $\lambda = 1$, a preliminary inspection of the plots of the Studentized residuals r_i against the fitted values, H and D for each of the three

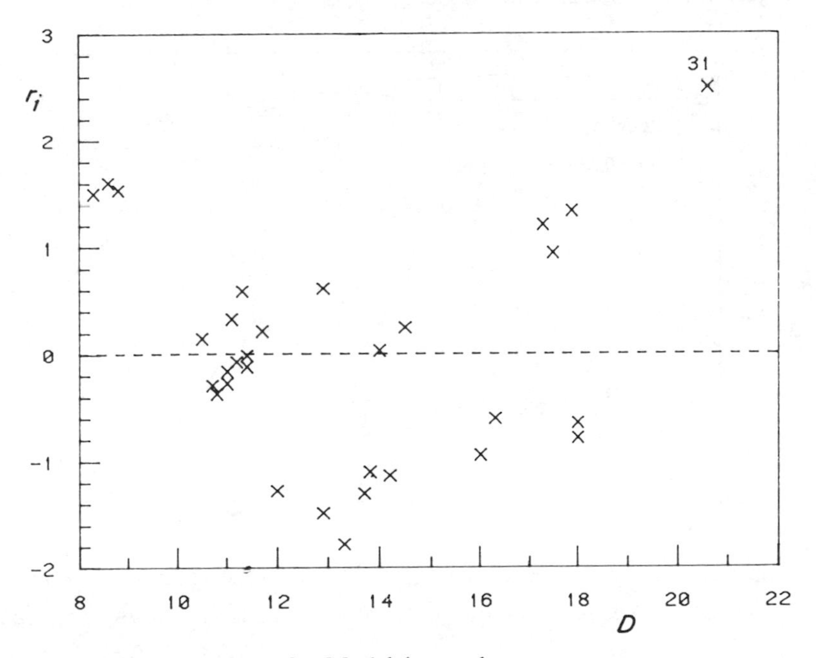

Figure 2.4.1 r_i versus D for Model 1, tree data

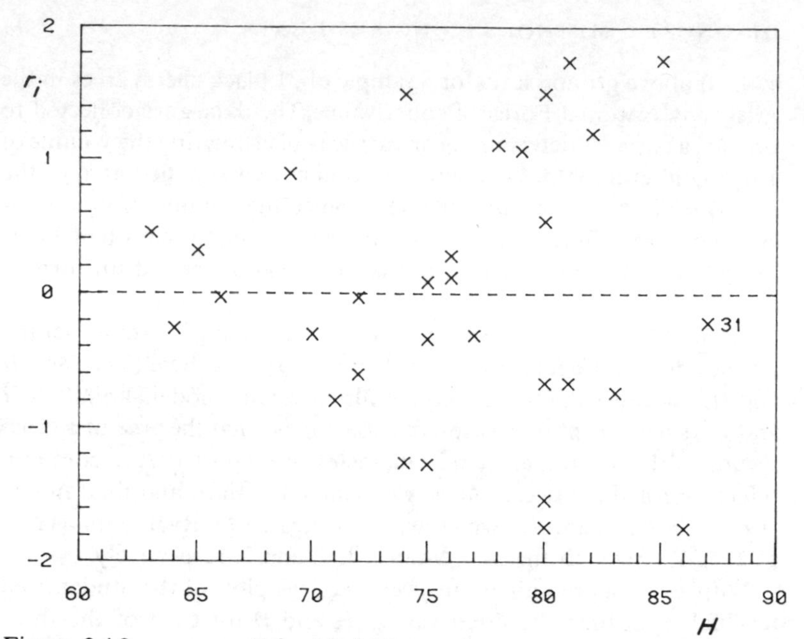

Figure 2.4.2 r_i versus H for Model 3, tree data

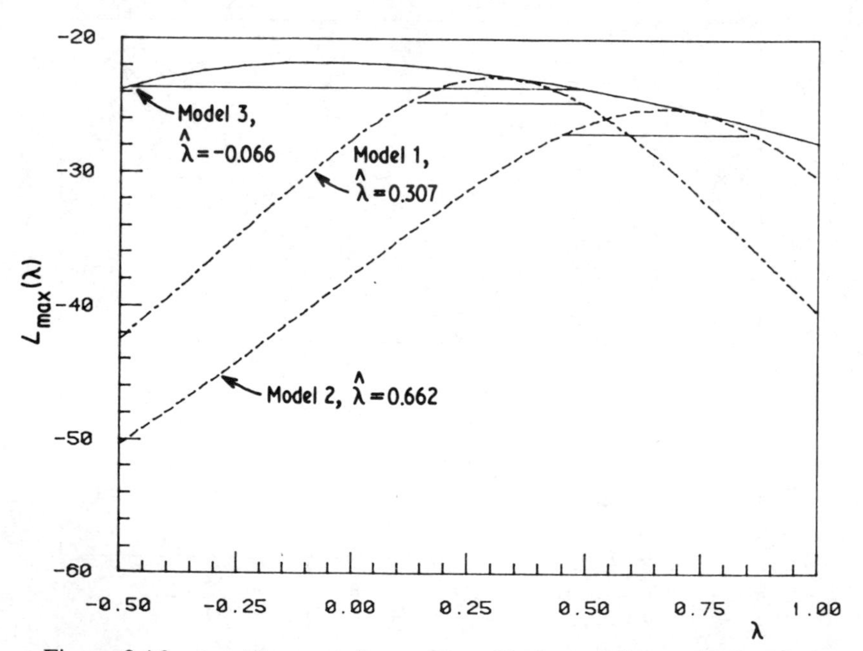

Figure 2.4.3 $L_{max}(\lambda)$ versus λ, tree data. Horizontal lines correspond to asymptotic 95% confidence intervals

models confirms that transformation is likely to be worthwhile. For Model 1 the plot of r_i versus D given in Fig. 2.4.1 shows a clear nonlinear trend, while for Model 3 the plot of r_i versus H given in Fig. 2.4.2 strongly suggests that the variability increases with H. Other plots yield similar conclusions, although some are a bit ambiguous.

Plots of $L_{\max}(\lambda)$ against λ and the approximate 95% confidence intervals from (2.4.12) for each of the three models are given in Fig. 2.4.3. The maximum likelihood estimate of λ indicates a different transformation for each model. (Maximum likelihood estimates were determined by golden section search; see Kennedy and Gentle, 1980, p. 432.) For Model 3, $\hat{\lambda} = -0.066$ and the suggested transformation is $(Vol)^{(0)} = \log(Vol)$, while the suggested transformations for Models 1 and 2 are $\hat{\lambda} = 1/3$ and $2/3$, respectively. Comparing Models 1 and 3 we see that the transformation $\lambda = 1/3$ is compatible with both likelihoods. Also, if $\lambda = 1/3$ is used to transform Vol in Model 3, then the term in D^2 is unnecessary and if $\lambda = 0$ is used to transform Model 3, then D^2 does contribute to the fit. (From (2.4.13), the F-statistics for D^2 are $F(0; Z) = 11.9$ and $F(1/3; Z) = 0.03$.) Based on this analysis, there is little reason to prefer Model 3 over the simpler $(Vol)^{(1/3)}$ on D and H. It is reassuring that the variables in the latter model are dimensionally compatible, a condition often overlooked in practice.

A comparison between Models 2 and 3 can be carried out in a manner analogous to that given above. The essential difference is that the transformation suggested by Model 2 does not seem compatible with the likelihood for Model 3. In this comparison, Model 3 may be preferable.

The residual mean squares in terms of $\mathbf{Z}^{(\lambda)}$ for Model 1 with $\lambda = 1/3$, Model 2 with $\lambda = 2/3$, and Model 3 with $\lambda = 0$ are 4.84, 5.62, and 4.68, respectively. Based on this and the previous analysis, Model 1 with $\lambda = 1/3$ is our preference from among those considered. Unfortunately, this transformation does not seem to correct all of the deficiencies noted earlier. The transformation successfully induces additivity, and the scatterplot of r_i versus H given in Fig. 2.4.4 indicates that the variance structure has been improved, although case 31 now stands out.

This example is intended to illustrate the use of the Box–Cox procedure and the kinds of results that can be expected. Certainly, other reasonable models can be formulated. For example, the relation between the volume, height, and diameter of a cylinder or cone, $Vol \propto D^2H$, suggests an additive model with all variables replaced by their logarithms. Transformations of the explanatory variables will be

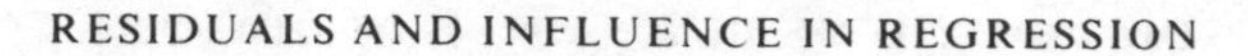

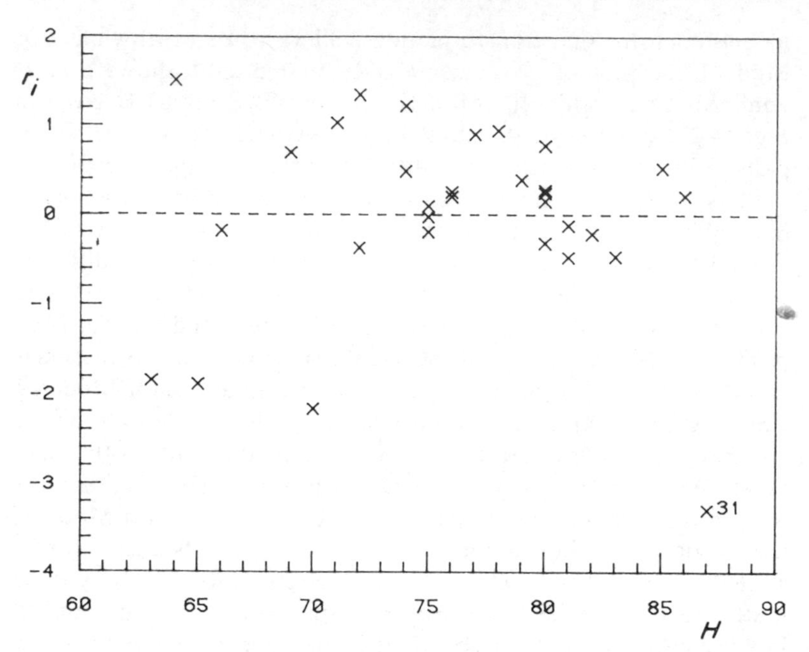

Figure 2.4.4 r_i versus H, for Model 1 with $\lambda = 1/3$, tree data

considered later in this section. For further discussion of transformations in this data set, see Atkinson (1982).□

2.4.3 DIAGNOSTIC METHODS

Andrews (1971a) demonstrates that the likelihood method for choosing a transformation is sensitive to outlying responses. Since the scale is subject to question prior to the application of this methodology, diagnostic procedures applied to the untransformed data may not yield reliable conclusions. An outlier in the untransformed data, for example, may be brought into line by a transformation. Carroll (1980) proposed a robust method obtained by replacing the likelihood function by an objective function that is less sensitive to outlying responses. Although Carroll's method is superior to the likelihood method in terms of robustness properties, it is still sensitive to outliers. Diagnostic support for the likelihood method is clearly important.

In this section we discuss two additional methods for assessing the need to transform the responses. These methods, due to Atkinson (1973, 1982) and Andrews (1971a), can be based on new explanatory variables constructed from the original data and have graphical counterparts that are useful for identifying anomalies. Because of this diagnostic ability and ease of calculation, these methods should prove valuable by themselves or as support for a likelihood analysis.

Atkinson's method

Atkinson's method is based on the score statistic $t_D(\lambda_0)$ for the hypothesis $\lambda = \lambda_0$. The score statistic does not require iteration and can be obtained using standard regression routines. To see how this is done, let $\mathbf{Z}^{(\lambda_0)} = (z_i^{(\lambda_0)})$ and

$$\mathbf{G}_z^{(\lambda_0)} = \left.\frac{\partial \mathbf{Z}^{(\lambda)}}{\partial \lambda}\right|_{\lambda = \lambda_0}$$

where $\mathbf{Z}^{(\lambda)}$ may correspond to any of the single parameter transformation families discussed previously. Apart from an unimportant sign change, the score statistic is equal to the usual t-statistic for the hypothesis $\phi = 0$ in the model

$$\mathbf{Z}^{(\lambda_0)} = \mathbf{X}\boldsymbol{\beta} + \phi \mathbf{G}_z^{(\lambda_0)} + \boldsymbol{\varepsilon} \tag{2.4.14}$$

Asymptotically, the null distribution of $t_D(\lambda_0)$ is standard normal, but its distribution in small samples is intractable since both $\mathbf{Z}^{(\lambda_0)}$ and $\mathbf{G}_z^{(\lambda_0)}$ are random variables with nonstandard distributions.

In this approach to the calculation of $t_D(\lambda_0)$, $\mathbf{G}_z^{(\lambda_0)}$ is regarded as a new explanatory variable which Box (1980) terms a constructed variable. The corresponding model (2.4.14) can be viewed as an approximation obtained by expanding $\mathbf{Z}^{(\lambda)}$ in a Taylor series about λ_0 (Atkinson, 1982). In this expansion, the coefficient of the constructed variable $\phi = \lambda_0 - \lambda$ and thus the least squares estimate $\hat{\phi}$ of ϕ provides a quick estimate $\tilde{\lambda}$ of λ,

$$\tilde{\lambda} = \lambda_0 - \hat{\phi} \tag{2.4.15}$$

Estimates obtained in this way will often be good approximations to $\hat{\lambda}$, but Atkinson (1982) demonstrates by example that adequate agreement cannot be guaranteed in general. Nonetheless, in the absence of special software $\tilde{\lambda}$ may prove useful.

Another adjunct to this method is the added variable plot for the constructed variable $G_z^{(\lambda_0)}$. Ideally, this plot should show a consistent and clear linear trend, indicating that the evidence for the transformation is spread evenly throughout the data. Outliers in an added variable plot may correspond to cases which are distorting the evidence for a transformation and thus require special attention. Substantial curvature may be an indication that a modification of the transformation family would permit a closer representation. Suppose, for example, that the folded power transformation family (2.4.4) yields a plot with a strong and consistent linear trend. The added variable plot for the power family (2.4.1) will likely show strong curvature since the constructed variables for the power and folded-power families are not linearly related.

Andrews' method

Like Atkinson's method, Andrews' method is based on a test of the hypothesis $\lambda = \lambda_0$. The test statistic is constructed by expanding $\mathbf{Y}^{(\lambda)}$ about λ_0,

$$\mathbf{Y}^{(\lambda_0)} \cong \mathbf{Y}^{(\lambda)} + (\lambda_0 - \lambda)\,\mathbf{G}_y^{(\lambda_0)}$$

where

$$\mathbf{G}_y^{(\lambda_0)} = \left.\frac{\partial \mathbf{Y}^{(\lambda)}}{\partial \lambda}\right|_{\lambda = \lambda_0}$$

Since $\mathbf{Y}^{(\lambda)} = \mathbf{X}\boldsymbol{\beta} + \boldsymbol{\varepsilon}$,

$$\mathbf{Y}^{(\lambda_0)} \cong \mathbf{X}\boldsymbol{\beta} + (\lambda_0 - \lambda)\mathbf{G}_y^{(\lambda_0)} + \boldsymbol{\varepsilon} \tag{2.4.16}$$

This model is similar to the model (2.4.14) used in the construction of the score statistic. However, (2.4.14) is based on the normalized transformed responses $z_i^{(\lambda_0)}$ whereas (2.4.16) is based on $y_i^{(\lambda_0)}$. In effect, Andrews' approach ignores the Jacobian of the transformation.

The statistic for Andrews' test is equal to the t-statistic for the hypothesis $\lambda_0 - \lambda = 0$ in the model

$$\mathbf{Y}^{(\lambda_0)} = \mathbf{X}\boldsymbol{\beta} + (\lambda_0 - \lambda)\,\hat{\mathbf{G}}_y^{(\lambda_0)} + \boldsymbol{\varepsilon} \tag{2.4.17}$$

where $\hat{\mathbf{G}}_y^{(\lambda_0)}$ is equal to $\mathbf{G}_y^{(\lambda_0)}$ evaluated at the fitted values from the null model $\mathbf{Y}^{(\lambda_0)} = \mathbf{X}\boldsymbol{\beta} + \boldsymbol{\varepsilon}$. It follows immediately from the work of

Milliken and Graybill (1970) that the t-statistic has a standard t-distribution with $n-p'-1$ degrees of freedom. That is, Andrews' test is exact.

As in Atkinson's method, the least squares estimate of $\lambda_0-\lambda$ from (2.4.17) can be used to obtain a quick estimate of λ and the added variable plot for the constructed variable $\hat{\mathbf{G}}_y^{(\lambda_0)}$ should be inspected for unusual features.

Application to the power family

It is informative to compare the constructed variables for the Andrews and Atkinson procedures for the power family in combination with the hypothesis of no transformation ($\lambda_0=1$). In this situation it is easily verified that

$$\hat{\mathbf{G}}_y^{(1)}=[\hat{y}_i\log(\hat{y}_i)-\hat{y}_i+1],$$

where $\hat{y}_i$ is the i-th fitted value from $\mathbf{Y}=\mathbf{X}\boldsymbol{\beta}+\boldsymbol{\varepsilon}$, and

$$\mathbf{G}_z^{(1)}=\{y_i\log[y_i/g(y)]-y_i+\log[g(y)]+1\}$$

The associated test statistics depend on these constructed variables only through the residuals $(\mathbf{I}-\mathbf{V})\mathbf{G}^{(1)}$ from the regression of $\mathbf{G}^{(1)}$ on $\mathbf{X}$. If $\mathbf{X}$ contains a constant column, $(\mathbf{I}-\mathbf{V})\mathbf{1}=\mathbf{0}$ and the constructed variables simplify to

$$\hat{\mathbf{G}}_y^{(1)}=[\hat{y}_i\log(\hat{y}_i)] \tag{2.4.18}$$

and

$$\mathbf{G}_z^{(1)}=(y_i\log[y_i/g(y)]-y_i) \tag{2.4.19}$$

Since $(\mathbf{I}-\mathbf{V})\hat{\mathbf{Y}}=\mathbf{0}$, it is clear that the approximation of $\mathbf{G}_z^{(1)}$ obtained by substituting $\hat{y}_i$ for y_i is equivalent to $\hat{\mathbf{G}}_y^{(1)}$. Thus, in this situation $\hat{\mathbf{G}}_y^{(1)}$ may be regarded as an approximate version of $\mathbf{G}_z^{(1)}$.

Although Andrews' method yields an exact test, there is evidence that this method has some loss of power relative to Atkinson's method (Atkinson, 1973). Andrews' method also has certain robustness properties that are not shared by Atkinson's method. With replication, for example, all cases in a single cell will have the same fitted value and consequently methods based on $\hat{\mathbf{Y}}$ will be less sensitive to a single outlier than those based on $\mathbf{Y}$ (Atkinson, 1982).

Tukey's test

Tukey's well-known single degree of freedom for nonadditivity is obtained using the constructed variable $\hat{\mathbf{G}}=(\hat{y}_i^2)$ obtained under the hypothesis of no transformation ($\lambda_0 = 1$). Andrews (1971a, b) and Atkinson (1982) discuss the relationship between Tukey's test and their respective methods. A discussion of Tukey's test applied to two-way tables is given in Section 2.5.

EXAMPLE 2.4.2. TREE DATA NO. 2. The score statistics $t_D(1)$ and the corresponding quick estimates $\tilde{\lambda}$ for each of the three models used for the tree data of Example 2.4.1 are given in Table 2.4.2. For comparison, the likelihood estimates are also given. In each situation, the need for a transformation is indicated by the score statistic and the agreement between $\hat{\lambda}$ and $\tilde{\lambda}$ seems adequate.

Table 2.4.2 *Transformation statistics, tree data*

	Full data			*Case 31 deleted*		
Model	$t_D(1)$	$\tilde{\lambda}$	$\hat{\lambda}$	$t_D(1)$	$\tilde{\lambda}$	$\hat{\lambda}$
1. H, D	7.41	0.394	0.307	6.84	0.252	0.234
2. H, D^2	3.25	0.686	0.662	3.55	0.537	0.550
3. H, D, D^2	3.18	0.134	−0.066	4.38	−0.169	−0.121

Figure 2.4.5 contains added variable plots for the constructed variables (2.4.19) in each model. In the plots corresponding to Models 1 and 2, case 31 stands out as a possible outlier and thus may be having an undue effect on the analysis. In the plot for Model 3, case 31 is not as noticeable. The effects of case 31 can be seen by removing it from the data and recomputing $\tilde{\lambda}$, $\hat{\lambda}$, and $t_D(1)$. These values are also given in Table 2.4.2. For each model the agreement between $\tilde{\lambda}$ and $\hat{\lambda}$ is still quite good and the score statistics indicate that transformations are still needed. Without case 31, however, the suggested transformations can change. For model 2, the suggested transformation is $\lambda = 2/3$ for the full data and $\lambda = 1/2$ for the reduced data. Either transformation may yield an adequate model since, as further analysis will show, they are compatible with the likelihood based on the full and reduced data. □

EXAMPLE 2.4.3. JET FIGHTERS NO. 3. In Example 2.3.4, rescaling of *FFD* to log (*FFD*), with *FFD* measured in months after January 1940 was done on logical grounds. We now consider the choice of scale for *FFD* more systematically.

Figure 2.4.6 gives a scatter plot of r_i versus fitted values for the regression of *FFD* on the other variables in addition to a constant. This

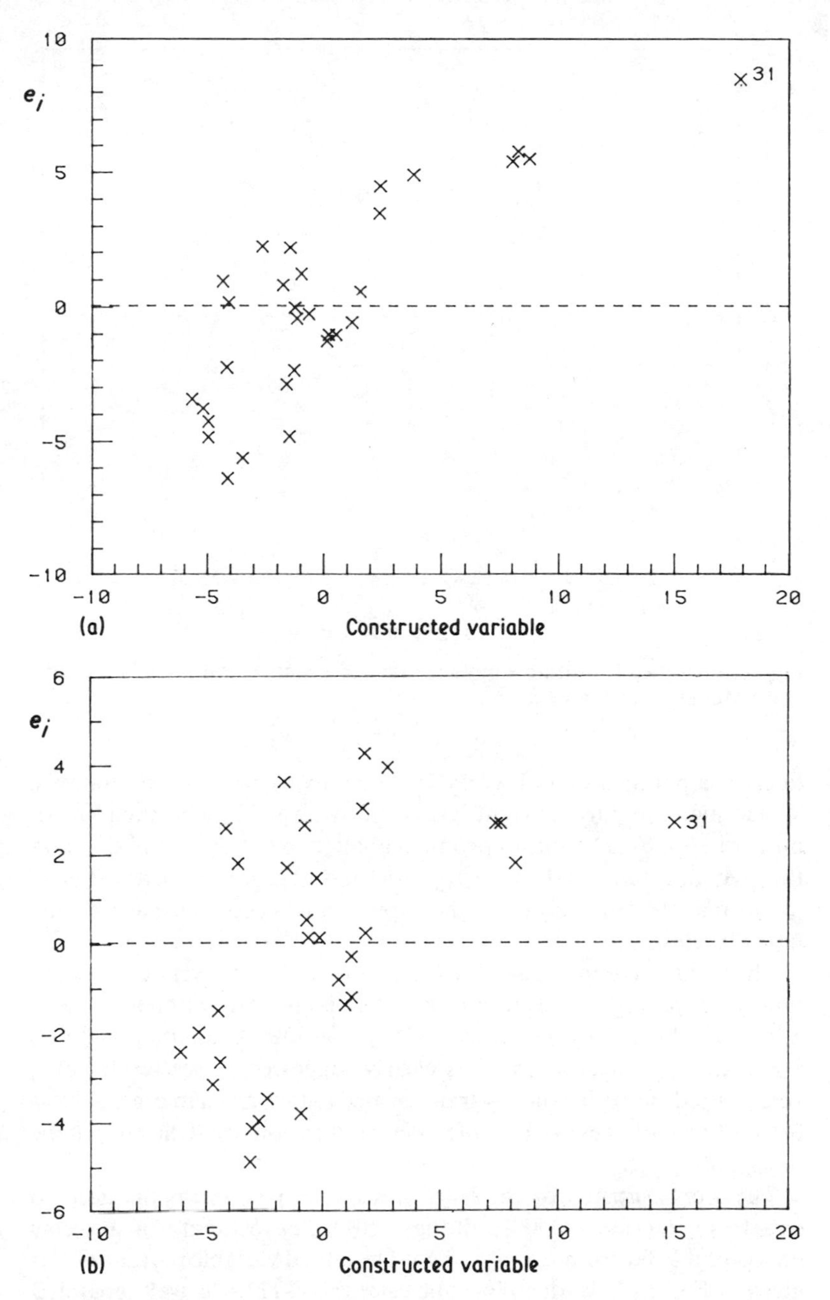

(Legend overleaf)

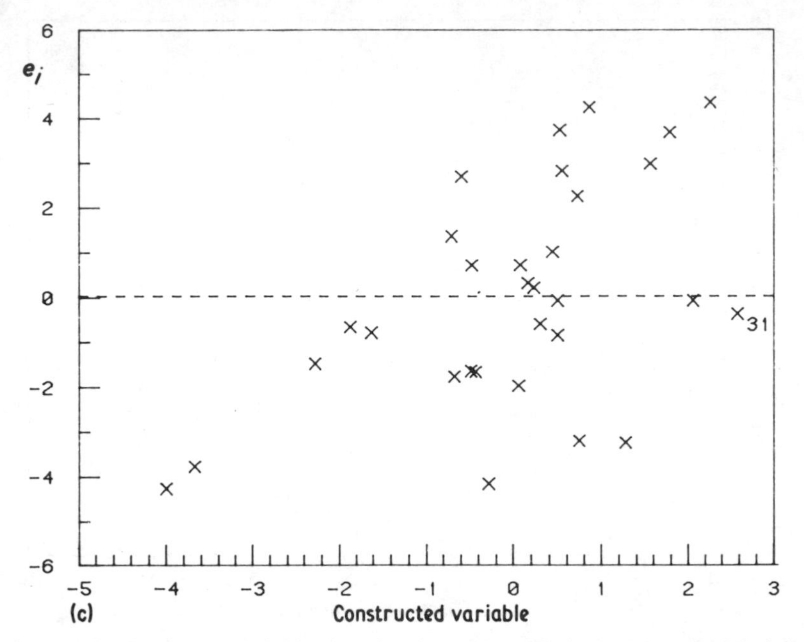

Figure 2.4.5 Added variable plots for the score statistic, tree data. (a) Model 1. (b) Model 2. (c) Model 3

figure is a paragon of ambiguity. It allows a variety of interpretations, spread increasing to the right, a slight downward bow in the plot, an outlier in the F-111A, or no problem at all, depending on the skill and the preconceptions of the investigator. Finding a pattern in a scatter of points may be a difficult task, and often renders plots such as this one nearly useless.

The solid curve in Figure 2.4.7 is a plot of $L_{\max}(\lambda)$ versus λ for the family of power transformations. The likelihood estimate of λ is $\hat{\lambda} = -0.024$ and the asymptotic 95% confidence interval excludes $\lambda = 1$; the log transformation is clearly suggested. A scatterplot of r_i versus fitted values for the log transformed data is given in Fig. 2.4.8. As before, this plot does not provide a clear indication of a deficiency in the model, although the F-111A still stands out.

The score statistic for the hypothesis $\lambda_0 = 1$ confirms the need to transform, $t_D(1) = -3.88$, although the quick estimate of λ seems unacceptably far from $\hat{\lambda}$, $\tilde{\lambda} = -0.54$. The added variable plot for $\mathbf{G}_z^{(1)}$ is given in Fig. 2.4.9. In this plot, one case, the F-111A, is well separated

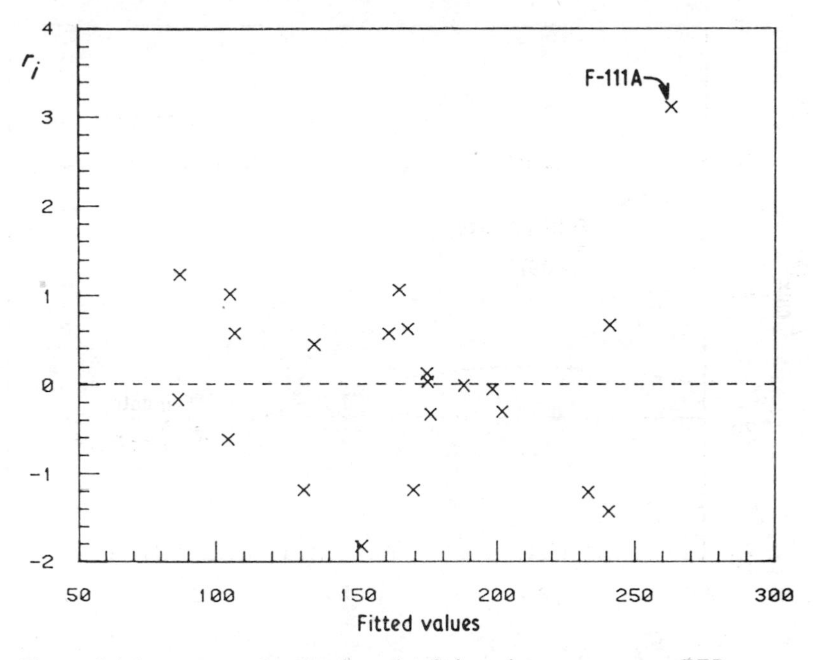

Figure 2.4.6 r_i versus fitted values, jet fighter data, response = *FFD*

from the rest and it appears that our conclusions may change if it were deleted. Figure 2.4.10 is the constructed variable plot for $\mathbf{G}_z^{(1)}$ after deletion of the F-111A. The linear feature of Fig. 2.4.9 is now completely absent, and the need to transform is less clear. Without the F-111A, $t_D(1) = -0.05$. Inferences based on the likelihood method are also very sensitive to the presence of this case. As shown by the dashed curve in Fig. 2.4.7, $\hat{\lambda}$ is close to 1 when the F-111A is deleted.

Although the evidence for the need to transform *FFD* is weak and depends heavily on the F-111A, the log transformation may be sensible for two reasons. First, as previously stated, *FFD* is a stand-in for technological level and *LFFD* is more palatable than *FFD*. Second, while the F-111A does seem to be different, it is the most recent aircraft in the data, and for that reason we may wish to modify a model to provide a better fit to it than we would for a plane developed 20 years earlier. □

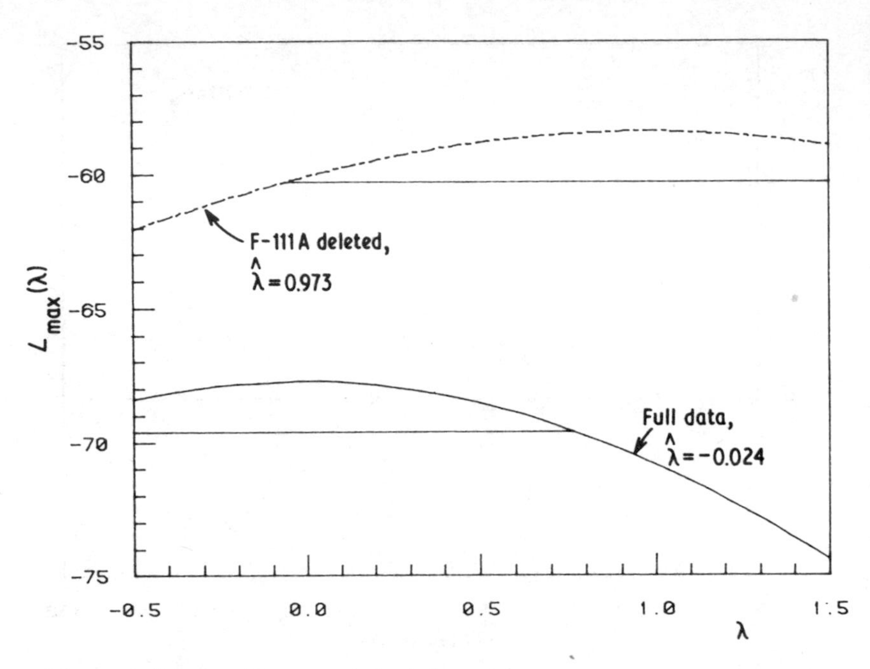

Figure 2.4.7 $L_{max}(\lambda)$ versus λ, jet fighter data

2.4.4 TRANSFORMING THE EXPLANATORY VARIABLES

Box and Tidwell (1962) suggest a general procedure to aid in the selection of transformations for the explanatory variables. A useful version of their procedure begins with the assumption that the response y_i can be written as

$$y_i = \beta_0 + \sum_{j=1}^{p} \beta_j x_{ij}^{(\lambda_j)} + \varepsilon_i, \qquad i = 1, 2, \ldots, n \tag{2.4.20}$$

where $x_{ij}^{(\lambda_j)}$ denotes the transformation of the j-th explanatory variable and the ε_is are (approximately) normal with zero mean and constant variance σ^2. Any of the single parameter transformation families discussed previously in this section may be used for $x_{ij}^{(\lambda_j)}$. (Extensions to multiple parameter transformation families are immediate.) Of course, we may also have $x_{ij}^{(\lambda_j)} = x_{ij}$, $i = 1, 2, \ldots, n$, for selected j.

As an alternative to the use of nonlinear methods, inferences about the λ_js can be based on an approximation to (2.4.20) obtained by

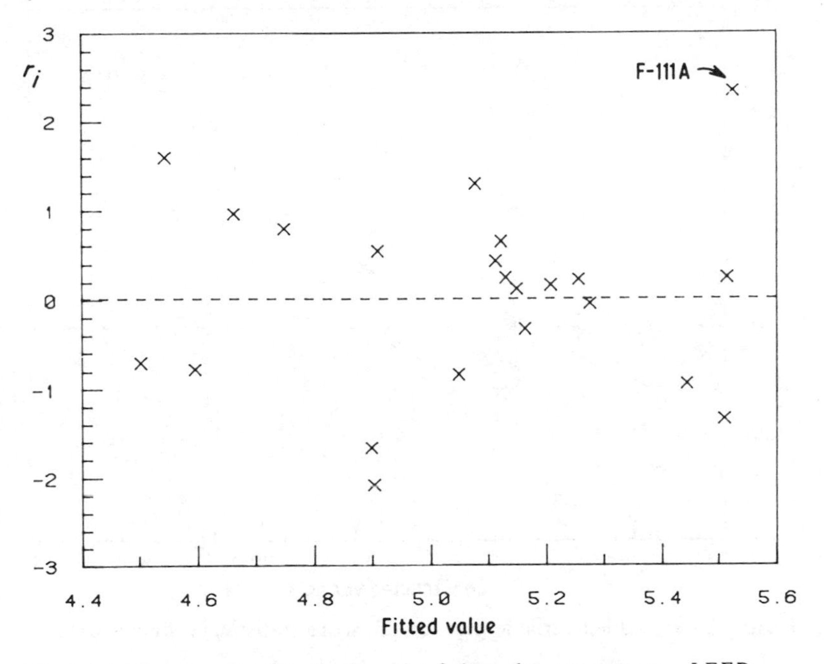

Figure 2.4.8 r_i versus fitted values, jet fighter data, response = *LFFD*

expanding $x_{ij}^{(\lambda_j)}$ about the hypothesized values $\lambda_{0j}, j = 1, 2, \ldots, p$,

$$y_i = \beta_0 + \sum_{j=1}^{p} \beta_j x_{ij}^{(\lambda_{0j})} + \sum_{j=1}^{p} \beta_j(\lambda_j - \lambda_{0j}) g_{ij}^{(\lambda_{0j})} + \varepsilon_{ij} \qquad (2.4.21)$$

where

$$g_{ij}^{(\lambda_{0j})} = \left. \frac{\partial}{\partial \lambda_j} x_{ij}^{(\lambda_j)} \right|_{\lambda_j = \lambda_{0j}}$$

In this model, the transformation parameter λ_j is related to the slope of the added variable plot for the constructed variable $g_{ij}^{(\lambda_{0j})}$. A linear trend in such a plot may be taken as an indication that $\lambda_j \neq \lambda_{0j}$; the absence of a linear trend indicates that either $\lambda_j = \lambda_{0j}$ or $\beta_j = 0$. As before, these plots can also be used to identify outlying cases that may be distorting the evidence for a transformation.

The approximate model (2.4.21) is still nonlinear in the parameters, but a quick estimate $\tilde{\lambda}_j$ of λ_j is

$$\tilde{\lambda}_j = \lambda_{0j} + \frac{\hat{\phi}_j}{\hat{\beta}_j} \qquad (2.4.22)$$

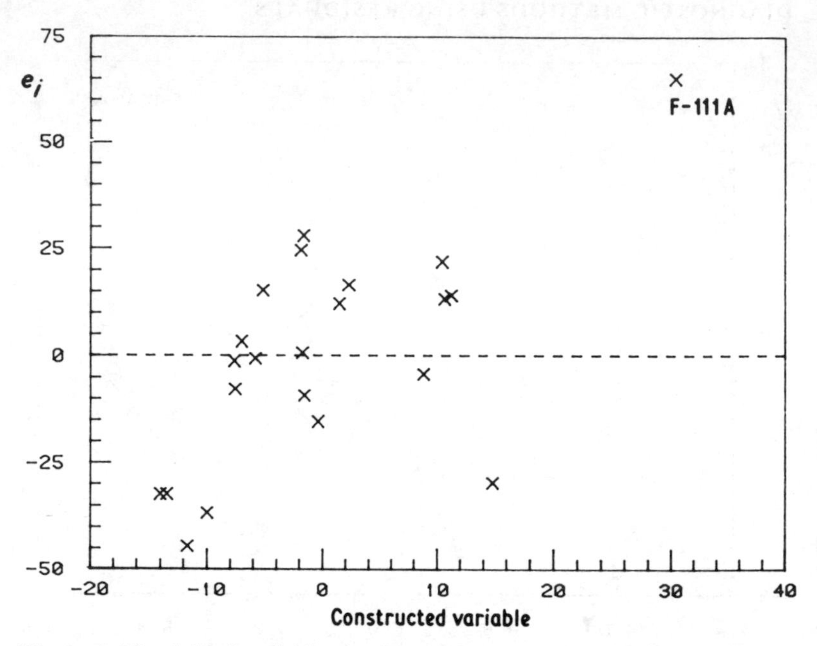

Figure 2.4.9 Added variable plot for the score statistic, jet fighter data

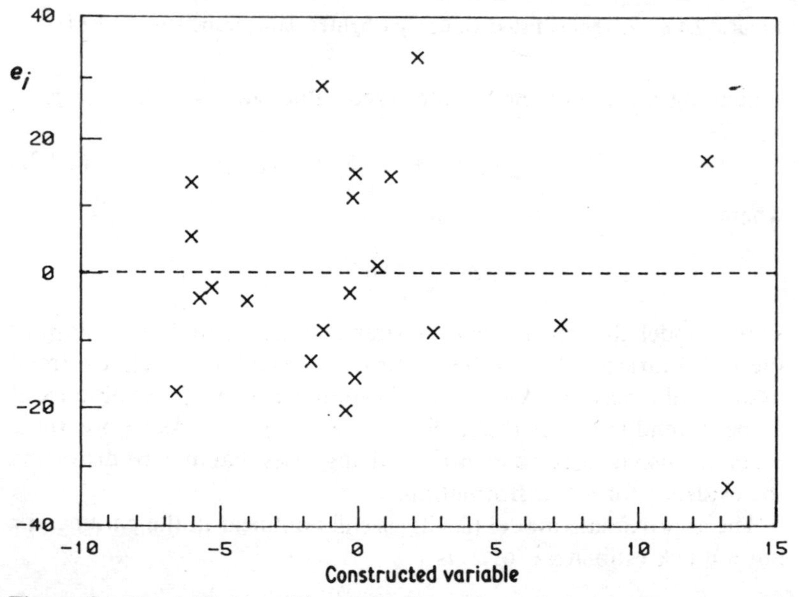

Figure 2.4.10 Added variable plot for the score statistic, jet fighter data with the F-111A deleted

where $\hat{\phi}_j$ is the least squares estimate of $\beta_j(\lambda_j - \lambda_{0j})$ from (2.4.21) and $\hat{\beta}_j$ is the least squares estimate of β_j from the null model

$$y_i = \beta_0 + \sum_j \beta_j x_{ij}^{(\lambda_{0j})} + \varepsilon_i$$

Further iteration using the $\tilde{\lambda}_j$s as starting values may be used to find estimates which further reduce the residual sum of squares for the original model (2.4.20). The quick estimate in combination with the added variable plot will usually suffice for diagnosing the need to transform. For the method to be effective, however, $\hat{\beta}_j$ must have a relatively small standard error. Substantial collinearity among the columns of **X**, for example, can result in unreasonable results, particularly for the quick estimates.

For illustration, consider the family of power transformations (2.4.1). The constructed variable is

$$g^{(\lambda_0)} = \frac{\lambda_0[x^{\lambda_0}\log(x)] - [x^{\lambda_0} - 1]}{\lambda_0^2}$$

which is equivalent to

$$g^{(\lambda_0)} = \frac{x^{\lambda_0}\log(x)}{\lambda_0}$$

since the projection of $(g(\lambda_0))$ onto the orthogonal complement of the space spanned by the remaining columns (variables) of **X** is all that matters. When the hypothesis is that of no transformation ($\lambda_0 = 1$) the constructed variable is simply $g^{(1)} = x\log(x)$ which can be easily computed with nearly any regression program.

EXAMPLE 2.4.4. TREE DATA NO. 3. We use the tree data described in Example 2.4.1 to illustrate the use of the Box–Tidwell procedure. Figure 2.4.11 gives the added variable plots for the constructed variables $D\log(D)$ and $H\log(H)$. The estimated coefficients with their estimated standard errors for the regressions of *Vol* on (D, H) and on $(D, H, D\log(D), H\log(H))$ are given in Table 2.4.3.

The plot for $D\log(D)$ in Fig. 2.4.11 shows a clear linear trend and thus indicates the need to transform diameter. From Table 2.4.3, the corresponding quick estimate of the power is $\tilde{\lambda}_D = 1 + 7.204/4.708 = 2.53$. In contrast, the plot for $H\log(H)$ shows no linear trend and thus a transformation of H is probably unnecessary. The suggested model, *Vol* on $D^{2.5}$ and H, is not far from the model *Vol* on D^2 and H that was used in Example 2.4.1. In fact, the latter model might be preferred based on ease of interpretation. Recall from Example 2.4.1,

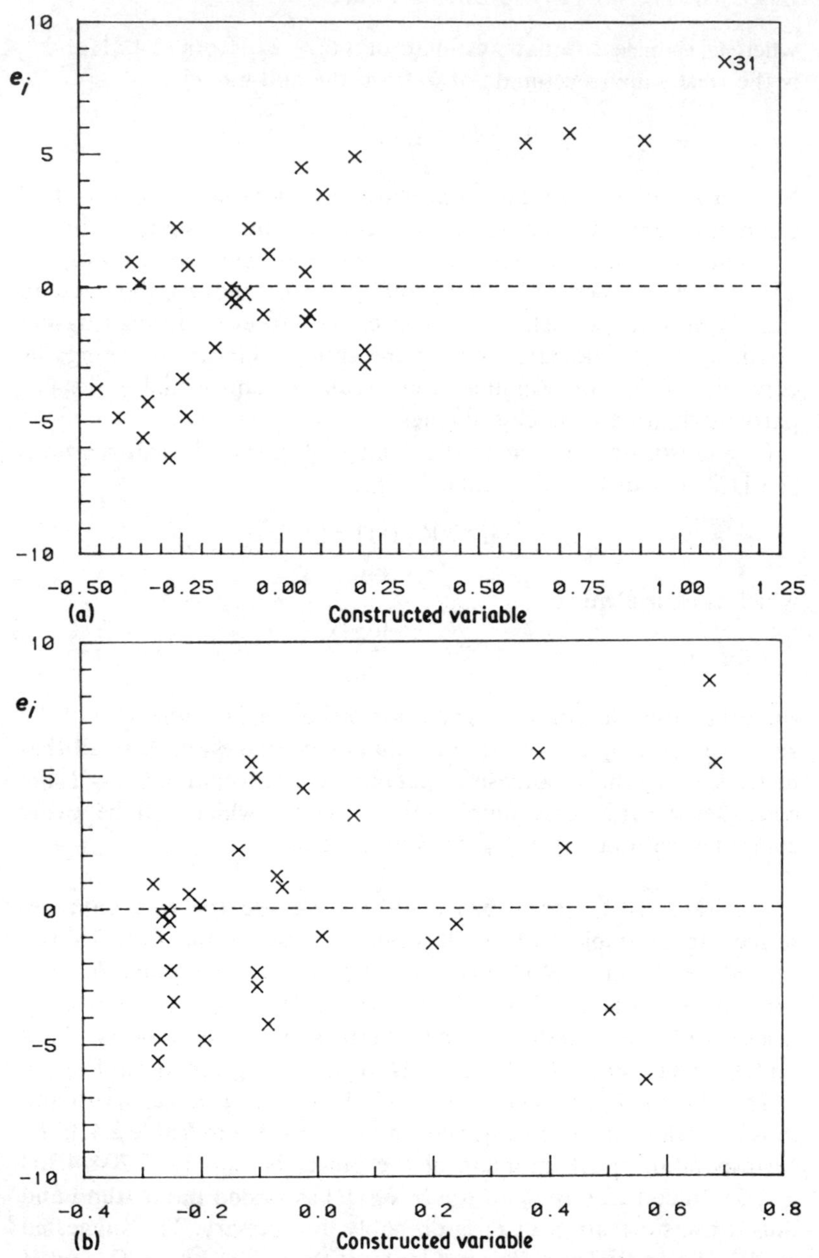

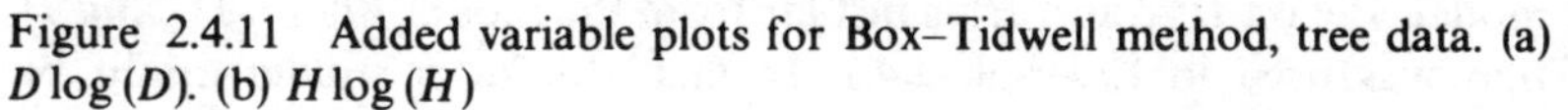

Figure 2.4.11 Added variable plots for Box–Tidwell method, tree data. (a) $D \log(D)$. (b) $H \log(H)$

Table 2.4.3 *Regression summaries for two models, tree data*

	Estimate	*s.e.*	*Estimate*	*s.e.*
Intercept	−57.987	8.638	65.567	124.718
D	4.708	0.264	−21.463	5.065
H	0.339	0.130	−1.757	9.363
$D \log(D)$			7.204	1.394
$H \log(H)$			0.405	1.762

however, that *Vol* on D^2 and H can be refined further by transforming *Vol*. However, for *Vol* on $D^{2.5}$ and H, the score statistic $t_D(1) = -1.44$ suggests that a transformation of *Vol* may not provide much improvement.

In some problems, iteration may provide substantially improved estimates of the λ_js for those variables requiring a transformation. In this example, the iterated estimate of λ_D remains close to 2.5.

Finally, the Box–Tidwell procedure applied to the model $(Vol)^{1/3}$ on D and H that was suggested in Example 2.4.1 may be used to argue that transformations of D and H after *Vol* are not likely to result in much additional improvement. Such sequential procedures should not be confused with methods for the simultaneous estimation of transformations for the response and explanatory variables. □

EXAMPLE 2.4.5. CLOUD SEEDING NO. 5. In the Box–Cox method of selecting a transformation for the responses, the scales of the explanatory variables are held fixed. This may not be appropriate for the cloud seeding data since the response Y and prewetness P are both measures of amount of rainfall. It seems sensible that these variables should be measured in the same scale.

To investigate the need for transforming Y and P simultaneously, we use the power family in combination with the model,

$$\mathbf{Y}^{(\lambda)} = \mathbf{X}_1 \boldsymbol{\beta}_1 + \beta_5 \mathbf{P}^{(\lambda)} + \beta_{15} \mathbf{A} \mathbf{P}^{(\lambda)} + \boldsymbol{\varepsilon} \tag{2.4.23}$$

where $\mathbf{X}_1$ is the 24×9 matrix of explanatory variables excluding prewetness and action × prewetness, $\mathbf{P}^{(\lambda)}$ is the 24-vector of transformed prewetness values, $\mathbf{A}$ is a 24×24 diagonal matrix with i-th diagonal element equal to 1 if the i-th day was seeded and 0 otherwise, and $\boldsymbol{\varepsilon}$ is $N(\mathbf{0}, \sigma^2 \mathbf{I})$. Following the discussion in Section 2.4.2, the log likelihood maximized over $\boldsymbol{\beta}$ and σ^2 for fixed λ is

$$L_{\max}(\lambda) = -\frac{n}{2} \log \{ \mathbf{Z}^{(\lambda)\mathrm{T}} [\mathbf{I} - \mathbf{V}^{(\lambda)}] \mathbf{Z}^{(\lambda)} / n \}$$

where $\mathbf{V}^{(\lambda)}$ is the usual projection matrix for model (2.4.23). The maximum likelihood estimate $\hat{\lambda}$ of λ can be read from a plot of $L_{\max}(\lambda)$ versus λ for a few selected values of λ. For each plotted point a new value of $\mathbf{P}^{(\lambda)}$ and thus a new value of $\mathbf{V}^{(\lambda)}$ has to be computed.

Figure 2.4.12 gives a plot of $L_{\max}(\lambda)$ versus λ along with the associated 95% asymptotic confidence interval. The maximum likelihood estimate is $\hat{\lambda} = 0.401$ and $\lambda = 1$ is well outside the confidence interval. Since Y and P are both measures of the volume of rainfall, the cube root transformation seems a sensible choice.

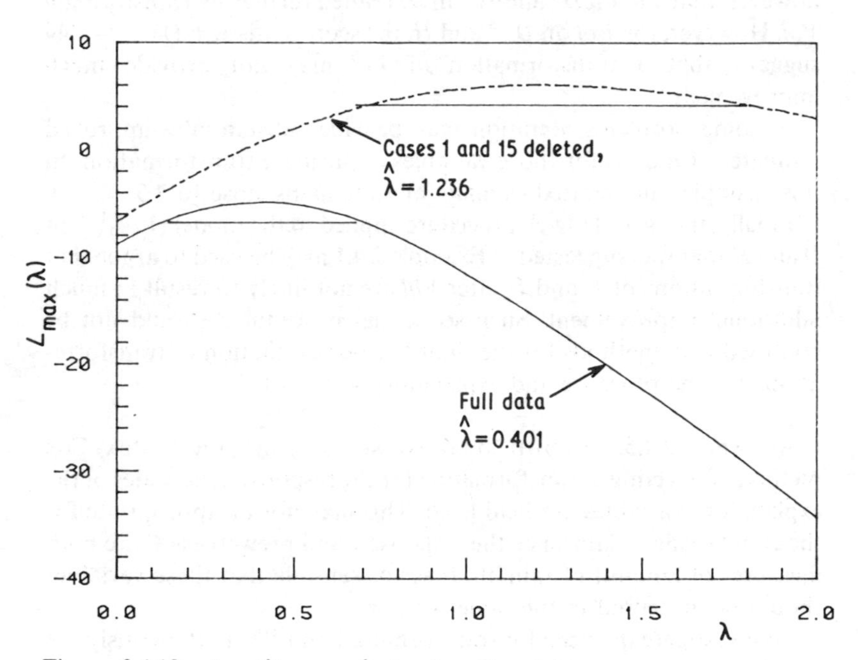

Figure 2.4.12 $L_{\max}(\lambda)$ versus λ, cloud seeding data

A diagnostic plot for this procedure can be obtained by using model (2.4.14) in combination with the Box–Tidwell method. Model (2.4.14) with $\lambda_0 = 1$ can be rewritten as

$$\mathbf{Z}^{(1)} = \mathbf{X}_1\boldsymbol{\beta}_1 + \beta_5\mathbf{P}^{(\lambda)} + \beta_{15}\mathbf{A}\mathbf{P}^{(\lambda)} + (1-\lambda)\mathbf{G}_z^{(1)} + \boldsymbol{\varepsilon}$$

Next, expanding $\mathbf{P}^{(\lambda)}$ about $\lambda = 1$ as in (2.4.21) and rearranging terms leads to

$$\mathbf{Z}^{(1)} = \mathbf{X}\boldsymbol{\beta} + (\lambda - 1)\,[\beta_5\mathbf{G}_P^{(1)} + \beta_{15}\mathbf{A}\mathbf{G}_P{}^{(1)} - \mathbf{G}_z^{(1)}] + \boldsymbol{\varepsilon}$$

where $\mathbf{X}$ is the matrix containing all untransformed explanatory variables and

$$\mathbf{G}_p^{(1)} = \left.\frac{\partial P^{(\lambda)}}{\partial \lambda}\right|_{\lambda=1} = P \log(P)$$

This form suggests that an added variable plot for the constructed variable

$$\mathbf{G} = (\hat{\beta}_5 \mathbf{G}_P^{(1)} + \hat{\beta}_{15} \mathbf{A} \mathbf{G}_P^{(1)} - \mathbf{G}_z^{(1)}), \tag{2.4.24}$$

where $\hat{\beta}_5$ and $\hat{\beta}_{15}$ are the least squares estimates of β_5 and β_{15} from the regression of $\mathbf{Z}^{(1)}$ on $\mathbf{X}$, may be a useful diagnostic.

Figure 2.4.13 gives the added variable plot for the constructed variable $\mathbf{G}$. There seems to be a strong linear trend in this plot, but cases 1 and 15 clearly stand apart and may be controlling our impression of the plot in addition to the results of the likelihood analysis. The least squares slope of the added variable plot is -0.762 which is an estimate of $\lambda - 1$. Thus a quick estimate of λ is $\tilde{\lambda} = 1 - 0.762 = 0.238$ which is not too far from the cube root transformation suggested previously.

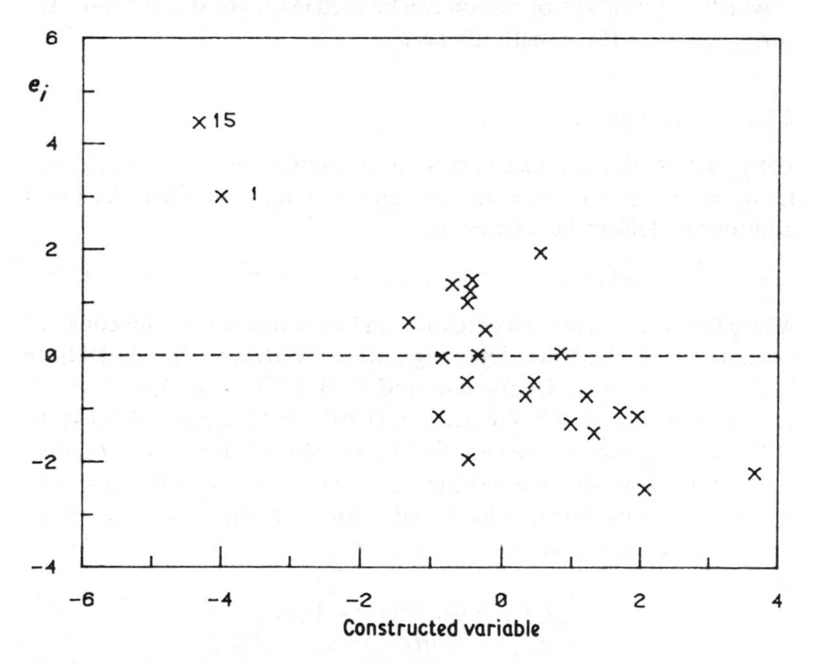

Figure 2.4.13 Added variable plot for (2.4.24), cloud seeding data

When cases 1 and 15 are removed, the evidence for a transformation disappears. The dashed curve in Fig. 2.4.12 is a plot of $L_{max}(\lambda)$ versus λ for the reduced data. The maximum likelihood estimate is now $\hat{\lambda} = 1.24$ and $\lambda = 1$ is well within the 95% asymptotic confidence interval. In the absence of additional information, the results of any analysis of these data, regardless of the transformation used, should be interpreted with caution. In future examples using the cloud seeding data, we will use the cube root transformation. □

2.5 Residual analysis in two-way tables

The linear model for the unreplicated two-way table is an example of the kind of model that could be studied using the diagnostic methods discussed earlier in this chapter. However, since the appearance of Tukey's (1949) one degree of freedom test of additivity, a body of methods that take advantage of the special structure of two-way tables has developed. These methods, which can often be generalized to higher-dimensional layouts, merit special study as examples of the ways in which additional information can be used in diagnostic methods. We survey some of these methods here.

2.5.1 OUTLIERS

Let y_{ij} denote the response in row i and column j of an $r \times c$ table, and let $\mu_{ij} = Ey_{ij}$, $\mu = \bar{\mu}_{..}$, $\alpha_i = \bar{\mu}_{i.} - \bar{\mu}_{..}$ and $\beta_j = \bar{\mu}_{.j} - \bar{\mu}_{..}$. Then the usual additive model can be written as

$$y_{ij} = \mu + \alpha_i + \beta_j + \varepsilon_{ij}, \qquad i = 1, 2, \ldots, r; j = 1, 2, \ldots, c \tag{2.5.1}$$

where the errors ε_{ij} are uncorrelated and have mean zero and constant variance σ^2. Methods for detecting outliers relative to this model have been investigated by Gentleman and Wilk (1975a, b), Daniel (1978), Draper and John (1980), Gentleman (1980), and Galpin and Hawkins (1981) among others. Barnett and Lewis (1978) give an informative discussion. Generally, the residuals $e_{ij} = (y_{ij} - \bar{y}_{i.} - \bar{y}_{.j} + \bar{y}_{..})$ from a fit of the additive model are reliable indicators of a single outlying cell. If, for example, a single outlying value of magnitude θ occurs in cell (1, 1) then

$$\begin{aligned} Ee_{11} &= \theta(r-1)(c-1)/rc, \\ Ee_{1j} &= -\theta(r-1)/rc, \qquad j \geq 2 \\ Ee_{i1} &= -\theta(c-1)/rc, \qquad i \geq 2 \end{aligned} \tag{2.5.2}$$

and

$$Ee_{ij} = \theta/rc, \qquad \text{otherwise}$$

Thus, the residual corresponding to the outlying cell will have the largest absolute expectation provided $r \geq 3$, $c \geq 3$. Since the residual variances are constant for this model, there is no essential distinction to be made between the ordinary residuals and the Studentized residuals. A formal, normal theory test for a single outlier can be based on the maximum normed residual

$$MNR = \max_{ij} |e_{ij}| / \left(\sum_{ij} e_{ij}^2 \right)^{1/2} \tag{2.5.3}$$

This statistic is equivalent to $\max_i |t_i|$ obtained from the mean shift outlier model described in Section 2.2.2. The 1%, 5% and 10% points of MNR for $r = 3(1)10$, $c = 3(1)10$, from Galpin and Hawkins (1981), are reproduced in Table 2.5.1.

When two or more outlying values are present the residuals will often lack noticeable peculiarities since the effects of multiple outliers can filter through the entire table of residuals in complicated ways (Gentleman and Wilk, 1975a; Daniel, 1978). Gentleman (1980) discusses methods for finding the k most likely outliers; that is, the k observations whose removal provides the greatest reduction in the residual sum of squares. This is equivalent to finding the k observations that maximize the multiple case Studentized residual.

2.5.2 NONADDITIVITY

When nonadditivity is suspected, a useful initial representation of the response is

$$y_{ij} = \mu + \alpha_i + \beta_j + \gamma_{ij} + \varepsilon_{ij}, \qquad i = 1, \ldots, r; j = 1, \ldots, c \tag{2.5.4}$$

where μ, α_i, β_j, and ε_{ij} are as previously defined and $\gamma_{ij} = \mu_{ij} - \bar{\mu}_{i.} - \bar{\mu}_{.j} + \bar{\mu}_{..}$. Of course, the usual additive model is obtained if $\gamma_{ij} = 0$ for all i and j. An equivalent condition for additivity is that all two-by-two contrasts of the form $\mu_{ij} - \mu_{i'j} - \mu_{ij'} + \mu_{i'j'}$ be equal to zero. Johnson and Graybill (1972a) exploit this fact to develop a method for estimating σ^2 in the presence of partial nonadditivity.

A variety of models and tests for nonadditivity can be obtained by imposing additional structure on the interaction terms γ_{ij}. The model associated with Tukey's test (1949) is perhaps the best known and is

Table 2.5.1 *Critical values for MNR = maximum normed residual in two-way tables. Starred values are exact. Source: Galpin and Hawkins (1981), reprinted with permission*

	c							
r	3	4	5	6	7	8	9	10
				$\alpha = 0.01$				
3	0.660 33*							
4	0.674 84*	0.665 11*						
5	0.664 34*	0.639 95*	0.607 97*					
6	0.645 97*	0.613 02*	0.577 74*	0.546 28*				
7	0.625 76*	0.587 67*	0.550 80*	0.519 01*	0.491 93			
8	0.605 84*	0.564 63*	0.527 07*	0.495 38	0.468 70	0.445 99		
9	0.586 96*	0.543 86*	0.506 11*	0.474 75	0.448 57	0.426 41	0.407 36	
10	0.569 35*	0.525 16*	0.487 50	0.456 58	0.430 94	0.409 31	0.390 79	0.374 69
				$\alpha = 0.05$				
3	0.648 10*							
4	0.645 12*	0.620 66*						
5	0.624 15*	0.589 71*	0.555 13*					
6	0.600 08*	0.560 79*	0.524 91*	0.494 59				
7	0.576 66*	0.535 13*	0.498 97	0.468 99	0.443 96			
8	0.554 98*	0.512 56*	0.476 60	0.447 15	0.422 73	0.402 13		
9	0.535 21*	0.492 65	0.457 12	0.428 27	0.404 47	0.384 47	0.367 36	
10	0.517 24	0.474 98	0.439 98	0.411 75	0.388 56	0.369 11	0.352 51	0.338 12

$\alpha = 0.10$

3	0.637 15*							
4	0.624 79*	0.594 44*						
5	0.599 44*	0.561 85*	0.527 11*					
6	0.573 03*	0.532 74*	0.497 58	0.468 38				
7	0.548 46*	0.507 47*	0.472 57	0.443 95	0.420 19			
8	0.526 26	0.485 52	0.451 15	0.423 21	0.400 13	0.380 70		
9	0.506 30	0.466 31	0.432 60	0.405 33	0.382 90	0.364 07	0.347 98	
10	0.488 32	0.449 34	0.416 33	0.389 73	0.367 90	0.349 62	0.334 02	0.320 51

obtained by setting

$$\gamma_{ij} = \phi\alpha_i\beta_j \tag{2.5.5}$$

for all i and j. This model may be viewed as a way of modeling a linear-by-linear interaction in latent variables associated with rows and columns and is sometimes called Tukey's concurrence model. Tukey's 1 df test of additivity is obtained using a standard procedure for converting a nonlinear model to a linear model (Milliken and Graybill, 1970): Replace α_i and β_j in γ_{ij} with their least squares estimates $\hat{\alpha}_i$ and $\hat{\beta}_j$ from the additive model (2.5.1) and then construct the usual F-statistic F_T for the hypothesis $\phi = 0$. Under the null hypothesis and normality, this statistic has an F-distribution with 1 and $(r-1)(c-1)-1$ df. Graphical aids useful in interpreting this test can be constructed using the methodology of Section 2.3. The power of Tukey's test and its robustness in non-normal situations have been investigated by Ghosh and Sharma (1963) and Yates (1972).

Mandel (1961) suggested two alternative structures for the interaction terms:

$$\gamma_{ij} = a_i\beta_j \tag{2.5.6}$$

and

$$\gamma_{ij} = \alpha_i b_j \tag{2.5.7}$$

for all i and j. The models associated with (2.5.6) and (2.5.7) are called the row and column regression models, respectively. These models can also be motivated by appealing to the notion of latent variables associated with rows (2.5.7) or columns (2.5.6) and, relative to (2.5.5), are more flexible approaches to nonadditivity. In the row model, a test of additivity is obtained by replacing β_j in (2.5.6) with its least squares estimate from (2.5.1) and then constructing the usual F-statistic F_{row} for the hypothesis $a_1 = a_2 = \ldots = a_r = 0$. Under the null hypothesis and normality, this statistic has an F-distribution with $r-1$ and $(r-1)(c-2)$ df. The analogous test for the column model is constructed in the same way.

It is important to remember that these tests for nonadditivity are obtained by approximating a nonlinear model with a linear model that can be handled using standard techniques. Under the alternative hypotheses, the statistics F_T and F_{row} do not have noncentral F-distributions, as would usually be the case in standard applications. In the presence of nonadditivity, the residual mean square resulting from a fit of the linearized version of the model is positively biased. This bias can be severe if ϕ or Σa_i^2 is large. Hence, these methods should not be regarded as being more than relatively straightforward ways of detecting nonadditivity. If nonadditivity is found, it may be wise to abandon the linearized version of the model in favor of more appropriate methods of analysis. For example, the data might be transformed to restore additivity.

In the approaches of Tukey and Mandel, it is necessary to assume the presence of main effects ($\alpha_i \neq 0$, $\beta_j \neq 0$) for the interactions to be present. This and the specific structures assumed for γ_{ij} place an often unwarranted limitation on the types of nonadditivity that will be detected by these techniques. Johnson and Graybill (1972b) proposed setting

$$\gamma_{ij} = \delta w_i u_j \tag{2.5.8}$$

where $\Sigma_i w_i = \Sigma_j u_j = 0$ and $\Sigma w_i^2 = \Sigma u_j^2 = 1$, as a more general structure for detecting nonadditivity. Essentially, this assumes that the interaction γ_{ij} is a quadratic function of latent variables that need not be related to the main effects. This form would be appropriate if the interactions do not occur systematically across the entire table, but do occur systematically in a subset of the full table or in only an isolated cell. For example, a single outlier in cell (1, 1) corresponds to $\mathbf{W}^{\mathrm{T}} = (w_i) = k_w(r-1, -1, -1, \ldots, -1)$ and $\mathbf{U}^{\mathrm{T}} = (u_i) = k_u(c-1, -1, -1, \ldots, -1)$ (see Equation (2.5.2)), where k_w and k_u are constants chosen to insure that $\mathbf{W}^{\mathrm{T}}\mathbf{W} = \mathbf{U}^{\mathrm{T}}\mathbf{U} = 1$. As a second illustration, consider a nonadditive table in which the subtables formed by the first $s \leq r-1$ rows and the last $r-s$ rows are additive, the nonadditivity being due solely to the difference between the subtables. Let $\mu_{1ij} = \mu_{ij}$ for $i = 1, 2, \ldots, s$, $j = 1, 2, \ldots, c$ and $\mu_{2ij} = \mu_{ij}$ for $i = s+1, \ldots, r$ and $j = 1, 2, \ldots, c$. Then $w_i = k_w(r-s)$ for $i = 1, 2, \ldots, s$ and $w_i = -k_w s$ for $i = s+1, \ldots, r$, and $\mathbf{U}^{\mathrm{T}} = k_u(\bar{\mu}_{1.j} - \bar{\mu}_{1..} - \bar{\mu}_{2.j} + \bar{\mu}_{2..})$. Generally, (2.5.8) can model any alternative situation in which the matrix with elements Ee_{ij} is of rank 1.

Since the interaction terms are not functions of the main effects, the method used previously for constructing an easy test of additivity ($\delta = 0$) does not apply. Instead, Johnson and Graybill derive the maximum likelihood estimators under the assumption of normality. Let $\mathbf{E} = (e_{ij})$ be the $r \times c$ matrix of residuals from the additive model. Then the maximum likelihood estimators of the parameters in the

Johnson–Graybill model are

$$\hat{\mu} = \bar{y}_{..}$$
$$\hat{\alpha}_i = \bar{y}_{i.} - \bar{y}_{..}$$
$$\hat{\beta}_j = \bar{y}_{.j} - \bar{y}_{..}$$
$$\hat{\delta}^2 = \text{largest eigenvalue of } \mathbf{E}^{\mathrm{T}}\mathbf{E}$$
$$\hat{\mathbf{W}} = \text{normalized eigenvector of } \mathbf{E}\mathbf{E}^{\mathrm{T}} \text{ associated with } \hat{\delta}^2$$
$$\hat{\mathbf{U}} = \text{normalized eigenvector of } \mathbf{E}^{\mathrm{T}}\mathbf{E} \text{ associated with } \hat{\delta}^2$$

and

$$\hat{\sigma}^2 = \left[\sum_i \sum_j e_{ij}^2 - \hat{\delta}^2\right] \Big/ rc$$

The likelihood ratio test statistic for $H\colon \delta = 0$ is a monotonic function of

$$\Delta = \hat{\delta}^2 \Big/ \sum_i \sum_j e_{ij}^2 \tag{2.5.9}$$

Large values of Δ furnish evidence against additivity. Johnson and Graybill (1972b) give the upper 1%, 5%, and 10% points of the null distribution of Δ. These are reproduced in Table 2.5.2.

In addition to providing a reasonably flexible test, the Johnson–Graybill approach provides a useful method for diagnosing more specific forms of nonadditivity. Plots of $\hat{w}_i$ versus $\hat{\alpha}_i$ or $\hat{u}_j$ versus $\hat{\beta}_j$ may suggest that the column or row regression models respectively, or perhaps Tukey's model, is appropriate. If $\hat{\mathbf{W}}$ and $\hat{\mathbf{U}}$ each contain a single relatively large value then this may be taken as an indication of an outlier in the cell corresponding to the coordinates of the large values. The signs of the elements of $\hat{\mathbf{W}}$ serve to identify a decomposition by rows of the full table into two subtables that may be more nearly additive. If the elements of the same sign are of the same order of magnitude then this might be taken as an indication that the subtables are additive. At the very least, such subtables require further inspection when nonadditivity is present. Similar comments apply to $\hat{\mathbf{U}}$.

For further discussion of the Johnson–Graybill approach and extentions to situations in which the rank of the matrix of the expected residuals is greater than 1, see Hegemann and Johnson (1976a) and Mandel (1971). Bradu and Gabriel (1978) present a graphical technique as an aid to determining an appropriate model. Hegemann and Johnson (1976b) compare the power of Tukey's test to that based on the Johnson–Graybill model. Their general conclusion is that if the row and column effects are large and the structure $\gamma_{ij} = \phi\alpha_i\beta_j$ is appropriate then Tukey's test will have the greater power. Otherwise, the Johnson–Graybill test is preferred.

EXAMPLE 2.5.1 AGRICULTURAL EXPERIMENT. We illustrate the methods suggested in this section using data on a two-way classification design from Carter, Collier, and Davis (1951). This data set was used by Johnson and Graybill (1972b) to illustrate the use of their test. The data, which are part of a larger experiment to determine the effectiveness of blast furnace slags as agricultural liming materials on three soil types, are presented in Table 2.5.3(a). The response is yield of corn in bushels per acre.

As a base, we first consider the fit obtained from the additive model (2.5.1). The ANOVA table and the normed residuals, $e_{ij}/(\sum_i \sum_j e_{ij}^2)^{1/2}$, are presented in Tables 2.5.3(c) and 2.5.3(b), respectively. The results in Table 2.5.3(c) suggest that the average soil effects are significant while the average treatment effects are not. The usual estimate of σ^2 from the additive model is 79.0. The pattern of the signs of the normed residuals in Table 2.5.3(b) might be taken as an indication that the additive model is not appropriate. The MNR occurs in cell (5, 3) and has the value 0.603 which has a p-value less than 0.05. Evidently, there is reason to suspect that the observation in cell (5, 3) does not conform to the assumed model; that is, either the model or the observation is wrong.

We next fit the Johnson–Graybill model. The maximum likelihood estimates of the interaction parameters are

$$\hat{\delta}^2 = 943.02$$

$$\hat{\mathbf{W}}^{\mathrm{T}} = (-0.476,\ -0.337,\ 0.086,\ 0.040,\ 0.767,\ -0.212,\ 0.131)$$

and

$$\hat{\mathbf{U}}^{\mathrm{T}} = (-0.206,\ -0.581,\ 0.787)$$

The maximum likelihood estimate of σ^2 is 0.21. This is a biased estimate of σ^2. Johnson and Graybill proposed an alternative estimator $\tilde{\sigma}^2$ of σ^2 that will be unbiased when additivity holds,

$$\tilde{\sigma}^2 = \frac{\left(\sum_{ij} e_{ij}^2 - \hat{\delta}^2\right)}{(r-1)(c-1) - E(\hat{\delta}^2/\sigma^2)}$$

where the expectation is taken under the hypothesis $\delta = 0$. Tables of $E(\hat{\delta}^2/\sigma^2)$ are available from Mandel (1971). If $\delta \neq 0$ then the estimate is no longer unbiased. For the problem at hand $E(\hat{\delta}^2/\sigma^2) = 8.94$ and

Table 2.5.2 *Upper percentage points for null distribution of* Δ *(Equation (2.5.9)). Starred values are exact. Source: Johnson and Graybill (1972b), reprinted with permission*

	c									
r	3	4	5	6	7	8	10	12	16	20
					$\alpha = 0.01$					
3	0.999 97*									
4	0.9975*	0.8930								
5	0.9883*	0.9303*	0.9004							
6	0.9743*	0.9082	0.7825	0.7194						
7	0.9587*	0.8619*	0.7325	0.7325	0.6457					
8	0.9429*	0.8446	0.7407	0.6470	0.6243	0.5809				
10	0.9135*	0.7575	0.6924	0.6516	0.5523	0.5044	0.4452			
12	0.8879*	0.7411	0.6514	0.5702	0.5170	0.4911	0.4372	0.3969		
16	0.8472*	0.6256	0.5788	0.5167	0.4560	0.4331	0.3628	0.3372	0.2837	
20	0.8164*	0.5966	0.5462	0.4955	0.4229	0.3945	0.3364	0.3095	0.2567	0.2249
32	0.7571*	0.5367	0.4788	0.4198	0.3700	0.3326	0.2818	0.2485	0.2049	0.1794
50	0.7089*	0.5043	0.4423	0.3722	0.3178	0.2864	0.2430	0.2139	0.1731	0.1496
100	0.6498*	0.4463	0.3771	0.3189	0.2720	0.2421	0.1977	0.1698	0.1359	0.1151
					$\alpha = 0.05$					
3	0.9994*									
4	0.9873*	0.8567								
5	0.9648*	0.8811*	0.8407							
6	0.9406*	0.8505	0.7294	0.6681						
7	0.9168*	0.8003*	0.6823	0.6703	0.5957					
8	0.8974*	0.7811	0.6815	0.5985	0.5733	0.5345				
10	0.8630*	0.7043	0.6361	0.5901	0.5096	0.4680	0.4143			

12	0.8357*	0.6936	0.5979	0.5242	0.4774	0.4501	0.4016	0.3665		
16	0.7950*	0.6295	0.5356	0.4760	0.4227	0.3991	0.3390	0.3064	0.2656	
20	0.7661*	0.6290	0.5054	0.4542	0.3932	0.3652	0.3139	0.2876	0.2408	0.2117
32	0.7127*	0.5349	0.4469	0.3894	0.3454	0.3105	0.2644	0.2335	0.1938	0.1702
50	0.6713*	0.5978	0.4127	0.3482	0.3002	0.2706	0.2296	0.2020	0.1643	0.1423
100	0.6218*	0.4610	0.3583	0.3024	0.2595	0.2311	0.1892	0.1629	0.1306	0.1116
					$\alpha = 0.10$					
3	0.9975*									
4	0.9743*	0.8349								
5	0.9429*	0.8458*	0.8021							
6	0.9135*	0.8130	0.6975	0.6398						
7	0.8879*	0.7631*	0.6548	0.6358	0.5687					
8	0.8660*	0.7435	0.6487	0.5725	0.5462	0.5098				
10	0.8308*	0.6749	0.6057	0.5570	0.4972	0.4489	0.3982			
12	0.8037*	0.6594	0.5695	0.5001	0.4563	0.4289	0.3830	0.3506		
16	0.7647*	0.6022	0.5131	0.4550	0.4137	0.3815	0.3268	0.2952	0.2562	
20	0.7376*	0.5737	0.4843	0.4329	0.3778	0.3502	0.3023	0.2765	0.2326	0.2049
32	0.6886*	0.5161	0.4306	0.3740	0.3329	0.2992	0.2556	0.2257	0.1881	0.1654
50	0.6512*	0.4887	0.3978	0.3362	0.2913	0.2624	0.2226	0.1960	0.1598	0.1386
100	0.6071*	0.4421	0.3490	0.2942	0.2533	0.2255	0.1850	0.1595	0.1279	0.1001

Table 2.5.3 *Agricultural data. Source: Carter et al.* (1951). (a) Data. (b) Normed residuals. (c) Analysis of variance

	(a) Data			*(b) Normed residuals*		
	Soil			*Soil*		
Treatment	1	2	3	1	2	3
a	11.1	32.6	63.3	−0.066	−0.299	0.365
b	15.3	40.8	65.0	−0.082	−0.185	0.268
c	22.7	52.1	58.8	0.023	0.046	−0.069
d	23.8	52.8	61.4	0.011	0.021	−0.032
e	25.6	63.1	41.1	0.158	0.445	−0.603
f	31.2	59.5	78.1	−0.082	−0.095	0.177
g	25.8	55.3	60.2	0.040	0.067	−0.107

(c) *Analysis of variance*

Source	*df*	*SS*	*MS*	*F*
Treatments	6	731.1	121.8	0.88
Soils	2	5696.3	2848.2	20.61
Residual	12	947.4	79.0	

$\tilde{\sigma}^2 = 1.43$. Both estimates of σ^2 are considerably smaller than that obtained from the additive model. The likelihood ratio test statistic for $\delta = 0$ has the value $\Delta = 0.9954$, with the corresponding p-value less than 0.01.

In an effort to understand the precise nature of the nonadditivity, we turn to an inspection of $\hat{\mathbf{W}}$ and $\hat{\mathbf{U}}$. First, plots of $\hat{w}_i$ versus $\hat{\alpha}_i$ and $\hat{u}_j$ versus $\hat{\beta}_j$ do not display a clear linear trend and thus neither Tukey's model nor the two versions of Mandel's approach is likely to provide an adequate explanation. The corresponding F-tests confirm this observation, as the three F-statistics are all less than 1. Second, an inspection of $\hat{\mathbf{W}}$ reveals that treatments a, b, and f seem to form an additive subtable; the elements of $\hat{\mathbf{W}}$ corresponding to these treatments are all negative and of the same order of magnitude. However, the remaining positive elements are not of the same order of magnitude; the element corresponding to treatment e is 19 times as large as the element corresponding to treatment d. The interpretation is that, while the subset formed by treatments c, d, e, and g may be more nearly additive than the full table, it does not seem to form an additive subset.

Similarly, inspection of $\hat{\mathbf{U}}$ suggests that the first two columns form an additive subtable. Inspecting both vectors simultaneously with a view towards detecting a single outlier isolates treatment e in the third column as a possible outlying cell. This again implicates cell (5, 3).

At this point a number of options for further analysis are available. One might, for example, replace the suspected outlier with a pseudo-value and reanalyze the data. However, this as well as many other techniques requires the specification of a model and at this point an appropriate model is unknown. A more useful procedure is to delete treatment e entirely and reanalyze the data for additivity.

With treatment e deleted, the estimate of σ^2 based on the additive model is 30.0 and the F-tests corresponding to the interactive components in the Tukey and Mandel models are again nonsignificant. Fitting the Johnson–Graybill model to the reduced data set yields the following estimates

$$\begin{aligned}
\hat{\delta}^2 &= 295.24 \\
\hat{\mathbf{W}}^{\mathrm{T}} &= (0.623, 0.373, -0.382, -0.300, 0.150, -0.463) \\
\hat{\mathbf{U}}^{\mathrm{T}} &= (-0.205, -0.582, 0.787) \\
\hat{\sigma}^2 &= 2.25 \\
\tilde{\sigma}^2 &= 15.15 \\
\Delta &= 0.9853
\end{aligned}$$

Both estimates of σ^2 have increased, and Δ still has a p-value of less than 0.01 indicating that some nonadditivity remains. Inspection of $\hat{\mathbf{W}}$ reveals that treatments a, b, and f again form an additive subtable and that treatments c, d, and g form a subtable that is more nearly additive than when treatment e was included. The interpretation of $\hat{\mathbf{U}}$ is the same as previously given. In short, it appears that the nonadditivity present in the reduced data set is due to the difference between the sets of treatment $\{a, b, f\}$ and $\{c, d, g\}$ in the third column. In retrospect, much the same conclusions might have been reached from an inspection of a plot of the data such as that given in Fig. 2.5.1. The response lines for treatments a, b, and f are nearly parallel as are those for $\{c, d, g\}$, while the response line for treatment e is anomalous. Of course, hindsight is usually more accurate than foresight and such visual inspections become difficult in larger tables.

A separate analysis of each set of treatments suggests that the data within a set are additive: for both treatment sets the tests of $\delta = 0$ have corresponding p-values greater than 0.05. Further, under the additive model, the estimates of σ^2 from the treatment sets $\{a, b, f\}$ and $\{c, d, g\}$

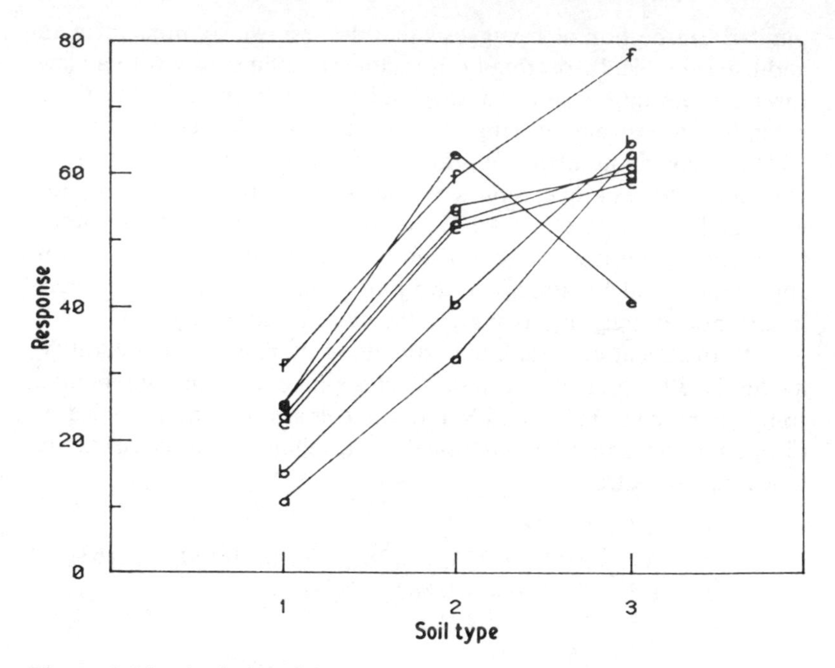

Figure 2.5.1 Agricultural data

are 9.23 and 1.01, respectively. These estimates are, of course, much smaller than the original estimate from the additive model. The ratio of the estimates, each being based on 4 df is 9.14 and this is approximately the 2.5% point of the appropriate F-distribution. This suggests that either nonadditivity is still present in treatment set $\{a, b, f\}$ or the variances of the treatment sets are different.

To this point, the analysis suggests that the nonadditivity in the data is due primarily to the differences between the treatment subtables $\{a, b, f\}$, $\{c, d, g\}$, and $\{e\}$. Depending on interest, the analysis of the treatment effects could be carried on in a variety of ways.

In addition to this analysis of the Johnson–Graybill model, a transformation to induce additivity could prove useful. Indeed, the entire Johnson–Graybill approach might have been overlooked in favor of the transformation methods of Section 2.4. As illustrated below, however, this may often be unwise since not all nonadditivity can be removed by a transformation.

A plot of $L_{\max}(\lambda)$ for the power family is given as Fig. 2.5.2. The maximum likelihood estimate is $\hat{\lambda} = 0.497$ and $\lambda = 1$ (no transform-

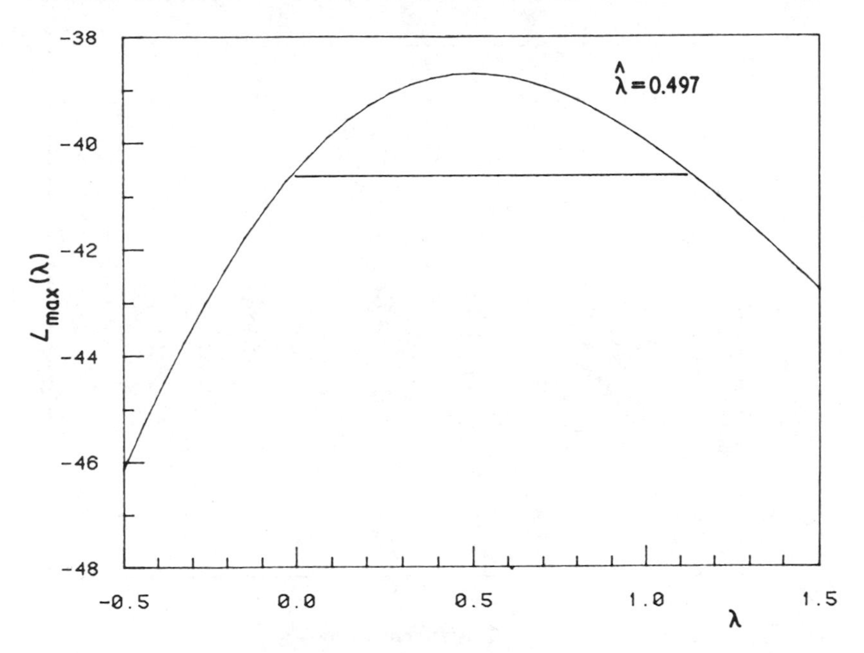

Figure 2.5.2 $L_{\max}(\lambda)$ versus λ, agricultural data

ation) is near one end of the asymptotic 95% confidence interval. There is only mild evidence of the need to transform. The score statistic is $t_D(1) = 1.64$ and the corresponding quick estimate is $\tilde{\lambda} = 0.087$ which suggests the log transformation. (In this example, the agreement between $\hat{\lambda}$ and $\tilde{\lambda}$ does not seem adequate.) Based on the transformed data with $\hat{\lambda} = 0.497$, $MNR = 0.582$ for cell (5, 3) with a p-value near 0.05, and $\Delta = 0.9828$ which has a p-value of less than 0.01.

While the transformed data are still nonadditive, it is possible that the results of the likelihood analysis are being distorted by the outlying cell. This suspicion is reinforced by the added variable plots for the score statistic given in Fig. 2.5.3. Another application of the likelihood method, this time without treatment e, gives $\hat{\lambda} = 0.397$ which is consistent with the cube root transformation. For cube root transformed data without treatment e, $\Delta = 0.9680$. Again, substantial nonadditivity remains. □

The exploratory approach used in the previous example can, if necessary, be formalized. Marasinghe and Johnson (1981a, b) provide

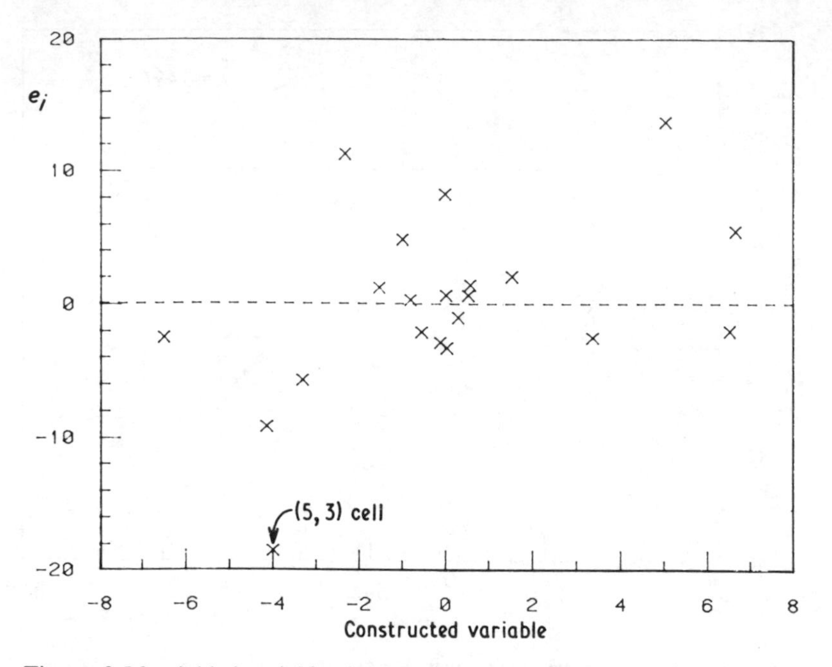

Figure 2.5.3 Added variable plot for the score statistic, agricultural data

likelihood ratio test statistics and associated critical values for the hypotheses $\mathbf{HW} = \mathbf{0}$, $\mathbf{GU} = \mathbf{0}$, and $\mathbf{HW} = \mathbf{GU} = \mathbf{0}$ where $\mathbf{H}$ and $\mathbf{G}$ are full rank matrices of contrasts. Besides providing formal tests, this material can help avoid the problems of overinterpretation that are inherent in any exploratory analysis.

CHAPTER 3

Assessment of influence

'To arrive inductively at laws of this kind, where one quantity *depends* on or *varies with* another, all that is required is a series of careful and exact measures in every different state of the *datum* and *quaesitum*. Here, however, the mathematical form of the law being of the highest importance, the greatest attention must be given to the *extreme cases* as well as to all those points where the one quantity changes rapidly with a small change of the other.'

HERSCHEL, *op. cit.*

The diagnostic methods presented in the last chapter are useful for finding general inadequacies in a model. A related question that cannot be easily addressed by those methods is that of stability, or the study of the variation in the results of an analysis when the problem formulation (see Fig. 1.2.1) is modified. If a case is deleted, for example, results based on the reduced data set can be quite different from those based on the complete data, as was suggested by many of the examples in Chapter 2. We call the study of the dependence of conclusions and inferences on various aspects of a problem formulation the study of influence.

The basic idea in influence analysis is quite simple. We introduce small perturbations in the problem formulation, and then monitor how the perturbations change the outcome of the analysis. The important questions in designing methods for influence analysis are the choices of the perturbation scheme, the particular aspect of an analysis to monitor, and the method of measurement. The possible answers for these separate questions can lead to a variety of different diagnostics. For example, diagnostics resulting from perturbation schemes applied to the data case by case can be quite different from those resulting from perturbation schemes applied to assumptions such as normality of errors.

In this chapter, we consider only one perturbation scheme in which the data are modified by deletion of cases, either one at a time or in groups. Case deletion diagnostics have found the greatest acceptance, and have been applied in many problems besides linear least squares regression. We will also limit our study to aspects of the analysis that can be summarized by the sample influence curve, to be described here at some length. Other approaches to the study of influence are described in later chapters.

3.1 Motivation

Not all cases in a set of data play an equal role in determining estimates, tests, and other statistics. For linear least squares, the results of the last chapter suggest that cases with v_{ii} near 1 or with large Studentized residuals will play a larger role. In some problems, the character of the regression may be determined by only a few cases while most of the data is essentially ignored. An extreme example of this is given in Fig. 3.1.1 for simple regression. If the one point separated from the others is moved, downweighted, or completely removed from the data, the resulting analysis may change substantially, as illustrated by the two regression lines computed with and without the separated point. While the change in the line can be anticipated from inspection of the

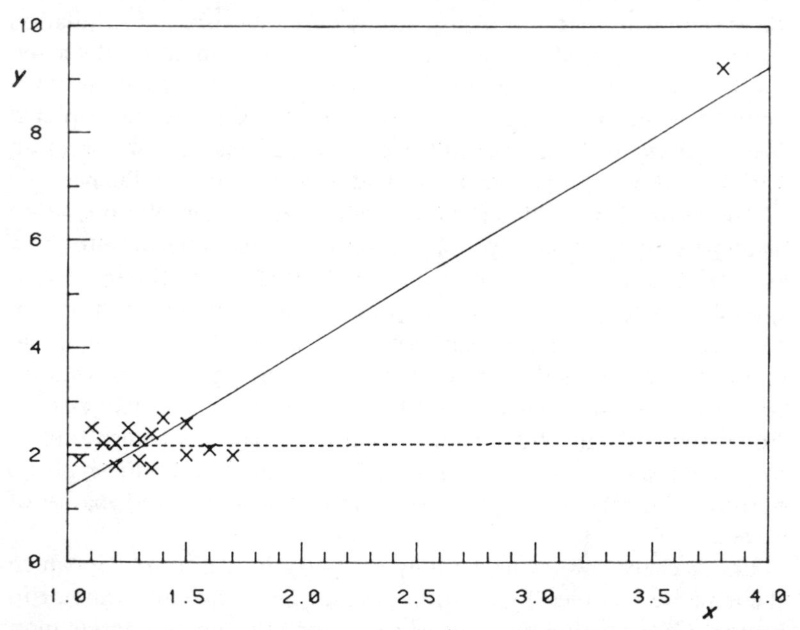

Figure 3.1.1 A simple regression scatter plot. —— regression of y on x, all data, $R^2 = 0.90$. ---- regression with the separated case removed, $R^2 < 0.01$

scatterplot, the change in other summaries such as R^2 can be startling. For the complete data in Fig. 3.1.1, $R^2 = 0.90$, while if the one separated case is removed R^2 is less than 0.01 (see Weisberg, 1980a, Example 3.3, for a discussion of the dependence of R^2 on the spread of the independent variables).

Table 3.1.1 contains a further example of a somewhat different character. In fitting the model $Ey = \beta_0 + \beta_1 x_1 + \beta_2 x_2$, case 4 may be considered an outlier because of its large Studentized residual, but it will have only modest influence on the estimates of the βs. Deletion of case 6, with $v_{66} = 1$, will result in a rank deficient model, so this case has a large influence. This example is deceptively simple, but the same conditions can occur in much larger problems if the role of case 6 is taken over by a small set of cases, and if the structure of x_1 and x_2 is made less obvious by a nonsingular linear transformation.

Table 3.1.1 *A hypothetical example*

Case	x_1	x_2	y	v_{ii}	e_i	r_i	t_i
1	1	1	3	0.4714	−0.1286	−0.3886	−0.34
2	1	2	4	0.2857	−0.1429	−0.3714	−0.33
3	1	3	5	0.1857	−0.1571	−0.3826	−0.34
4	1	4	7	0.1714	0.8286	2.0000	∞
5	1	5	7	0.2429	−0.1857	−0.4689	−0.42
6	0	6	8	1.0000	0	Undefined	Undefined
7	1	7	9	0.6429	−0.2143	−0.7878	−0.74

$\hat{\beta}_0 = 1.914$ ($t = 2.64$); $\hat{\beta}_1 = 0.200$ ($t = 0.37$); $\hat{\beta}_2 = 1.015$ ($t = 10.76$).
$\hat{\sigma}^2 = 0.2071$; df $= 4$; $R^2 = 0.97$

The ability to find influential cases can benefit the analyst in at least two ways. First, the study of influence yields information concerning reliability of conclusions and their dependence on the assumed model. For example, the usefulness of the complete data regression in Fig. 3.1.1 is highly dependent on the validity of the separated case. Alternatively, if deletion of an influential case from a data set changes the sign of an estimated parameter, relevant inference concerning that parameter may be in doubt. Second, we shall see that cases in the p-dimensional observation space that are far removed from other cases will tend to have, on the average, a relatively large influence on the analysis. This, in turn, may indicate areas in the observation space with inadequate coverage for reliable estimation and prediction.

The techniques developed here are not intended to provide rules for the rejection of data, as influential cases are not necessarily undesirable. Often, in fact, they can provide more important information than most other cases.

The emphasis in this chapter is on detecting influential cases rather than on how to deal with them once they are found, since final judgments must necessarily depend on context, making global recommendations impossible. Some of the possible actions can be given, however. If the influential cases correspond to gross measurement errors, recording or keypunching errors, or inappropriate experimental conditions, then they should be deleted or, if possible, corrected. If an influential case cannot be traced to conditions that justify its removal and the model is known to be approximately correct, a formal outlier test might be useful, although such tests cannot be expected to be powerful. Collecting more data or reporting the results of separate analyses with and without the cases in question are two additional possibilities that are often appropriate. Finally, in situations where predictions are important it may be possible to circumvent partially the effects of influential cases by isolating stable regions, or regions where the influence is minimal and unimportant.

In the next three sections of this chapter we review some of the results concerning the influence curve. Sample versions of the influence curve provide justification for the basic tools used for finding influential cases.

3.2 The influence curve

Let T_n be a vector-valued statistic of length k based on an independent and identically distributed sample $\mathbf{z}_1, \mathbf{z}_2, \ldots, \mathbf{z}_n$ from the cdf F defined on R^m. Of interest is the assessment of the change in T_n when some specific aspect of the problem is slightly changed. A first step in such an assessment is to find a *statistical functional* T that maps (a subset of) the set of all cdf's onto R^k such that, if $\hat{F}$ is the empirical cdf based on $\mathbf{z}_1, \mathbf{z}_2, \ldots, \mathbf{z}_n$, then $T(\hat{F}) = T_n$. If such a functional exists, then we can study the properties of T_n by examining the behavior of $T(F)$ or $T(\hat{F})$ when F or $\hat{F}$ is perturbed.

As a simple example, consider $m = k = 1$ and $T_n = n^{-1} \Sigma z_i = \bar{z}$. The corresponding statistical functional is

$$T(F) = \int z \, \mathrm{d}F(z)$$

so clearly $T(\hat{F}) = \int z\,\mathrm{d}\hat{F} = \bar{z}$. This estimator would be called robust if 'small' changes in F or $\hat{F}$ do not produce wild fluctuations in $T(F)$ or $T(\hat{F})$.

To aid in our study, we make use of the influence curve (Hampel, 1968, 1974; nearly parallel work appears in Tukey, 1970, and Andrews, Bickel, Hampel, Huber, Rogers and Tukey, 1972), and, in particular, various finite sample versions derived from it. For the most part, our approach to the influence curve is heuristic; for rigorous treatments, interested readers are urged to consult recent books by Huber (1981) and Serfling (1981).

Let $\delta_{\mathbf{z}}$ denote the cdf giving mass 1 to $\mathbf{z}$ in R^m. The vector-valued influence curve $IC_{T,F}(\mathbf{z})$ of T at F is defined pointwise by

$$IC_{T,F}(\mathbf{z}) = \lim_{\varepsilon \to 0_+} \frac{T[(1-\varepsilon)F + \varepsilon\delta_{\mathbf{z}}] - T(F)}{\varepsilon} \tag{3.2.1}$$

provided the limit exists for all $\mathbf{z}$ in R^m. Thus, the influence curve is just the ordinary right-hand derivative, evaluated at $\varepsilon = 0$, of $T[(1-\varepsilon)F + \varepsilon\delta_{\mathbf{z}}]$. It gives a measure of the influence on the statistical functional T of adding an observation at $\mathbf{z}$ as $n \to \infty$. For notational simplicity, the dependence of the influence curve on F and T will be supressed when no confusion is likely to result.

The original use of the influence curve and related notions exploited by von Mises (1947, 1964) and expanded upon by Reeds (1976) is in determining asymptotic properties of an estimator. Hampel (1968) and Andrews *et al.* (1972) use influence curves to compare estimators and to suggest robust modifications of existing estimation techniques. For example, M-estimators are modified versions of maximum likelihood estimates that have desirable properties for the corresponding influence curves. The main use in this work is anticipated by Devlin, Gnanadesikan and Kettenring (1975), Pregibon (1979, 1981), Cook and Weisberg (1980), and Hinkley (1977): The influence curve is used to monitor the influence of individual cases on estimates.

The following introductory example illustrates the use of the influence curve and suggests specific procedures for special purposes.

EXAMPLE 3.2.1. THE SAMPLE AVERAGE. The influence curve for $\mu = T(F) = \int z\,\mathrm{d}F$ (with $k = m = 1$) can be computed directly from (3.2.1) to be

$$IC(z) = \lim_{\varepsilon \to 0} \frac{(1-\varepsilon)\mu + \varepsilon z - \mu}{\varepsilon} = z - \mu \tag{3.2.2}$$

An undesirable property of the sample average is that its influence curve is unbounded; that is, small changes in F can produce large changes in the estimator.

The influence curve can also be used in a number of ways to see how individual cases affect the sample average. Suppose that a single additional case z were added, giving a sample of size $n+1$ and the new sample cdf $\hat{F}_+ = n\hat{F}/(n+1) + (1/(n+1))\delta_z$. It easily follows that

$$T(\hat{F}_+) = T(\hat{F}) + \frac{1}{n+1}(z - T(\hat{F}))$$

or

$$\bar{z}_+ = \bar{z} + \frac{1}{n+1}(z - \bar{z}) \tag{3.2.3}$$

where $\bar{z}_+ = T(\hat{F}_+)$. For a fixed sample size n, $\bar{z}_+ - \bar{z}$ increases linearly as z deviates from $\bar{z}$. This gives the influence of a single future case on the current sample average $\bar{z}$ and only indirectly reflects the influence of z_i, $i = 1, 2, \ldots, n$ on $\bar{z}$. Equation (3.2.3) is related to the sensitivity curves suggested by Tukey (1970).

The influence of the i-th case z_i on $\bar{z}$ may be determined by removing z_i from the sample and proceeding as before,

$$\bar{z} = \bar{z}_{(i)} + \frac{1}{n}(z - \bar{z}_{(i)}), \qquad i = 1, 2, \ldots, n$$

where $\bar{z}_{(i)}$ denotes the sample average computed without the i-th case. This describes a collection of n influence curves obtained by deleting each case in turn. The influence of z_i on $\bar{z}$ is obtained by evaluating the i-th curve at $z = z_i$. This results in the n case statistics

$$\bar{z} - \bar{z}_{(i)} = \frac{1}{n}(z_i - \bar{z}_{(i)}), \qquad i = 1, 2, \ldots, n,$$

which can be expressed more informatively by writing $(z_i - \bar{z}_{(i)})$ in terms of the full sample average,

$$\bar{z} - \bar{z}_{(i)} = (z_i - \bar{z})/(n-1) \tag{3.2.4}$$

Thus, the influence of a single case depends on the sample size and the full sample residuals. Any case with a sufficiently large residual will be influential for the sample average. □

3.3 The influence curve in the linear model

The first step in finding the influence curve for the least squares estimator of $\boldsymbol{\beta}$ in model (2.1.1) is to construct the appropriate functional T. Following Hinkley (1977), let the $(p'+1)$-vector $(\mathbf{x}^T, y)$ have a joint cdf F with

$$E_F\left\{\begin{pmatrix}\mathbf{x}\\ y\end{pmatrix}(\mathbf{x}^T, y)\right\} = \begin{pmatrix}\boldsymbol{\Sigma}(F) & \gamma(F)\\ \gamma^T(F) & \tau(F)\end{pmatrix} \tag{3.3.1}$$

By allowing $\mathbf{x}$ to have design measure (3.3.1) will also describe problems with $\mathbf{X}$ fixed. The functional corresponding to the least squares estimator of $\boldsymbol{\beta}$ is

$$T(F) = \boldsymbol{\Sigma}^{-1}(F)\gamma(F) \tag{3.3.2}$$

assuming, of course, that $\boldsymbol{\Sigma}$ is nonsingular. Next, let $\delta_{\mathbf{z}} = \delta_{(\mathbf{x}, y)}$ be the cdf that places mass 1 at $(\mathbf{x}^T, y)$. The p'-dimensional influence curve as a function of $(\mathbf{x}^T, y)$ is defined pointwise by (3.2.1). An explicit formula is obtained by writing

$$\begin{aligned}\boldsymbol{\Sigma}((1-\varepsilon)F+\varepsilon\delta_{\mathbf{z}}) &= (1-\varepsilon)\left(\boldsymbol{\Sigma}(F)+\frac{\varepsilon}{1-\varepsilon}\mathbf{x}\mathbf{x}^T\right)\\ \gamma((1-\varepsilon)F+\varepsilon\delta_{\mathbf{z}}) &= (1-\varepsilon)\gamma(F)+\varepsilon y\mathbf{x}\end{aligned} \tag{3.3.3}$$

From (3.3.3) updating $\boldsymbol{\Sigma}^{-1}(F)$ to $\boldsymbol{\Sigma}^{-1}((1-\varepsilon)F+\varepsilon\delta_{\mathbf{z}})$ is equivalent to adding a new case at $\mathbf{x}$ with weight $\varepsilon/(1-\varepsilon)$. Using Appendix A.2,

$$\begin{aligned}\boldsymbol{\Sigma}^{-1}((1-\varepsilon)F+\varepsilon\delta_{\mathbf{z}}) = \frac{1}{1-\varepsilon}\Big[&\boldsymbol{\Sigma}^{-1}(F) - \frac{\varepsilon}{1-\varepsilon}\boldsymbol{\Sigma}^{-1}(F)\times\\ &\mathbf{x}\left(1+\frac{\varepsilon}{1-\varepsilon}\mathbf{x}^T\boldsymbol{\Sigma}^{-1}(F)\mathbf{x}\right)^{-1}\mathbf{x}^T\boldsymbol{\Sigma}^{-1}(F)\Big]\end{aligned} \tag{3.3.4}$$

Substituting for $T((1-\varepsilon)F+\varepsilon\delta_{\mathbf{z}})$ in the definition of the influence curve, simplifying, and taking the limit gives

$$IC_{\boldsymbol{\beta},F}(\mathbf{x}, y) = \boldsymbol{\Sigma}^{-1}(F)\mathbf{x}(y - \mathbf{x}^T T(F)) \tag{3.3.5}$$

If interest centers on a set of q independent linear combinations of the elements of $\boldsymbol{\beta}$, then it is more appropriate to consider the influence curve for these combinations. Let $\boldsymbol{\psi} = \mathbf{Z}\boldsymbol{\beta}$, where $\mathbf{Z}$ is a $q \times p'$ rank q matrix. It is easily shown that the influence curve for $\boldsymbol{\psi} = \mathbf{Z}\boldsymbol{\beta} = \mathbf{Z}T(F)$ is

$$IC_{\boldsymbol{\psi},F}(\mathbf{x}, y) = \mathbf{Z} \times IC_{\boldsymbol{\beta},F}(\mathbf{x}, y) \tag{3.3.6}$$

As with the influence curve for the sample average, the influence curve for linear least squares regression is unbounded in each component as $y - \mathbf{x}^{\mathrm{T}}T(F)$ becomes large. This observation has led to the development of robust regression methods that generally bound influence by downweighting cases with large residuals. In addition, however, the componentwise influence can grow large, even if $y - \mathbf{x}^{\mathrm{T}}T(F)$ is small, if $\mathbf{x}$ is far from $E_F(\mathbf{x})$ and substantially in a direction of an eigenvector corresponding to a small eigenvalue of $\Sigma(F)$. Robust regression methods may also be highly influenced by such cases, as discussed in Chapter 5.

For the influence curve to provide a useful diagnostic procedure in regression, (3.3.6) must be modified by replacing $(\mathbf{x}, y)$ by $(\mathbf{x}_i, y_i)$, $i = 1, 2, \ldots, n$, and by replacing parameters by statistics. Although (3.3.6) is a useful theoretical diagnostic, as Hampel (1974), Mallows (1975), and others have pointed out, it describes an estimation technique with respect to a theoretical sampling population F. In any finite sample situation, more information relevant to the specific problem can be obtained by removing dependence upon F and using an asymptotically equivalent finite sample version, like those in Example 3.2.1, that corresponds directly to the observed data.

3.4 Sample versions of the influence curve

Several finite sample versions of the influence curve that depend on an observed sample have been suggested. Two of these, which shall be called the empirical influence curve (*EIC*) and the sample influence curve (*SIC*), have received the greatest attention, and will be discussed most completely here; both are discussed by Mallows (1975). They will be presented as a continuation of the previous section on least squares estimation of $\boldsymbol{\beta}$, but the ideas are general and the application to other situations should be clear.

3.4.1 EMPIRICAL INFLUENCE CURVES

In general, the *EIC* is obtained by substituting the sample cdf $\hat{F}$ for F in the influence curve. For linear models, using (3.3.5) and setting $\hat{\boldsymbol{\beta}} = T(\hat{F})$ gives

$$EIC(\mathbf{x}, y) = n(\mathbf{X}^{\mathrm{T}}\mathbf{X})^{-1}\mathbf{x}(y - \mathbf{x}^{\mathrm{T}}\hat{\boldsymbol{\beta}}) \tag{3.4.1}$$

and

$$EIC_i = EIC(\mathbf{x}_i, y_i) = n(\mathbf{X}^{\mathrm{T}}\mathbf{X})^{-1}\mathbf{x}_i e_i \tag{3.4.2}$$

where, as usual, $e_i = y_i - \mathbf{x}_i^{\mathrm{T}}\boldsymbol{\beta}$. The *EIC* is appealing on several grounds, as it appears to be an exact analogy to the influence curve. It measures the effects of an infinitesimal perturbation of $\hat{F}$ at $\mathbf{x}_i$, and corresponds to the infinitesimal jackknife method of Jaeckel (1972; see also Miller, 1974). The *EIC* pretends that an infinitely large sample has been used to obtain $\hat{F}$, and it measures the instantaneous rate of change in the estimator as a single case at $\mathbf{x}$ is added to the data.

A second sample version of the influence curve can be constructed to display the influence of the i-th case on the computed estimate of $\boldsymbol{\beta}$. The general idea is to substitute the sample cdf with the i-th case deleted for F in the influence curve and then evaluate the resulting *EIC* at the i-th case. This is analogous to the treatment in Example 3.2.1.

Let $\hat{F}_{(i)}$ denote the sample cdf with the i-th case deleted. For least squares estimators of $\boldsymbol{\beta}$, substitution of $\hat{F}_{(i)}$ for F in (3.3.5) yields an empirical influence curve with the i-th case deleted,

$$EIC_{(i)}(\mathbf{x}, y) = (n-1)(\mathbf{X}_{(i)}^{\mathrm{T}}\mathbf{X}_{(i)})^{-1}\mathbf{x}(y - \mathbf{x}^{\mathrm{T}}\hat{\boldsymbol{\beta}}_{(i)}) \qquad (3.4.3)$$

where $\hat{\boldsymbol{\beta}}_{(i)} = T(\hat{F}_{(i)})$ and

$$\mathbf{X}_{(i)}^{\mathrm{T}}\mathbf{X}_{(i)}/(n-1) = \int \mathbf{x}\mathbf{x}^{\mathrm{T}}\,\mathrm{d}\hat{F}_{(i)}$$

This represents n *EIC*s, one for each $i = 1, 2, \ldots, n$. The influence of the i-th case is determined by evaluating (3.4.3) at $(\mathbf{x}_i, \mathbf{y}_i)$,

$$EIC_{(i)}(\mathbf{x}_i, y_i) = (n-1)(\mathbf{X}_{(i)}^{\mathrm{T}}\mathbf{X}_{(i)})^{-1}\mathbf{x}_i(y_i - \mathbf{x}_i^{\mathrm{T}}\hat{\boldsymbol{\beta}}_{(i)}) \qquad (3.4.4)$$

Using the relationships in Appendix A.2, this can be more informatively expressed in terms of the full sample,

$$EIC_{(i)}(\mathbf{x}_i, y_i) = (n-1)(\mathbf{X}^{\mathrm{T}}\mathbf{X})^{-1}\mathbf{x}_i e_i/(1 - v_{ii})^2 \qquad (3.4.5)$$

The interpretation of $EIC_{(i)}$ is analogous to that for *EIC*. It should be remembered, however, that $EIC_{(i)}$ is the result of the evaluation of n separate influence curves.

3.4.2 SAMPLE INFLUENCE CURVES

Both the *EIC* and $EIC_{(i)}$ are constructed under the fiction that infinitely large samples have been used to obtain $\hat{F}$ and $\hat{F}_{(i)}$, $i = 1, 2, \ldots, n$. The sample size n in (3.4.2) and (3.4.5) appears as a result of the covariance structure and does not necessarily reflect the effects of a finite sample. When investigating the influence of individual cases on computed statistics, a more explicit dependence on n is desirable, or else important finite sample characteristics can be obscured. A more desirable sample version of the influence curve can be obtained by setting $F = \hat{F}$ and taking $\varepsilon = -1/(n-1)$ in the definition of the influence curve (3.2.1). Evaluating at $\mathbf{z}_i^{\mathrm{T}} = (\mathbf{x}_i^{\mathrm{T}}, y_i)$ and $\varepsilon = -1/(n-1)$, we find $(1-\varepsilon)\hat{F} + \varepsilon\delta_{\mathbf{z}_i} = \hat{F}_{(i)}$. The *sample influence curve* is then

$$
\begin{aligned}
SIC_i &= -(n-1)\,(T(\hat{F}_{(i)}) - T(\hat{F})) \\
&= (n-1)\,(\hat{\boldsymbol{\beta}} - \hat{\boldsymbol{\beta}}_{(i)}) \\
&= \frac{(n-1)\,(\mathbf{X}^{\mathrm{T}}\mathbf{X})^{-1}\mathbf{x}_i e_i}{1 - v_{ii}}
\end{aligned}
\tag{3.4.6}
$$

which is proportional to the change in the estimate of $\boldsymbol{\beta}$ when a case is deleted.

The essential difference between these three sample versions of the influence curve appears in the power of the $(1 - v_{ii})$ term in the denominators, while the numerators are essentially the same when evaluated at the sample points. Recall that remote rows of $\mathbf{X}$ will tend to have $1 - v_{ii}$ small. The $EIC_{(i)}$ will be most sensitive to cases with v_{ii} large, while EIC will be least sensitive. The SIC lies between these versions in terms of relative weight given to v_{ii}.

Any of these sample versions of the influence curve for $\boldsymbol{\beta}$ may be transformed to a sample influence curve for $\boldsymbol{\psi} = \mathbf{Z}\boldsymbol{\beta}$ by multiplying on the left by $\mathbf{Z}$; see (3.3.6).

An alternative and perhaps more immediately revealing derivation of EIC_i, $EIC_{(i)}$, and SIC_i can be obtained from a related perturbation scheme (Pregibon, 1979, 1981; Belsley *et al.*, 1980). Let all cases have error variance σ^2, except for case i with $\mathrm{var}(\varepsilon_i) = \sigma^2/w_i$, $w_i > 0$. Then, using Appendix A.2, the weighted least squares estimator of $\boldsymbol{\beta}$ as a function of w_i can be written as

$$
\hat{\boldsymbol{\beta}}(w_i) = \hat{\boldsymbol{\beta}} - \frac{(\mathbf{X}^{\mathrm{T}}\mathbf{X})^{-1}\mathbf{x}_i(1 - w_i)e_i}{1 - (1 - w_i)v_{ii}}
\tag{3.4.7}
$$

Differentiating (3.4.7) with respect to w_i yields

$$
\Delta\hat{\boldsymbol{\beta}}(w_i) = \frac{\mathrm{d}}{\mathrm{d}w_i}\hat{\boldsymbol{\beta}}(w_i) = \frac{(\mathbf{X}^{\mathrm{T}}\mathbf{X})^{-1}\mathbf{x}_i e_i}{(1 - (1 - w_i)v_{ii})^2}
\tag{3.4.8}
$$

The EIC_i, apart from the multiplier n, is found by evaluating $\Delta\hat{\boldsymbol{\beta}}(w_i)$ at $w_i = 1$ and thus describes the rate of change in the estimator as w_i deviates from 1. Similarly, the $EIC_{(i)}$ is found by evaluating $\Delta\hat{\boldsymbol{\beta}}(w_i)$ as $w_i \to 0$, and it measures the rate of change in the estimator as the i-th case is deleted. The SIC is a compromise between EIC and $EIC_{(i)}$ since

$$
(n-1)^{-1} SIC_i = \int_0^1 \Delta\hat{\boldsymbol{\beta}}(w_i)\,\mathrm{d}w_i
\tag{3.4.9}
$$

is the average gradient over the whole interval.

EXAMPLE 3.4.1. CLOUD SEEDING NO. 6. Figure 3.4.1 contains a graph of the estimate of β_{14} from the cloud seeding model (2.4.23) with $\lambda = 1/3$ as the weight for case 2 is varied from 0 to 1. The comments made above concerning the three empirical influence curves are clear. The EIC corresponds to the derivative at $w_i = 1$, which seems too small, while $EIC_{(i)}$, the derivative as $w_i \to 0$, seems too large. The SIC, which corresponds to the slope of the line joining $\hat{\boldsymbol{\beta}}(0)$ and $\hat{\boldsymbol{\beta}}(1)$, appears to provide a more satisfactory summary of this curve.□

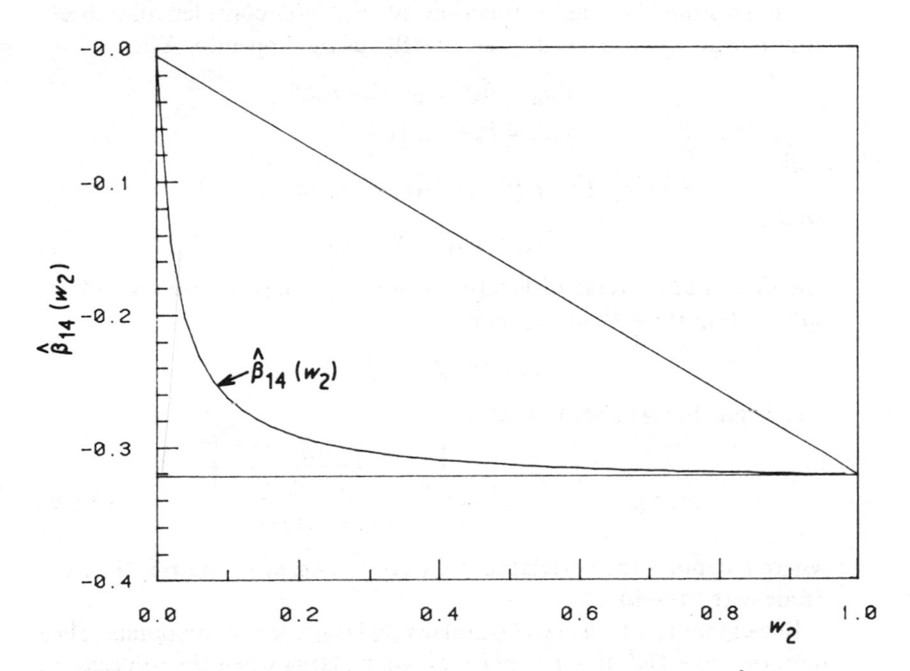

Figure 3.4.1 $\hat{\beta}_{14}(w_2)$ versus w_2, cloud seeding data. Note: s.e. $(\hat{\beta}_{14}(1)) = 0.177$, s.e. $(\hat{\beta}_{14}(0)) = 0.324$

EXAMPLE 3.4.2. PARTIAL F-TESTS. Partial F-tests for the hypothesis that the individual coefficients of $\boldsymbol{\beta}$ are zero are commonly used to simplify a linear model. When using this procedure, it is not uncommon to find that retention of a particular coefficient depends on the presence of a single case. This behavior seems particularly prevalent when the model contains polynomial terms. The influence of individual cases on the partial F-tests can be seen from the SIC for the associated F-statistics (Cook, 1979).

Let $\hat{\beta}_k$ denote the k-th component of $\hat{\boldsymbol{\beta}}$ and define

$$\tau_k = \hat{\beta}_k/\hat{\sigma}\sqrt{(b_k)}$$

where b_k is the k-th diagonal element of $(\mathbf{X}^T\mathbf{X})^{-1}$. The partial F-statistic for the hypothesis that $\beta_k = 0$ is $F_k = \tau_k^2$. Further, let $\hat{\beta}_{k(i)}$, $\tau_{k(i)}$, $b_{k(i)}$, $\hat{\sigma}_{(i)}^2$, and $F_{k(i)}$ denote the analogous quantities based on the data set without the i-th case.

Characteristics of the $SIC_i = (n-1)[F_k - F_{k(i)}]$ are most easily seen by expressing $F_{k(i)}$ as a function of F_k. We consider the three components comprising $F_{k(i)}$ separately: using Appendix A.2,

$$\hat{\beta}_{k(i)} = \hat{\beta}_k - c_{ki}e_i/(1-v_{ii}),$$
$$b_{k(i)} = b_k + c_{ki}^2/(1-v_{ii}),$$

and

$$\hat{\sigma}_{(i)}^2 = [(n-p')\hat{\sigma}^2 - e_i^2/(1-v_{ii})]/(n-p'-1)$$

where

$$c_{ki} = \mathbf{d}_k^T(\mathbf{X}^T\mathbf{X})^{-1}\mathbf{x}_i$$

and $\mathbf{d}_k$ is a unit vector of length p' with 1 in the k-th position. After substituting these three forms into

$$F_{k(i)} = \hat{\beta}_{k(i)}^2/\hat{\sigma}_{(i)}^2 b_{k(i)}$$

a little algebra will verify that

$$F_{k(i)} = \frac{(n-p'-1)r_i^2}{(n-p'-r_i^2)} \frac{\left[\frac{\tau_k}{r_i} - \rho\left(\frac{v_{ii}}{1-v_{ii}}\right)^{1/2}\right]^2}{[1+\rho^2 v_{ii}/(1-v_{ii})]} \tag{3.4.10}$$

where ρ denotes the correlation between $\hat{\beta}_k$ and $\mathbf{x}_i^T\hat{\boldsymbol{\beta}}$, and r_i is the i-th Studentized residual.

Recall that $v_{ii}/(1-v_{ii})$ will be relatively large for remote points. The term $(n-p'-1)r_i^2/(n-p'-r_i^2) = t_i^2$ will be large when the i-th case is an outlier, and under the null hypothesis it has an $F(1, n-p'-1)$ distribution.

It seems clear from inspection of Equation (3.4.10) that almost anything can happen to the partial F-statistics when a case is removed. Two general observations seem particularly interesting, however: Suppose that the deleted case appears to be an outlier (r_i^2 is large) and that $\rho(v_{ii}/(1-v_{ii}))^{1/2}$ is negligible; empirical investigations indicate that typically ρ is not negligible by itself. Then,

$$F_{k(i)} \cong F_k\left(\frac{n-p'-1}{n-p'-r_i^2}\right) > F_k$$

Deleting a case with $r_i^2 > 1$ in a dense region will tend to increase all partial F-statistics.

Next, consider the deletion of a point that fits the model quite well ($r_i^2 \leq 1$). Then,

$$F_k - F_{k(i)} \cong \frac{\{\rho\tau_k[v_{ii}/(1-v_{ii})]^{1/2} + r_i\}^2}{1 + \rho^2 v_{ii}/(1-v_{ii})} - r_i^2$$

and we can generally expect all partial F-statistics greater than one to decrease when a conforming point which has v_{ii} large is deleted.□

3.5 Applications of the sample influence curve

The sample influence curve defined at (3.4.6) has natural appeal as the basis for diagnostic techniques that locate influential cases. We recall again its basic properties: It is computed from observed data and apart from constants it is interpreted as the change in a statistic when a case is deleted. Also, for many problems including linear least squares regression, the *SIC*, or approximations thereof, can be easily computed. We shall see that the sample influence curve has other desirable properties derived from geometrical considerations and from extensions to the study of the influence of groups of cases.

In the remainder of this chapter, methods for finding influential cases are developed from the sample influence curve; methods based on the *EIC* or $EIC_{(i)}$ can be developed similarly. To be most useful, such methods should allow the cases to be ordered on the basis of influence. For linear least squares, the *SIC* for $\hat{\boldsymbol{\beta}}$ is a p'-dimensional vector and there is no natural ordering of multidimensional vectors. Even in the case $p' = 2$ where a scatterplot of the *SIC* can be constructed and inspected, there is no natural way to construct a complete ordering of the points on the basis of influence. It is necessary, therefore, to use a norm to characterize influence and order cases. A norm may be regarded as a function which maps the *SIC* into R^1. Of course, there is a natural ordering (less than) for points in R^1. The choice of a norm to characterize the *SIC* is a crucial part of the study of influence.

Norms can be usefully defined from properties of a model. We call such norms *external*. Alternatively, they can be defined without reference to the model by considering the n values of the sample influence curve as a multivariate sample, and applying an *internal* norm. After a study of characterizing norms for the influence of a single case, we turn to multiple case influence measures, which are straightforward generalizations of the one-at-a-time statistics. Other norms for influence are discussed in Chapters 4 and 5.

3.5.1 EXTERNAL SCALING

In linear least squares regression, the sample influence curve for $\boldsymbol{\beta}$ is given by

$$(n-1)^{-1} SIC_i = \hat{\boldsymbol{\beta}} - \hat{\boldsymbol{\beta}}_{(i)} \tag{3.5.1}$$

Since SIC_i is a p'-vector, it is useful to consider norms $D_i(\mathbf{M}, c)$ determined by a symmetric, positive (semi-)definite $p' \times p'$ matrix $\mathbf{M}$ and a positive scale factor c:

$$\begin{aligned} D_i(\mathbf{M}, c) &= (n-1)^2 \frac{(SIC_i)^{\mathrm{T}} \mathbf{M} (SIC_i)}{c} \\ &= \frac{(\hat{\boldsymbol{\beta}}_{(i)} - \hat{\boldsymbol{\beta}})^{\mathrm{T}} \mathbf{M} (\hat{\boldsymbol{\beta}}_{(i)} - \hat{\boldsymbol{\beta}})}{c} \end{aligned} \tag{3.5.2}$$

Contours of constant $D_i(\mathbf{M}, c)$ are ellipsoids of dimension equal to the rank of $\mathbf{M}$. The contours may be viewed as being centered at $\hat{\boldsymbol{\beta}}$ or $\hat{\boldsymbol{\beta}}_{(i)}$, both interpretations being used in what follows.

This general norm has a useful interpretation in terms of linear combinations of the elements of $\boldsymbol{\beta}$. Let $\mathbf{z}$ denote an arbitrary $p' \times 1$ vector, $k = ((n-1)^2 c)^{-1}$ and, assuming that $\mathbf{M}$ is positive definite, let

$$\mathscr{E} = \{\mathbf{z} | \mathbf{z}^{\mathrm{T}} \mathbf{M}^{-1} \mathbf{z} \leq k\}$$

As a function of $\mathbf{z}$, the SIC for $\mathbf{z}^{\mathrm{T}} \boldsymbol{\beta}$ is

$$SIC_i(\mathbf{z}) = (n-1) \mathbf{z}^{\mathrm{T}} (\hat{\boldsymbol{\beta}} - \hat{\boldsymbol{\beta}}_{(i)})$$

and

$$\max_{\mathbf{z} \in \mathscr{E}} [SIC_i(\mathbf{z})]^2 = D_i(\mathbf{M}, c)$$

The maximum is attained in the direction of $\mathbf{M}(\hat{\boldsymbol{\beta}} - \hat{\boldsymbol{\beta}}_{(i)})$. Thus, $D_i(\mathbf{M}, c)$ can be interpreted as the maximum over $\mathbf{z}$ of the squared sample influence curves for $\mathbf{z}^{\mathrm{T}} \boldsymbol{\beta}$ when $\mathbf{z}$ is constrained to lie within the ellipsoid $\mathscr{E}$. Of course, the ordering over i of these maxima will not change if $k > 0$ is allowed to be arbitrary, but independent of i.

Clearly, the character of $D_i(\mathbf{M}, c)$ is determined by $\mathbf{M}$ and c, which may be chosen to reflect specific concerns. In what follows, we discuss both internal and external norms. The inner-product matrix $\mathbf{M}$ is nonstochastic for external norms in linear least squares. For internal norms, which are developed in Section 3.5.2, **M** is stochastic even for linear least squares.

A form of $D_i(\mathbf{M}, c)$ that reveals the effects of varying **M** and c is obtained by using Appendix A.2 to express $\hat{\beta}_{(i)}$ in terms of the full sample,

$$
\begin{aligned}
D_i(\mathbf{M}, c) &= \frac{e_i \mathbf{x}_i^{\mathrm{T}} (\mathbf{X}^{\mathrm{T}}\mathbf{X})^{-1} \mathbf{M} (\mathbf{X}^{\mathrm{T}}\mathbf{X})^{-1} \mathbf{x}_i e_i}{c(1 - v_{ii})^2} \\
&= r_i^2 \frac{\hat{\sigma}^2}{c} \frac{\mathbf{x}_i^{\mathrm{T}} (\mathbf{X}^{\mathrm{T}}\mathbf{X})^{-1} \mathbf{M} (\mathbf{X}^{\mathrm{T}}\mathbf{X})^{-1} \mathbf{x}_i}{(1 - v_{ii})} \\
&= r_i^2 \frac{\hat{\sigma}^2}{c} P_i(\mathbf{M}) \qquad (3.5.3)
\end{aligned}
$$

where r_i is the i-th Studentized residual and $P_i(\mathbf{M})$ is defined implicitly in this expression. By the nature of the regression problem, **M** and c should be chosen to make $D_i(\mathbf{M}, c)$ invariant under changes of scale and nonsingular linear transformations of the rows of **X**. In particular, c should be chosen so that $\hat{\sigma}^2/c$ is scale free. While there are many ways to achieve this, two stand out as obvious candidates: Choose $c = k\hat{\sigma}^2$ or $k\hat{\sigma}_{(i)}^2$, where $k > 0$ is a known constant that does not depend on **X**. The former choice was suggested by Cook (1977a, 1979) and Cook and Weisberg (1980), while the latter choice has been suggested by Belsley *et al.* (1980) and Atkinson (1981, 1982). If $c = k\hat{\sigma}^2$, then $r_i^2 \hat{\sigma}^2/c = r_i^2/k$. On the other hand, when $c = k\hat{\sigma}_{(i)}^2$ it follows from (2.2.8) that $r_i^2 \hat{\sigma}^2/c = t_i^2/k$ where, as before, t_i^2 is the i-th externally Studentized residual.

For either of these choices for c, the stochastic part of $D_i(\mathbf{M}, c)$ depends only on r_i^2. Since the null distribution of r_i^2 does not depend on **X**, or on the actual values of the parameters, it is reasonable to ask how the influence of the i-th case can be altered when the fit, as measured at the i-th case by r_i^2, is fixed. With r_i^2 fixed, it is clear from (3.5.3) that influence is a monotonically increasing function of $P_i(\mathbf{M})$. If $P_i(\mathbf{M})$ is large, the observed value of r_i^2 must be small for the case to be uninfluential. However, $Er_i^2 = 1$ under a correct model and thus cases with large $P_i(\mathbf{M})$ will typically be influential.

If **M** is nonstochastic, then so is $P_i(\mathbf{M})$, and its magnitude depends on the location of $\mathbf{x}_i^{\mathrm{T}}$ relative to the distance measures determined by the inner-product matrices $(\mathbf{X}^{\mathrm{T}}\mathbf{X})^{-1}\mathbf{M}(\mathbf{X}^{\mathrm{T}}\mathbf{X})^{-1}$ and $(\mathbf{X}^{\mathrm{T}}\mathbf{X})^{-1}$. It can be expected that cases with large v_{ii} will have $P_i(\mathbf{M})$ large also. However, this need not necessarily follow since by choice of **M** the numerator of $P_i(\mathbf{M})$ can be small even if v_{ii} is large.

We view $P_i(\mathbf{M})$ as the *potential*, relative to **M**, for the i-th case to be influential. Potential is important since it can be used to describe and detect configurations of the rows of $\mathbf{X}$ that are likely to produce highly influential cases.

$\mathbf{M} = \mathbf{X}^{\mathrm{T}}\mathbf{X}$, $c = p'\hat{\sigma}^2$

Although the class of invariant norms is large, one stands out by appeal to usual confidence ellipsoid arguments. A $(1-\alpha) \times 100\%$ confidence ellipsoid for $\boldsymbol{\beta}$ based on $\hat{\boldsymbol{\beta}}$ is given by the set of all $\boldsymbol{\beta}^*$ such that

$$\frac{(\boldsymbol{\beta}^* - \hat{\boldsymbol{\beta}})^{\mathrm{T}}(\mathbf{X}^{\mathrm{T}}\mathbf{X})(\boldsymbol{\beta}^* - \hat{\boldsymbol{\beta}})}{p'\hat{\sigma}^2} \leq F(1-\alpha; p', n-p') \tag{3.5.4}$$

This ellipsoid is centered at $\hat{\boldsymbol{\beta}}$, with contours determined by the eigenvalues and eigenvectors of $(\mathbf{X}^{\mathrm{T}}\mathbf{X})$; $p'\hat{\sigma}^2$ is a scale factor used to assign proper values to contours. Reference to (3.5.4) suggests setting $\mathbf{M} = \mathbf{X}^{\mathrm{T}}\mathbf{X}$ and $c = p'\hat{\sigma}^2$ in (3.5.2), to give

$$D_i \equiv D_i(\mathbf{X}^{\mathrm{T}}\mathbf{X}, p'\hat{\sigma}^2) = \frac{(\hat{\boldsymbol{\beta}}_{(i)} - \hat{\boldsymbol{\beta}})^{\mathrm{T}}(\mathbf{X}^{\mathrm{T}}\mathbf{X})(\hat{\boldsymbol{\beta}}_{(i)} - \hat{\boldsymbol{\beta}})}{p'\hat{\sigma}^2} \tag{3.5.5}$$

This measure, first proposed by Cook (1977a), gives the squared distance from $\hat{\boldsymbol{\beta}}$ to $\hat{\boldsymbol{\beta}}_{(i)}$ relative to the fixed geometry of $\mathbf{X}^{\mathrm{T}}\mathbf{X}$. By exploiting the similarity to (3.5.4), values of $D_i(\mathbf{X}^{\mathrm{T}}\mathbf{X}, p'\hat{\sigma}^2)$ can be converted to a familiar probability scale by comparing computed values to the $F(p', n-p')$ distribution. For example, if $D_i(\mathbf{X}^{\mathrm{T}}\mathbf{X}, p'\hat{\sigma}^2)$ equals the 0.50 value of the corresponding F distribution, then deletion of the i-th case would move the estimate of $\boldsymbol{\beta}$ to the edge of a 50% confidence ellipsoid relative to $\hat{\boldsymbol{\beta}}$. However, $D_i(\mathbf{X}^{\mathrm{T}}\mathbf{X}, p'\hat{\sigma}^2)$ is *not* distributed as F; this comparison is used only for converting D_i to a familiar scale (Cook, 1977b).

Figure 3.5.1 illustrates the measure $D_i(\mathbf{X}^{\mathrm{T}}\mathbf{X}, p'\hat{\sigma}^2)$ for a problem with $p' = 2$ and no intercept. The figure is derived from a linear approximation to a nonlinear problem to be discussed in Example 5.1.1. The elliptical contours correspond to $D_i(\mathbf{X}^{\mathrm{T}}\mathbf{X}, p'\hat{\sigma}^2) = \text{constant}$. Although contours of constant influence are elliptical, the $(\hat{\beta}_{1(i)}, \hat{\beta}_{2(i)})$ often tend to have nonelliptical scatter. In the figure, they generally fall along a curve, with the exception of the one clearly influential case in the lower-left corner.

Alternatively, D_i can be rewritten as

$$D_i = \frac{(\hat{\mathbf{Y}}_{(i)} - \hat{\mathbf{Y}})^{\mathrm{T}}(\hat{\mathbf{Y}}_{(i)} - \hat{\mathbf{Y}})}{p'\hat{\sigma}^2} \tag{3.5.6}$$

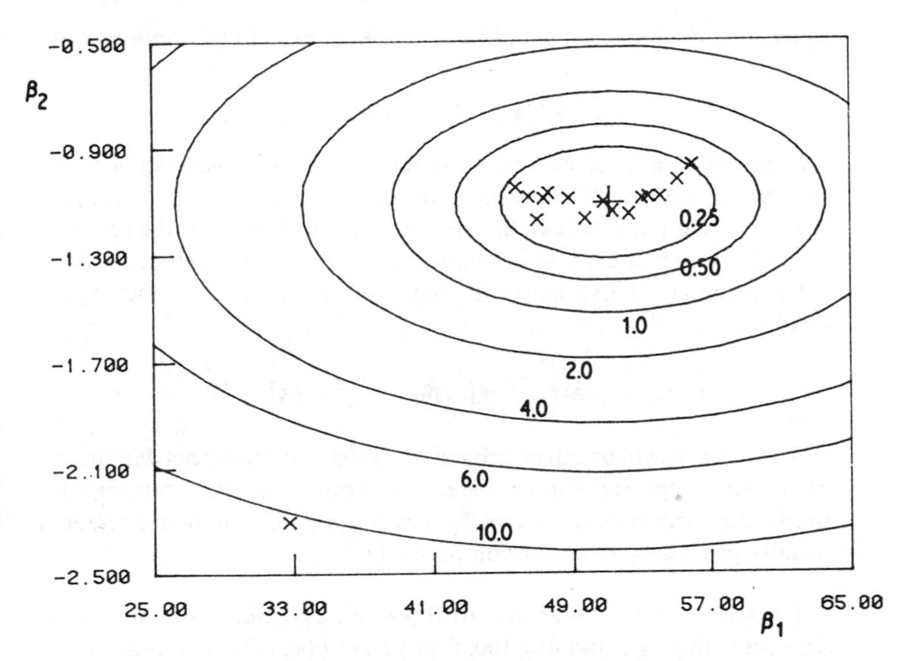

Figure 3.5.1 $\hat{\beta}_{2(i)}$ versus $\hat{\beta}_{1(i)}$ in a model with no intercept. + indicates the full data estimate $(\hat{\beta}_1, \hat{\beta}_2)$. Ellipses are contours of constant D_i with values shown

where $\hat{\mathbf{Y}}_{(i)} = \mathbf{X}\hat{\boldsymbol{\beta}}_{(i)}$. For problems where prediction is of more interest than estimation, D_i may be viewed as the usual Euclidean distance between $\hat{\mathbf{Y}}$ and $\hat{\mathbf{Y}}_{(i)}$. Clearly, any norm in the p'-dimensional estimation space may be regarded as a norm in the n-dimensional observation space provided that $\mathbf{M}$ is of the form $\mathbf{M} = \mathbf{X}^{\mathrm{T}}\mathbf{B}\mathbf{X}$.

A computationally convenient and revealing form for D_i is obtained by substituting $\mathbf{M} = \mathbf{X}^{\mathrm{T}}\mathbf{X}$ into (3.5.3) (Cook, 1977a)

$$D_i = \frac{1}{p'} r_i^2 \frac{v_{ii}}{1 - v_{ii}} \tag{3.5.7}$$

Apart from the constant p', D_i is the product of a random term r_i^2 and the potential $P_i(\mathbf{X}^{\mathrm{T}}\mathbf{X}) = v_{ii}/(1 - v_{ii})$, which is a monotonic function of v_{ii}. For linear least squares, computation and examination of the v_{ii} has become common practice (see, for example, Hoaglin and Welsch, 1978), and this is sensible if $\mathbf{M} = \mathbf{X}^{\mathrm{T}}\mathbf{X}$ is used to define a norm. The potential itself can be given several interesting interpretations. Cook (1977a) noted that $v_{ii}/(1-v_{ii}) = \text{var}(\hat{y}_i)/\text{var}(e_i)$. Weisberg (1980a) pointed out that

$$\mathbf{x}_i^{\mathrm{T}}(\mathbf{X}_{(i)}^{\mathrm{T}}\mathbf{X}_{(i)})^{-1}\mathbf{x}_i = v_{ii}/(1-v_{ii})$$

so the potential is a distance relative to the ellipsoids defined by $(\mathbf{X}_{(i)}^{\mathrm{T}}\mathbf{X}_{(i)})^{-1}$. Huber (1981) noted the relationship $\hat{y}_i = (1-v_{ii})\mathbf{x}_i^{\mathrm{T}}\hat{\boldsymbol{\beta}}_{(i)} + v_{ii}y_i$ so that potential can be interpreted as a function of the relative weight of y_i in determining $\hat{y}_i$. Finally, $v_{ii}/(1-v_{ii})$ is proportional to the total change in the variance of prediction at $\mathbf{x}_1, \ldots, \mathbf{x}_n$ when $\mathbf{x}_i$ is deleted,

$$v_{ii}/(1-v_{ii}) = \left[\sum_j \text{var}(\mathbf{x}_j^{\mathrm{T}}\hat{\boldsymbol{\beta}}_{(i)}) - \sum_j \text{var}(\mathbf{x}_j^{\mathrm{T}}\hat{\boldsymbol{\beta}})\right]\Big/\sigma^2$$

The i-th case will be called influential if D_i is large; the exact definition of large will depend on the problem, but D_i greater than 1, corresponding to distances between $\hat{\beta}$ and $\hat{\beta}_{(i)}$ beyond a 50% confidence region, usually provides a basis for comparison.

EXAMPLE 3.5.1. COMBINATIONS OF r_i^2, v_{ii}. Suppose that in a data set with $p' = 3$ and $n = 100$, four pairs of (e_i, v_{ii}) occur as given in Table 3.5.1. For each of these four cases $D_i = 3.0$, so deletion of any one of the four would move the estimate of $\boldsymbol{\beta}$ to the edge of a 95% confidence region about $\hat{\boldsymbol{\beta}}$, and each would be called highly influential. However, the reasons for the influence in the four cases are not the same. Cases 3 and 4 appear to be outliers given the extreme values for r_i, while for the other two cases the influence is apparent because of the potential; the large values of v_{ii} indicate that these cases are relatively far removed from the bulk of the data. □

EXAMPLE 3.5.2. RAT DATA. In an experiment to investigate the amount of a drug retained in the liver of a rat, 19 rats were

Table 3.5.1 *Residuals and* v_{ii} *for four hypothetical cases*

	e_i	v_{ii}	r_i	D_i
1	0.6325	0.9000	1.000	3.0
2	1.732	0.7500	1.732	3.0
3	9.000	0.2500	5.196	3.0
4	19.087	0.1000	9.000	3.0

randomly selected, weighed, placed under light ether anesthesia and given an oral dose of the drug. The dose an animal received was determined as approximately 40 mg of the drug per kilogram of body weight, since liver weight is known to be strongly related to body weight and it was felt that large livers would absorb more of a given dose than smaller livers. After a fixed length of time each rat was sacrificed, the liver weighed, and the percent of the dose in the liver determined.

The experimental hypothesis was that, for the method of determining the dose, there is no relationship between the percentage of the dose in the liver (Y) and the body weight (X_1), liver weight (X_2), and relative dose (X_3).

The data and sample correlations are given in Tables 3.5.2 and 3.5.3(a). As had been expected, the sample correlations between the response and the explanatory variables are all small, and none of the simple regressions of dose on any of the explanatory variables is significant, all having t-values less than 1 as shown in Table 3.5.3(b). However, the regression of Y on X_1, X_2, and X_3 gives a different and

Table 3.5.2 *Rat data. Source: Weisberg* (1980*a*)

X_1-*Body weight* (*g*)	X_2-*Liver weight* (*g*)	X_3-*Relative dose*	Y
176	6.5	0.88	0.42
176	9.5	0.88	0.25
190	9.0	1.00	0.56
176	8.9	0.88	0.23
200	7.2	1.00	0.23
167	8.9	0.83	0.32
188	8.0	0.94	0.37
195	10.0	0.98	0.41
176	8.0	0.88	0.33
165	7.9	0.84	0.38
158	6.9	0.80	0.27
148	7.3	0.74	0.36
149	5.2	0.75	0.21
163	8.4	0.81	0.28
170	7.2	0.85	0.34
186	6.8	0.94	0.28
146	7.3	0.73	0.30
181	9.0	0.90	0.37
149	6.4	0.75	0.46

Table 3.5.3 *Rat data.* (*a*) *Sample correlations.*

X_1-Body weight (g)	1.000			
X_2-Liver weight (g)	0.500	1.000		
X_3-Relative dose	0.990	0.490	1.000	
Y	0.151	0.203	0.338	1.000
	Body weight	Liver weight	Dose	Y

Table 3.5.3 *Rat data.* (*b*) *Regression summary, t-values in parentheses*

	Model including			
Coefficient	X_1	X_2	X_3	(X_1, X_2, X_3)
Intercept	0.196	0.220	0.133	0.266
	(0.89)	(1.64)	(0.63)	(1.37)
β_1 (rat weight)	0.0008			−0.0212
	(0.63)			(−2.67)
β_2 (liver weight)		0.0147		0.0143
		(0.86)		(0.83)
β_3 (dose)			0.235	4.178
			(0.96)	(2.74)

contradictory result: two of the explanatory variables, X_1 and X_3, have significant t-tests, with $p < 0.05$ in both cases, indicating that the two measurements combined are apparently useful indicators of Y. If X_2 is dropped from the model, the same phenomenon appears. The analysis so far might lead to the conclusion that a combination of dose and rat weight is associated with the response.

Figure 3.5.2 gives plots of r_i, v_{ii}, and D_i against case number for the model Y on X_1, X_2, X_3. The r_i do not display any unusual features as they are all less than 2, without obvious trends or patterns. However, inspecting the D_i, we locate a possible cause: case 3 has $D_3 = 0.93$; no other case has D_i bigger than 0.27, suggesting that case 3 alone may have large enough influence to induce the anomaly. The value of $v_{33} = 0.85$ indicates that the problem with this case is that the vector $\mathbf{x}_3$ is different from the others.

When case 3 is deleted, and the model is refit, the t-values for the coefficients of X_1, X_2, and X_3 are all substantially less than 1 in absolute value, so the anomalous result of a significant pair of regressors can be attributed to case 3 alone. Of course, this could have been anticipated from the discussion given in Example 3.4.2.

The reason for the influence of case 3 must now be studied. Inspection of the data indicates that this rat, with weight 190 g, was reported to have received a full dose of 1.00, which was a larger dose than it should have received according to the rule for assigning doses (for example, rat 8 with a weight of 195 g received a lower dose of 0.98). A number of causes for the result found in the first analysis are possible: (1) the dose or weight recorded for case 3 is in error or (2) the regression fit in the second analysis is not appropriate except in the

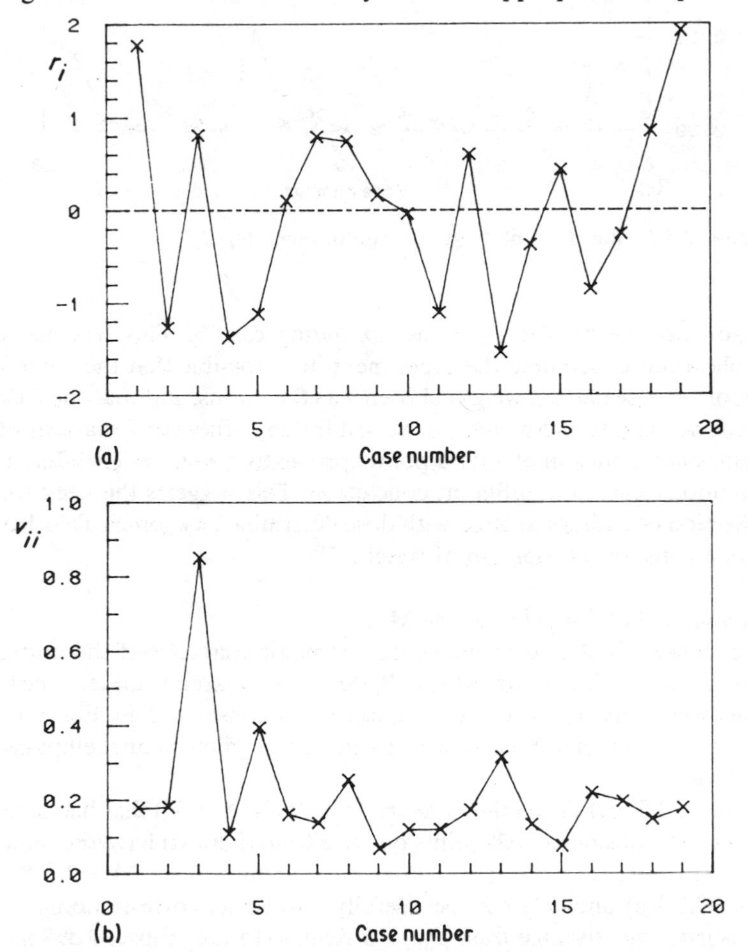

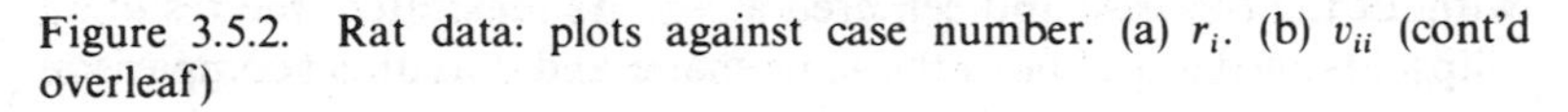

Figure 3.5.2. Rat data: plots against case number. (a) r_i. (b) v_{ii} (cont'd overleaf)

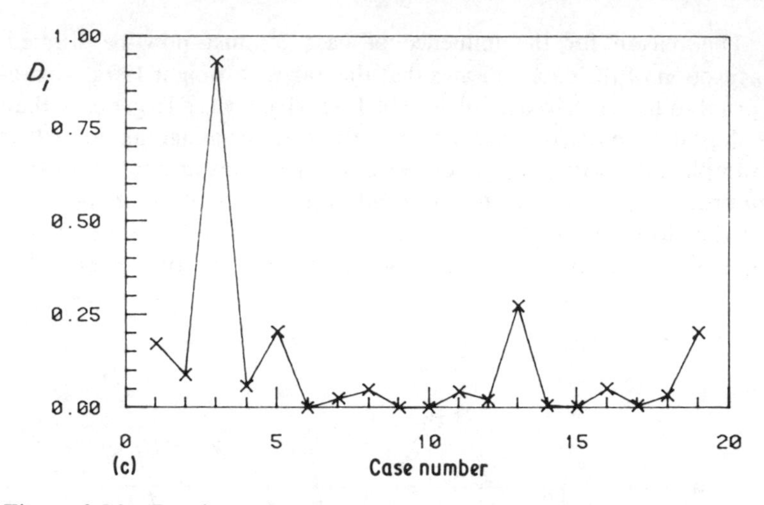

Figure 3.5.2 Rat data: plot against case number. (c) D_i

region defined by the 18 points excluding case 3. This has many implications concerning the experiment. It is possible that the combination of dose and rat weight chosen was fortuitous, and that the lack of relationship found would not persist for any other combinations of them, since inclusion of a data point apparently taken under different conditions leads to a different conclusion. This suggests the need for collection of additional data, with dose determined by some rule other than a constant proportion of weight.□

Alternative full rank choices for **M**, *c*

The choice of (**M**, c) determines the geometric character of the norm. The class of (**M**, c) for which $D_i(\mathbf{M}, c)$ is invariant under linear transformations is large, but the examples considered in Fig. 3.5.3 depict four obvious choices corresponding to p'-dimensional elliptical contours.

Figure 3.5.3(a) shows the measure $D_i = D_i(\mathbf{X}^{\mathrm{T}}\mathbf{X}, p'\hat{\sigma}^2)$ that has been previously considered. All points on the ellipsoid drawn have the same value for the characterizing norm. Measures using $\mathbf{M} = \mathbf{X}_{(i)}^{\mathrm{T}}\mathbf{X}_{(i)}$ (Figs. 3.5.3(b) and (c)) can be usefully viewed as corresponding to measuring the distance *from* $\hat{\boldsymbol{\beta}}_{(i)}$ *to* $\hat{\boldsymbol{\beta}}$ relative to the ellipsoid defined without the i-th case and centered at $\hat{\boldsymbol{\beta}}_{(i)}$. As illustrated, the resulting ellipsoids need not all be of the same shape, and thus direct comparison

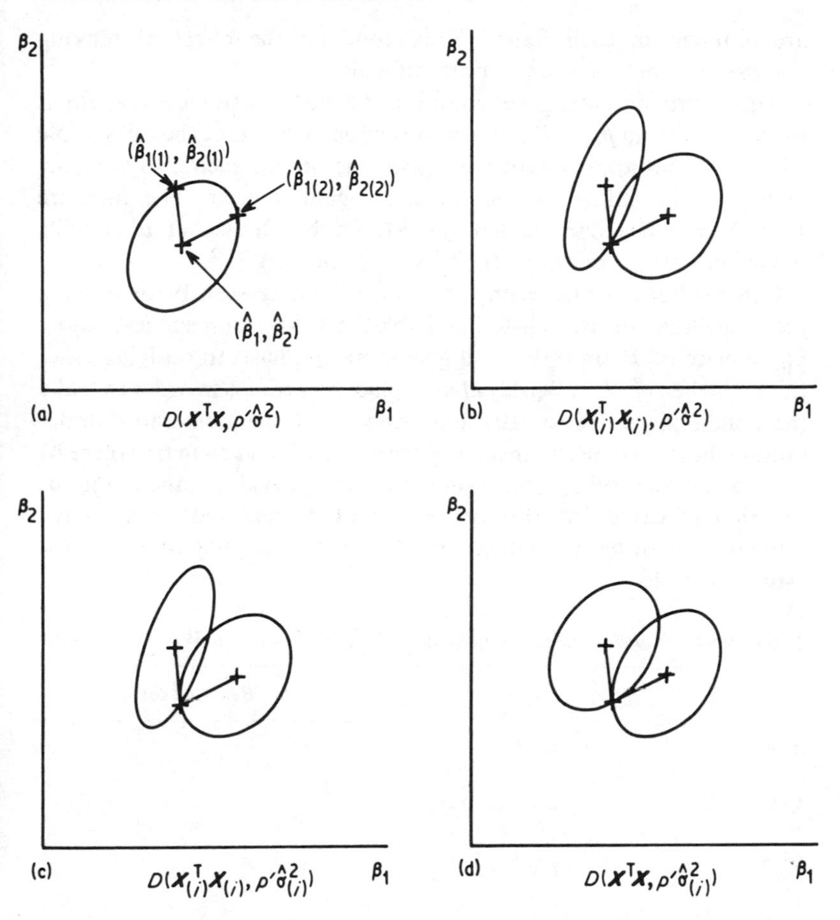

Figure 3.5.3 Graphical comparison of four norms: (a) $D_i(\mathbf{X}^T\mathbf{X}, p'\hat{\sigma}^2)$. (b) $D_i(\mathbf{X}_{(i)}^T \mathbf{X}_{(i)}, p'\hat{\sigma}^2)$. (c) $D_i(\mathbf{X}_{(i)}^T \mathbf{X}_{(i)}, p'\hat{\sigma}_{(i)}^2)$. (d) $D_i(\mathbf{X}^T\mathbf{X}, p'\hat{\sigma}_{(i)}^2)$

of the norm from case-to-case is questionable. From Fig. 3.5.3(b), for example, deletion of case 2 appears to lead to more nearly circular contours than did deletion of case 1 and, while $D_1(\mathbf{X}^T\mathbf{X}, p'\hat{\sigma}^2) = D_2(\mathbf{X}^T\mathbf{X}, p'\hat{\sigma}^2)$, the relationship between $D_1(\mathbf{X}_{(1)}^T\mathbf{X}_{(1)}, p'\hat{\sigma}^2)$ and $D_2(\mathbf{X}_{(2)}^T\mathbf{X}_{(2)}, p'\hat{\sigma}^2)$ is uncertain, as either may be larger.

In Fig. 3.5.3(b), $c = p'\hat{\sigma}^2$ while $c = p'\hat{\sigma}_{(i)}^2$ in Fig. 3.5.3(c). These two figures look alike and they have the same contours of constant value, but the values assigned to the contours are different, as the scale factors are different in each figure. This, too, has the effect of making comparisons between cases more difficult.

The fourth measure, graphed in Fig. 3.5.3(d), can be viewed again as the distance from $\hat{\boldsymbol{\beta}}_{(i)}$ to $\hat{\boldsymbol{\beta}}$ using ellipsoids determined by the full sample $\mathbf{M} = \mathbf{X}^{\mathrm{T}}\mathbf{X}$, but applying different scale factors for each i, so comparability of the values of the norm is again unclear. The measure $D_i(\mathbf{X}^{\mathrm{T}}\mathbf{X}, \hat{\sigma}^2_{(i)})$ has been called $(DFFITS)^2$ by Belsley *et al.* (1980). Atkinson (1981) discusses $[D_i(\mathbf{X}^{\mathrm{T}}\mathbf{X}, p'\hat{\sigma}^2_{(i)}/(n-p'))]^{1/2}$.

Other differences between these norms can be seen by examining their algebraic forms, as listed in Table 3.5.4. Atkinson suggests using $\hat{\sigma}^2_{(i)}$ in place of $\hat{\sigma}^2$ since this will give more emphasis to outlying cases ($t_i^2 > r_i^2$ when $r_i^2 > 1$). Belsley *et al.* replace $\hat{\sigma}^2$ with $\hat{\sigma}^2_{(i)}$ in order to make the denominator statistically independent of the numerator under normal theory. We prefer measures based on a fixed geometry where $\mathbf{M}$ and c do not depend on i since such measures provide an unambiguous ordering of cases. In addition, $\hat{\sigma}^2$ could be replaced by a robust estimator in order to reduce the effects of outlying cases on the estimated scale.

Table 3.5.4 *Normed influence measures. Source: Cook and Weisberg* (1980)

$\mathbf{M}$	c	*Reduced form*
$\mathbf{X}^{\mathrm{T}}\mathbf{X}$	$p'\hat{\sigma}^2$	$\frac{1}{p'} r_i^2 \frac{v_{ii}}{1-v_{ii}}$
$\mathbf{X}^{\mathrm{T}}\mathbf{X}$	$p'\hat{\sigma}^2_{(i)}$	$\frac{1}{p'} t_i^2 \frac{v_{ii}}{1-v_{ii}}$
$\mathbf{X}^{\mathrm{T}}_{(i)}\mathbf{X}_{(i)}$	$p'\hat{\sigma}^2$	$\frac{1}{p'} r_i^2 v_{ii}$
$\mathbf{X}^{\mathrm{T}}_{(i)}\mathbf{X}_{(i)}$	$p'\hat{\sigma}^2_{(i)}$	$\frac{1}{p'} t_i^2 v_{ii}$
$[\mathrm{diag}(\mathbf{X}^{\mathrm{T}}\mathbf{X})^{-1}]^{-1}$	$p'\hat{\sigma}^2_{(i)}$	$\frac{1}{p'} t_i^2 \frac{\mathbf{x}_i^{\mathrm{T}}(\mathbf{X}^{\mathrm{T}}\mathbf{X})^{-1}\mathbf{M}(\mathbf{X}^{\mathrm{T}}\mathbf{X})^{-1}\mathbf{x}_i}{1-v_{ii}}$
$\mathbf{I}$	$p'\hat{\sigma}^2$	$\frac{1}{p'} r_i^2 \frac{\mathbf{x}_i^{\mathrm{T}}(\mathbf{X}^{\mathrm{T}}\mathbf{X})^{-2}\mathbf{x}_i}{1-v_{ii}}$

Lower-dimensional norms

If $\mathbf{M}$ is chosen to have rank $q < p'$, contours of constant $D_i(\mathbf{M}, c)$ are q-dimensional ellipsoids. In particular, if $\mathbf{Z}$ is a $q \times p'$ rank q matrix such that $\boldsymbol{\psi} = \mathbf{Z}\boldsymbol{\beta}$ is of interest, then the norm with

$$\mathbf{M} = \mathbf{Z}^{\mathrm{T}}(\mathbf{Z}(\mathbf{X}^{\mathrm{T}}\mathbf{X})^{-1}\mathbf{Z}^{\mathrm{T}})^{-1}\mathbf{Z}; \qquad c = q\hat{\sigma}^2 \tag{3.5.8}$$

is an invariant norm corresponding to q-dimensional ellipsoidal confidence contours for $\boldsymbol{\psi}$ based on $\hat{\boldsymbol{\psi}} = \mathbf{Z}\hat{\boldsymbol{\beta}}$.

Suppose a subset of the elements of $\boldsymbol{\beta}$, say the last q, is of interest. Partition $\mathbf{X} = (\mathbf{X}_1, \mathbf{X}_2)$, where $\mathbf{X}_2$ is $n \times q$. If $\mathbf{Z} = (\mathbf{0}, \mathbf{I}_q)$,

$$(\mathbf{X}^{\mathrm{T}}\mathbf{X})^{-1}\mathbf{M}(\mathbf{X}^{\mathrm{T}}\mathbf{X})^{-1} = (\mathbf{X}^{\mathrm{T}}\mathbf{X})^{-1} - \begin{bmatrix} (\mathbf{X}_1^{\mathrm{T}}\mathbf{X}_1)^{-1} & \mathbf{0} \\ \mathbf{0} & \mathbf{0} \end{bmatrix} \tag{3.5.9}$$

Substituting this choice of $\mathbf{M}$ into the general form (3.5.3) for $D_i(\mathbf{M}, c)$ and simplifying yields

$$D_i^* \equiv D_i(\mathbf{M}, c) = \frac{r_i^2}{q}\frac{v_{ii} - u_{ii}}{1 - v_{ii}} = \frac{r_i^2}{q}\frac{w_{ii}^*}{1 - v_{ii}} \tag{3.5.10}$$

where $\mathbf{U} = \mathbf{X}_1(\mathbf{X}_1^{\mathrm{T}}\mathbf{X}_1)^{-1}\mathbf{X}_1^{\mathrm{T}}$, and $\mathbf{W}^* = \mathbf{V} - \mathbf{U}$ projects onto the columns of $(\mathbf{I} - \mathbf{U})\mathbf{X}_2$. The u_{ii} can be obtained from the projection matrix for $\mathbf{X}_1$, and the w_{ii}^* obtained by subtraction. The potential $P_i(\mathbf{M})$ for this measure is $w_{ii}^*/(1 - v_{ii})$ which will tend to be relatively small if the i-th row of $(\mathbf{I} - \mathbf{U})\mathbf{X}_2$ is sufficiently close to zero or to the sample average if the constant is not in $\mathbf{X}_1$.

Two special cases of lower-dimensional norms are of some interest. If we set $q = 1$, then the measure concentrates on a single coefficient of the parameter vector. If $c = \hat{\sigma}_{(i)}^2$ is used to replace $c = \hat{\sigma}^2$, the resulting measure is called $DFBETAS_{ij}$ by Belsley *et al.* (1980). The potential when $q = 1$ will be small if the i-th residual from the regression of $\mathbf{X}_2$ on $\mathbf{X}_1$ is small. In the general situation with $q = 1$, suppose $\boldsymbol{\psi} = \mathbf{z}^{\mathrm{T}}\boldsymbol{\beta}$. If $c = \hat{\sigma}^2$, the norm becomes (Cook, 1977a)

$$D_i(\mathbf{M}, c) = p' D_i \rho^2(\mathbf{x}_i^{\mathrm{T}}\hat{\boldsymbol{\beta}}, \mathbf{z}^{\mathrm{T}}\hat{\boldsymbol{\beta}}) \tag{3.5.11}$$

where $\rho(.,.)$ is the correlation. The maximum $p'D_i$ of this norm for fixed i occurs at $\mathbf{z} = \mathbf{x}_i$.

If $q = p$, $c = p\hat{\sigma}^2$, and the intercept is excluded, then

$$\mathbf{M} = \begin{pmatrix} 0 & \mathbf{0} \\ \mathbf{0} & \mathscr{X}^{\mathrm{T}}\mathscr{X} \end{pmatrix} \tag{3.5.12}$$

and

$$D_i(\mathbf{M}, c) = \frac{r_i^2}{p}\frac{v_{ii} - 1/n}{1 - v_{ii}} \tag{3.5.13}$$

When the intercept is not of interest, this last measure may be preferable to the more usual $D_i = D_i(\mathbf{X}^{\mathrm{T}}\mathbf{X}, p'\hat{\sigma}^2)$. We will continue to use D_i since modification of results for (3.5.13) is straightforward. When $\mathbf{x}_i = \bar{\mathbf{x}}$, $v_{ii} = 1/n$, and measure (3.5.13) is zero. Relative to this measure, observations at $\bar{\mathbf{x}}$ have no influence.

EXAMPLE 3.5.3. CLOUD SEEDING NO. 7. For the cloud seeding data, the coefficients for the seeding effect and interactions are of primary interest, so the choice of

$$\boldsymbol{\psi}^{\mathrm{T}} = (\beta_1, \beta_{13}, \beta_{14}, \beta_{15}, \beta_{16})$$

is suggested. The distance measure based on ellipsoids for $\boldsymbol{\psi}$, can be computed from (3.5.10). For the data using $Y^{(1/3)}$ and $P^{(1/3)}$, the values of D_i and D_i^{ψ} are given in Table 3.5.5 (the other columns in this table refer to a later example). The ordering of cases on influence is similar for the two measures. Case 2 is the most influential, but D_2 is over three times the size of D_2^{ψ}.□

Predictions

The diagnostics considered thus far measure the influence of individual cases in terms of their effects on the estimation of selected linear combinations of the elements of $\boldsymbol{\beta}$. The general measure $D_i(\mathbf{M}, c)$, however, is applicable in situations where prediction rather than estimation is the primary goal.

Let $\mathbf{X}_f$ be a $q \times p'$ matrix and suppose we wish to predict the q-vector of future values

$$\mathbf{Y}_f = \mathbf{X}_f\boldsymbol{\beta} + \boldsymbol{\varepsilon}_f \tag{3.5.14}$$

where $\boldsymbol{\varepsilon}_f$ is independent of the vector of errors $\boldsymbol{\varepsilon}$ in (2.1.1) and $\mathrm{Var}(\boldsymbol{\varepsilon}_f) = \sigma^2\mathbf{I}$. A point prediction for $\mathbf{Y}_f$ is $\hat{\mathbf{Y}}_f = \mathbf{X}_f\hat{\boldsymbol{\beta}}$, and

$$\mathrm{Var}(\hat{\mathbf{Y}}_f - \mathbf{Y}_f) = \sigma^2\mathbf{X}_f(\mathbf{X}^{\mathrm{T}}\mathbf{X})^{-1}\mathbf{X}_f^{\mathrm{T}} + \sigma^2\mathbf{I} \tag{3.5.15}$$

A $(1-\alpha)\times 100\%$ normal theory prediction region for $\mathbf{Y}_f$ is given by the collection of all q-vectors $\mathbf{Y}^*$ such that

$$\frac{(\hat{\mathbf{Y}}_f - \mathbf{Y}^*)^{\mathrm{T}}[\mathbf{X}_f(\mathbf{X}^{\mathrm{T}}\mathbf{X})^{-1}\mathbf{X}_f^{\mathrm{T}} + \mathbf{I}]^{-1}(\hat{\mathbf{Y}}_f - \mathbf{Y}^*)}{q\hat{\sigma}^2} \le F(1-\alpha; q, n-p') \tag{3.5.16}$$

The sample influence curve for the point prediction $\hat{\mathbf{Y}}_f$ is proportional to $(\hat{\mathbf{Y}}_f - \hat{\mathbf{Y}}_{f(i)}) = \mathbf{X}_f(\hat{\boldsymbol{\beta}} - \hat{\boldsymbol{\beta}}_{(i)})$. This in combination with (3.5.16)

Table 3.5.5 *Influence statistics, cloud seeding data*

Case	e_i	r_i	v_{ii}	t_i	D_i	D_i^{ψ}	δ_1^0	δ_i^1	J_i
1	0.578	0.522	0.552	0.507	0.030	0.031	0.703	0.774	0.172
2	−0.301	−1.143	0.975	−1.158	4.555	1.377	0.816	1.000	62.189
3	−0.645	−0.638	0.626	−0.622	0.062	0.043	0.547	0.605	0.192
4	1.287	0.940	0.314	0.935	0.037	0.024	0.700	0.492	0.073
5	−0.872	−0.686	0.411	−0.672	0.030	0.012	0.396	0.355	0.056
6	0.303	0.300	0.627	0.289	0.014	0.011	0.291	0.510	0.103
7	−2.861	−2.573	0.548	−3.528	0.729	0.714	0.911	0.935	0.220
8	2.438	1.683	0.233	1.828	0.078	0.097	0.727	0.480	0.061
9	0.070	0.055	0.412	0.053	0.000	0.000	0.009	0.008	0.001
10	−1.545	−1.165	0.358	−1.183	0.069	0.077	0.493	0.367	0.059
11	0.747	0.554	0.334	0.538	0.014	0.012	0.284	0.157	0.032
12	−0.951	−0.630	0.167	−0.615	0.007	0.008	0.349	0.052	0.025
13	0.138	0.106	0.387	0.102	0.001	0.001	0.032	0.026	0.004
14	−0.203	−0.135	0.170	−0.130	0.000	0.000	0.008	0.003	0.001
15	2.165	1.620	0.347	1.742	0.126	0.154	0.709	0.637	0.082
16	0.045	0.035	0.374	0.033	0.000	0.000	0.002	0.001	0.000
17	−0.881	−1.071	0.753	−1.078	0.318	0.411	0.900	0.980	0.724
18	0.303	0.601	0.907	0.586	0.320	0.399	0.409	0.974	2.326
19	0.071	0.050	0.261	0.048	0.000	0.000	0.002	0.001	0.000
20	0.645	0.638	0.626	0.622	0.062	0.071	0.547	0.605	0.192
21	−0.495	−0.345	0.249	−0.333	0.004	0.003	0.086	0.023	0.008
22	1.125	0.817	0.307	0.806	0.027	0.032	0.425	0.272	0.044
23	1.151	1.080	0.585	1.088	0.149	0.185	0.744	0.835	0.212
24	−2.313	−1.938	0.479	−2.209	0.315	0.303	0.910	0.908	0.165

suggests the norm $D_i(\mathbf{M}, c)$ with $c = q\hat{\sigma}^2$ and

$$\mathbf{M} = \mathbf{X}_f^{\mathrm{T}}[\mathbf{X}_f(\mathbf{X}^{\mathrm{T}}\mathbf{X})^{-1}\mathbf{X}_f^{\mathrm{T}} + \mathbf{I}]^{-1}\mathbf{X}_f \tag{3.5.17}$$

If a single prediction is of interest then $q = 1$, $\mathbf{X}_f = \mathbf{x}_f^{\mathrm{T}}$ and

$$D_i(\mathbf{M}, c) = p' D_i \rho^2(\mathbf{x}_i^{\mathrm{T}}\hat{\boldsymbol{\beta}}, \mathbf{x}_f^{\mathrm{T}}\hat{\boldsymbol{\beta}}) v_f/(1 + v_f) \tag{3.5.18}$$

where $v_f = \mathbf{x}_f^{\mathrm{T}}(\mathbf{X}^{\mathrm{T}}\mathbf{X})^{-1}\mathbf{x}_f$ and ρ^2 is defined following (3.5.11). Thus, the norm of the sample influence curve for a single prediction is simply the analogous norm (3.5.11) for estimation, reduced by the factor $v_f/(1 + v_f)$. Clearly, $p'D_i$ provides an upper bound for predictive as well as estimative influence.

A drawback to the use of $D_i(\mathbf{M}, c)$ for prediction is the requirement that $\mathbf{X}_f$ be specified *a priori*. If a model is to be used primarily for prediction, $\mathbf{X}_f$ may not be known during the development of the model. A possible solution to this problem is to construct $\mathbf{X}_f$ by choosing points that in some sense cover the region of interest. Coverage could be reflected both in terms of the location of the points and their density. From this point of view, a useful default is the choice $\mathbf{X}_f = \mathbf{X}$; that is, consider the predictions at the cases used to construct the model. Then,

$$\mathbf{M} = \mathbf{X}^{\mathrm{T}}[\mathbf{I} + \mathbf{V}]^{-1}\mathbf{X} = \tfrac{1}{2}(\mathbf{X}^{\mathrm{T}}\mathbf{X})$$

since $[\mathbf{I} + \mathbf{V}]^{-1} = \mathbf{I} - \frac{1}{2}\mathbf{V}$. When $\mathbf{X}_f = \mathbf{X}$,

$$D_i(\mathbf{M}, c) = \frac{p'}{2n} D_i \tag{3.5.19}$$

A second possible solution to the problem of an unspecified $\mathbf{X}_f$ is to set $q = 1$ and, for each i, choose $\mathbf{x}_f$ to maximize (3.5.18). Let $v_{fi} = \mathbf{x}_i^{\mathrm{T}}(\mathbf{X}^{\mathrm{T}}\mathbf{X})^{-1}\mathbf{x}_f$ and rewrite (3.5.18) as

$$D_i(\mathbf{M}, c) = \frac{r_i^2 v_{fi}^2}{(1 - v_{ii})(1 + v_f)}$$

Thus, maximizing $D_i(\mathbf{M}, c)$ by choice of $\mathbf{x}_f$ is equivalent to maximizing $v_{fi}^2/(1 + v_f)$ (see Appendix A.3). It follows that

$$\max_{\mathbf{x}_f}\left(\frac{v_{fi}^2}{1 + v_f}\right) = v_{ii} - \frac{1}{n+1} \tag{3.5.20}$$

and therefore

$$\max_{\mathbf{x}_f}[D_i(\mathbf{M}, c)] = r_i^2 \frac{v_{ii} - 1/(n+1)}{1 - v_{ii}} \tag{3.5.21}$$

For large n, this is essentially $p'D_i$.

3.5.2 INTERNAL SCALING

In the linear least squares problem, use of an external norm that corresponds to using confidence contours to order the values of the sample influence curve is straightforward and appealing. These norms are based on fixed metrics that do not depend on the observed behavior of the sample versions of the influence curve. Of course, they do depend on the expected behavior of the data in so far as $(\mathbf{X}^T\mathbf{X})^{-1}$ or the related inner-product matrices accurately reflect the variance of $\hat{\boldsymbol{\beta}}$. In contrast, internal norms are based on a matrix that derives from the observed values of the appropriate sample version of the influence curve. Internal norms may be constructed to be robust with respect to variations in the model or methods of analysis that would necessitate different external norms. If, for example, the model were altered to have $\mathrm{Var}(\mathbf{e}) = \sigma^2\mathbf{W}^{-1}$, where $\mathbf{W}$ is known, then to be consistent with previous rationale the inner-product matrix $\mathbf{X}^T\mathbf{X}$ for an external norm should be changed to $\mathbf{X}^T\mathbf{W}\mathbf{X}$.

We present two methods for internal scaling. In the first, the n values $\hat{\boldsymbol{\beta}} - \hat{\boldsymbol{\beta}}_{(i)}$ are treated as an unstructured p'-dimensional sample, and a multivariate outlier technique is used to order the values. Other methods for ordering a multivariate sample are given in Andrews (1972), Gnanadesikan (1977), and Barnett and Lewis (1978, Chapter 6). In the second, we consider the norms $D_i(\mathbf{M}, c)$, where $\mathbf{M}$, and c are chosen through use of the jackknife method.

Ordering using a multivariate outlier statistic

One method that is particularly well suited for study of the n values of the *SIC* is Wilks' (1963) criterion for detecting a single outlier in a multivariate sample. Let $\mathbf{b}_1, \mathbf{b}_2, \ldots, \mathbf{b}_n$ be p'-vectors, and define $\bar{\mathbf{b}} = n^{-1}\Sigma\,\mathbf{b}_i$ and $\boldsymbol{\Delta} = \Sigma\,(\mathbf{b}_i - \bar{\mathbf{b}})(\mathbf{b}_i - \bar{\mathbf{b}})^T$. Wilks' criterion selects $\mathbf{b}_i$ as a possible outlier if *i minimizes*

$$\frac{|\boldsymbol{\Delta}_{(i)}|}{|\boldsymbol{\Delta}|} \tag{3.5.22}$$

Since $|\boldsymbol{\Delta}|$ is proportional to the square of the volume of a p'-dimensional ellipsoid, minimizing this ratio is equivalent to choosing $\mathbf{b}_i$ to minimize the volume remaining after $\mathbf{b}_i$ is deleted, so in some sense $\mathbf{b}_i$ must be far from the other vectors $\mathbf{b}_j$, $j \neq i$.

The results on determinants in Appendix A.2 can be used to simplify the ratio (3.5.22). One finds that minimizing (3.5.22) is equivalent to maximizing the distance

$$\delta_i = (\mathbf{b}_i - \bar{\mathbf{b}})^{\mathrm{T}}\boldsymbol{\Delta}^{-1}(\mathbf{b}_i - \bar{\mathbf{b}}) \tag{3.5.23}$$

over $i = 1, 2, \ldots, n$. For linear least squares, explicit formulae for the δ_i can be obtained for any of the empirical versions of the influence function discussed earlier in this chapter. It is useful here to discuss the *EIC* and the *SIC* separately.

For the empirical influence curve (3.4.2), it is sufficient to take $\mathbf{b}_i = (\mathbf{X}^{\mathrm{T}}\mathbf{X})^{-1}\mathbf{x}_i e_i$ and thus $\bar{\mathbf{b}} = \mathbf{0}$. The inner product matrix, say $\boldsymbol{\Delta}_0$, is then

$$\boldsymbol{\Delta}_0 = \sum \mathbf{b}_j\mathbf{b}_j^{\mathrm{T}} = (\mathbf{X}^{\mathrm{T}}\mathbf{X})^{-1}\left[\sum e_j^2\mathbf{x}_j\mathbf{x}_j^{\mathrm{T}}\right](\mathbf{X}^{\mathrm{T}}\mathbf{X})^{-1} \tag{3.5.24}$$

The quantity $(n/(n-p'))\boldsymbol{\Delta}_0$ is a robust estimate of Var $(\hat{\boldsymbol{\beta}})$ obtained using the weighted jackknife method proposed by Hinkley (1977). Substituting into (3.5.23), the corresponding normed measure is

$$\begin{aligned}\delta_i^0 &= (\mathbf{b}_i - \bar{\mathbf{b}})^{\mathrm{T}}\boldsymbol{\Delta}_0^{-1}(\mathbf{b}_i - \bar{\mathbf{b}}) \\ &= e_i^2\mathbf{x}_i^{\mathrm{T}}\left[\sum_j e_j^2\mathbf{x}_j\mathbf{x}_j^{\mathrm{T}}\right]^{-1}\mathbf{x}_i, \qquad i = 1, 2, \ldots, n\end{aligned} \tag{3.5.25}$$

The statistic δ_i^0 can be computed by first defining $\mathbf{W}$ to be an $n \times n$ diagonal matrix with diagonals e_j^2, $j = 1, \ldots, n$. The δ_i^0 are then the diagonal elements of $\mathbf{W}^{1/2}\mathbf{X}(\mathbf{X}^{\mathrm{T}}\mathbf{W}\mathbf{X})^{-1}\mathbf{X}^{\mathrm{T}}\mathbf{W}^{1/2}$, the projection on the columns of $\mathbf{W}^{1/2}\mathbf{X}$.

For the sample influence curve defined at (3.4.6), we can take $\mathbf{b}_i = (\mathbf{X}^{\mathrm{T}}\mathbf{X})^{-1}\mathbf{x}_i e_i/(1 - v_{ii})$. Since $\bar{\mathbf{b}}$ is not zero, the form (3.5.23) does not simplify. The cross product matrix, say $\boldsymbol{\Delta}_1$, is

$$\begin{aligned}\boldsymbol{\Delta}_1 &= \sum(\mathbf{b}_j - \bar{\mathbf{b}})(\mathbf{b}_j - \bar{\mathbf{b}})^{\mathrm{T}} \\ &= (\mathbf{X}^{\mathrm{T}}\mathbf{X})^{-1}\left(\sum\frac{e_j^2}{(1-v_{jj})^2}\mathbf{x}_j\mathbf{x}_j^{\mathrm{T}} - n\bar{\mathbf{z}}\bar{\mathbf{z}}^{\mathrm{T}}\right)(\mathbf{X}^{\mathrm{T}}\mathbf{X})^{-1}\end{aligned} \tag{3.5.26}$$

where $n\bar{\mathbf{z}} = \Sigma\mathbf{x}_j e_j/(1 - v_{jj})$. The matrix $(n-1)\boldsymbol{\Delta}_1/n$ is the estimate of Var $(\hat{\boldsymbol{\beta}})$ obtained from the usual, unweighted jackknife (Miller, 1974). Although the corresponding internally scaled measure δ_i^1 can be computed exactly, some desirable algebraic simplification is possible if the usually small correction for the center $\bar{\mathbf{z}}$ is neglected. Setting $\bar{\mathbf{z}} = \mathbf{0}$ and substituting into (3.5.23), the resulting measure is

$$\delta_i^1 \simeq \frac{e_i^2}{(1-v_{ii})^2}\mathbf{x}_i^{\mathrm{T}}\left[\sum\frac{e_j^2}{(1-v_{jj})^2}\mathbf{x}_j\mathbf{x}_j^{\mathrm{T}}\right]^{-1}\mathbf{x}_i \tag{3.5.27}$$

As with the *EIC*, this measure can be computed as the diagonal elements of the projection on the columns of $\mathbf{W}^{1/2}\mathbf{X}$, where the $n \times n$ diagonal matrix $\mathbf{W}$ has diagonal elements $e_i^2/(1-v_{ii})^2$.

Jackknife method

The jackknife can be used to provide an alternative internal scaling method for empirical influence curves. In the most frequently used version of the jackknife, estimates are obtained by averaging n analyses, each obtained by deleting one case at a time from the data (Miller, 1974, provides a review). In many problems, jackknife estimates of parameters and variances have desirable properties. For example, Hinkley (1977) suggests $n\mathbf{\Delta}_0/(n-p')$ as an alternative estimator of $\text{Var}(\hat{\beta})$ that is robust against nonconstant error variances. This, in turn, suggests the use of $J_i \equiv D_i(\mathbf{\Delta}_0^{-1}, p'n/(n-p'))$ as an alternative to $D_i = D_i(\mathbf{X}^{\mathrm{T}}\mathbf{X}, p'\hat{\sigma}^2)$. The interpretation of J_i is the same as that of D_i, except that the metric should now be more robust. The statistic J_i, for $i = 1, 2, \ldots, n$, is given by

$$J_i \equiv D_i(\mathbf{\Delta}_0^{-1}, p'n/(n-p')) = \frac{n-p'}{np'} \frac{e_i^2 \mathbf{x}_i^{\mathrm{T}} [\sum e_j^2 \mathbf{x}_j \mathbf{x}_j^{\mathrm{T}}]^{-1} \mathbf{x}_i}{(1-v_{ii})^2}$$

$$= \frac{n-p'}{np'} \frac{\delta_i^0}{(1-v_{ii})^2} \tag{3.5.28}$$

J_i provides an interesting compromise between δ_i^0 and δ_i^1. In addition, the interpretation of J_i as a robust version of D_i has some appeal. A drawback of J_i, at least for linear least squares, is that its computation will generally require a second pass through the data to obtain δ_i^0, while D_i is computed directly from r_i and v_{ii}.

EXAMPLE 3.5.4. CLOUD SEEDING NO. 8. Table 3.5.5 lists the values of δ_i^0, δ_i^1, and J_i, as well as D_i and D_i^ψ as discussed in Example 3.5.3 for the cloud seeding data in the cube root scale. The statistics show reasonable agreement, although $J_2 = 62.189$ and $\delta_2^1 = 1.000$ are remarkably large, stressing the role of case 2 more clearly. Also, $J_{18} = 2.326$ suggests further interest in case 18. The *EIC* measure δ_i^0 pays less attention to the v_{ii}, and the δ_i^0 are large for cases with large r_i^2. □

3.5.3 GRAPHICAL AIDS

The study of influence can be augmented by a number of graphical displays. The most elementary are plots of the statistics r_i, v_{ii}, and

$D_i(\mathbf{M}, c)$ against case number. As illustrated earlier in Example 3.5.2, these plots provide a quick method of finding cases with large residuals, high potential, and high influence. They will be especially effective if the sample size is too large to make examination of lists of statistics useful, or if the ordering of cases is meaningful.

Atkinson (1981) has suggested that influence for an entire sample can be assessed by a display of the $[D_i(\mathbf{M}, c)]^{1/2}$ in a half-normal plot, with a simulated envelope added as described in Section 2.3.4. High influence cases will appear as isolated points at the far right of the graph. If no cases are influential this plot should be approximately straight. If part of the plot falls outside the simulated envelope, then some evidence is given that the assumptions used to compute the envelope, usually normality, independence, and constant variance, do not apply.

A third graphical aid for the assessment of influence is the added variable plot discussed in Section 2.3.2. Using the notation of that section, the added variable plot is a graph of the residuals obtained when $\mathbf{X}_k$ is deleted from the model, $(\mathbf{I} - \mathbf{U}_k)\mathbf{Y}$, against the residuals from the regression of $\mathbf{X}_k$ on the other Xs, $(\mathbf{I} - \mathbf{U}_k)\mathbf{X}_k$. In some ways these plots can be interpreted as a plot of y versus x in simple linear regression. Individual or groups of cases that stand apart from the rest of the cases should be investigated further. Their influence on the coefficient in question can be determined by deleting them, either individually or in groups, and recomputing the regression. Often, it will be found that such cases are influential.

While these plots are undoubtedly useful in trying to understand influence, they must be interpreted and used with some care since their use does not correspond to any standard case-by-case diagnostic method. When any case is omitted from the data, the projection matrix $\mathbf{U}_k$ changes and the entire character of the plot can change. In addition, these plots can fail to identify highly influential cases. If the i-th diagonal element of $\mathbf{U}_k$ is large, the corresponding elements of $(\mathbf{I} - \mathbf{U}_k)\mathbf{Y}$ and $(\mathbf{I} - \mathbf{U}_k)\mathbf{X}_k$ will tend to be small and thus the plotted point may not exhibit unusual characteristics, while the corresponding case could substantially influence $\hat{\beta}_k$.

EXAMPLE 3.5.5. CLOUD SEEDING NO. 9. Figure 3.5.4(a) is a half normal plot of $D_i^{1/2}$ for the cloud seeding data in the cube root scale. The relatively wide envelope at the right of the plot suggests that an influential case is very likely given the particular array of Xs in this data; one such influential case is observed. Aside from this one point, the plot

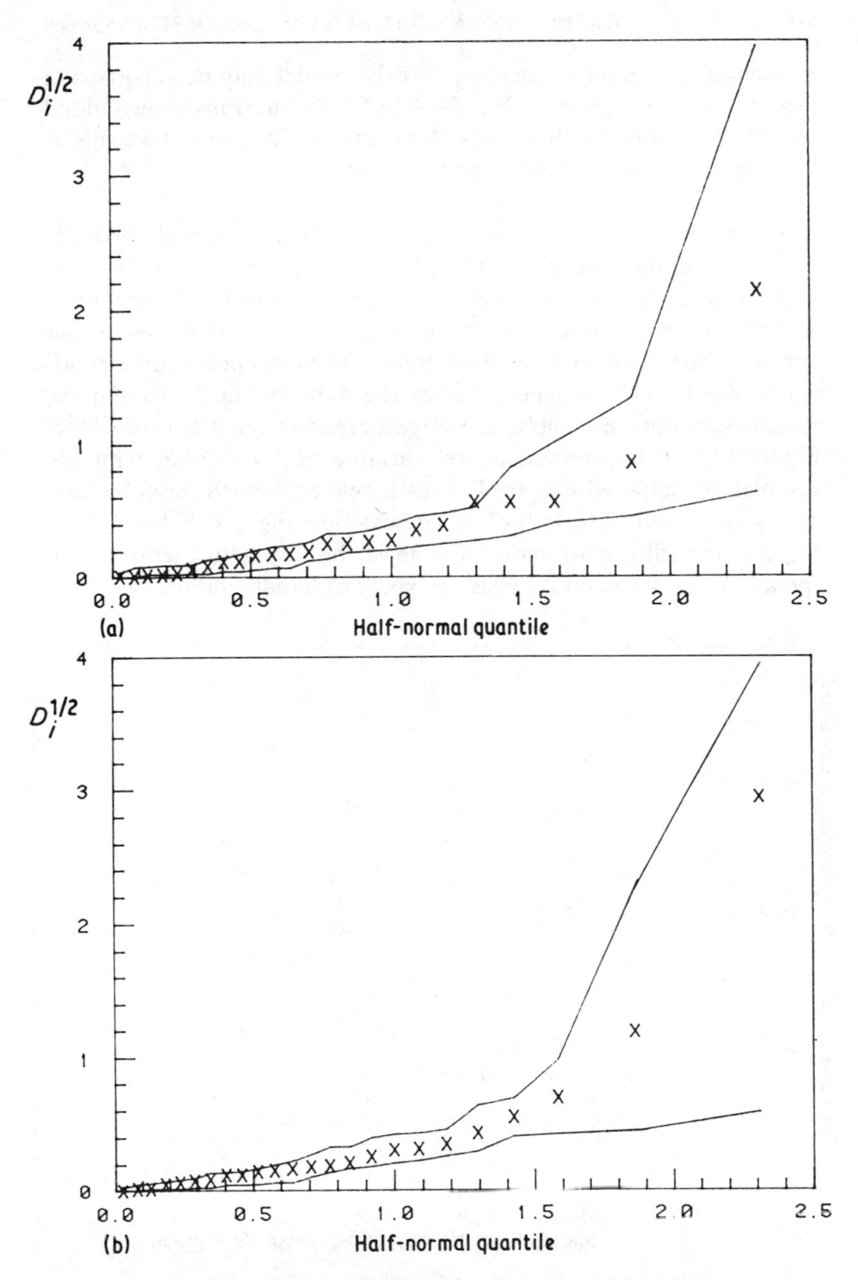

Figure 3.5.4 Half-normal plot of $D_i^{1/2}$, cloud seeding data. (a) *Y*, *P*, *AP* transformed via cube root transformation. (b) Untransformed data

is essentially straight, suggesting that the model may be adequate. In contrast, the plot given in Fig. 3.5.4(b) for the untransformed cloud seeding data is generally curved. Atkinson (1982) would take this as evidence of the need to transform the data. □

EXAMPLE 3.5.6. JET FIGHTERS NO. 4. Added variable plots for the jet fighter data are given in Figs. 2.3.8–2.3.10. In the plot for *SLF*, Fig. 2.3.10, it appears that the F-111A suppresses the usefulness of this variable since, if this case were deleted, the remaining cases would appear to show a slight systematic trend. When the point corresponding to the F-111A is removed from the data in Fig. 2.3.10, but the residuals are not recomputed, the slope increases from 0.0837 to 0.1156. Figure 3.5.5 is the correct added variable plot for *SLF*, with the residuals recomputed after the F-111A is deleted from the original data. The slope fitted here is 0.1386, so just deleting the F-111A from Fig. 2.3.10 results in an underestimate of the slope. Furthermore, the spread in Fig. 3.5.5 and 2.3.10 is markedly different, and the two plots

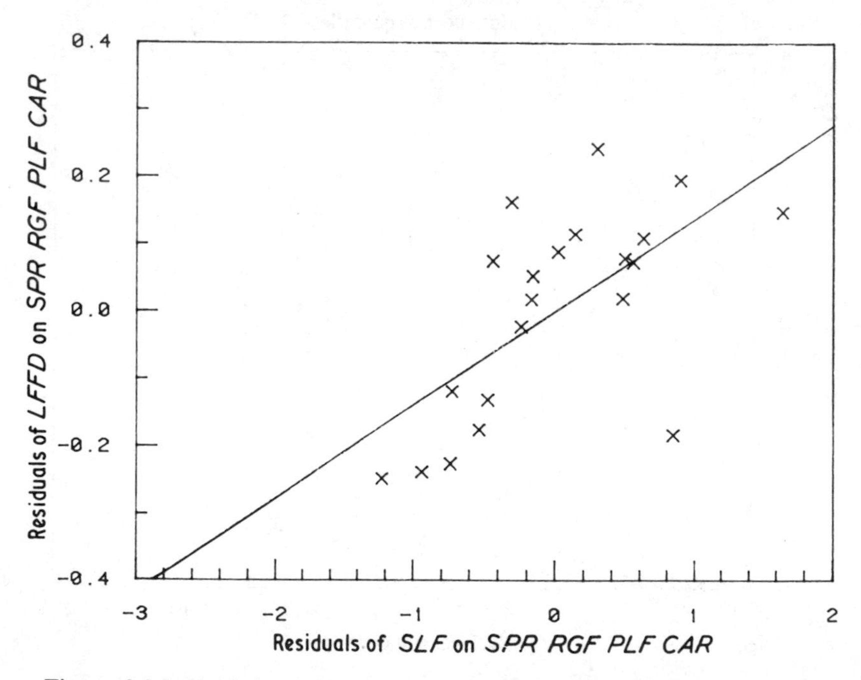

Figure 3.5.5 Added variable plot for *SLF* computed *without* the F-111A, jet fighter data

suggest different conclusions concerning SLF. From Fig. 3.5.5, SLF is more clearly important. □

3.6 Multiple cases

Both the derivation of the influence curves, and the diagnostic procedures developed from them, have concentrated on the effects of individual cases on estimates. For theoretical use of influence curves to study estimators, it can be expected that a study of pointwise influence will suffice. Additionally, in many practical data analytic problems, consideration of cases one at a time will provide the analyst with most of the information needed concerning the influence of cases on the fitted model. However, it can happen that a group of cases will be influential *en bloc*, but this influence can go undetected when cases are examined individually. This is illustrated with Fig. 3.6.1. If point C or D is deleted, the fitted regression will change very little. If both are deleted, the estimates of parameters may be very different. Conversely, if A or B is deleted the fitted line will change but if both are deleted, the fitted line will stay about the same.

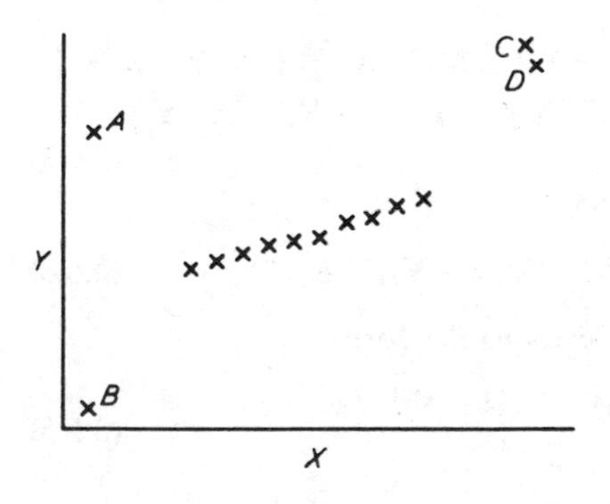

Figure 3.6.1 Illustration of joint influence. Source: Cook and Weisberg (1980)

The generalization of the influence curve and its empirical versions to multiple case problems is straightforward. Let I be an m-vector of indices of selected cases, $\mathrm{I}^{\mathrm{T}} = (i_1, i_2, \ldots, i_m)$, $1 \le i_j \le n$, and continue the earlier notation so that the subscript '(I)' means 'with the m cases indexed by I deleted,' while 'I' without parentheses will mean that only the cases indexed by I are remaining. For linear least squares, one obvious generalization of the sample influence curve is

$$SIC_{\mathrm{I}} = (n-m)(\hat{\beta} - \hat{\beta}_{(\mathrm{I})})$$

There are $\binom{n}{m}$ possible sets of cases at which the sample influence curve can be evaluated.

The (externally) normed measure $D_I(\mathbf{X}^T\mathbf{X}, p'\hat{\sigma}^2)$ is

$$D_I \equiv D_I(\mathbf{X}^T\mathbf{X}, p'\hat{\sigma}^2) = \frac{(\hat{\boldsymbol{\beta}}_{(I)} - \hat{\boldsymbol{\beta}})^T(\mathbf{X}^T\mathbf{X})(\hat{\boldsymbol{\beta}}_{(I)} - \hat{\boldsymbol{\beta}})}{p'\hat{\sigma}^2} \tag{3.6.1}$$

as given by Cook and Weisberg (1980). The other externally normed measures discussed in the last section are similarly defined for multiple cases. The geometric interpretation of these measures is identical to that for $m = 1$. An influential subset for estimating $\boldsymbol{\beta}$ will correspond to a large D_I.

As might be expected D_I can be expressed in multidimensional analogues of the r_i and v_{ii}. The results are obtained by first expressing $\hat{\boldsymbol{\beta}}_{(I)}$ as a function of $\hat{\boldsymbol{\beta}}$. Following Bingham (1977),

$$\begin{aligned}\hat{\boldsymbol{\beta}}_{(I)} &= (\mathbf{X}_{(I)}^T\mathbf{X}_{(I)})^{-1}\mathbf{X}_{(I)}^T\mathbf{Y}_{(I)} \\ &= (\mathbf{X}^T\mathbf{X} - \mathbf{X}_I^T\mathbf{X}_I)^{-1}(\mathbf{X}^T\mathbf{Y} - \mathbf{X}_I^T\mathbf{Y}_I)\end{aligned} \tag{3.6.2}$$

The inverse in (3.6.2) is computed using the basic formula in Appendix A.2 to give

$$\begin{aligned}\hat{\boldsymbol{\beta}}_{(I)} &= [(\mathbf{X}^T\mathbf{X})^{-1} + (\mathbf{X}^T\mathbf{X})^{-1}\mathbf{X}_I^T(\mathbf{I} - \mathbf{V}_I)^{-1}\mathbf{X}_I(\mathbf{X}^T\mathbf{X})^{-1}][\mathbf{X}^T\mathbf{Y} - \mathbf{X}_I^T\mathbf{Y}_I] \\ &= \hat{\boldsymbol{\beta}} - (\mathbf{X}^T\mathbf{X})^{-1}\mathbf{X}_I^T[-(\mathbf{I} - \mathbf{V}_I)^{-1}\mathbf{X}_I\hat{\boldsymbol{\beta}} + (\mathbf{I} + (\mathbf{I} - \mathbf{V}_I)^{-1}\mathbf{V}_I)\mathbf{Y}_I]\end{aligned} \tag{3.6.3}$$

Since $(\mathbf{I} - \mathbf{V}_I)^{-1} = \mathbf{I} + (\mathbf{I} - \mathbf{V}_I)^{-1}\mathbf{V}_I$,

$$\hat{\boldsymbol{\beta}}_{(I)} = \hat{\boldsymbol{\beta}} - (\mathbf{X}^T\mathbf{X})^{-1}\mathbf{X}_I^T(\mathbf{I} - \mathbf{V}_I)^{-1}\mathbf{e}_I \tag{3.6.4}$$

Finally, substituting into (3.6.1) leads to the form

$$D_I = \frac{\mathbf{e}_I^T(\mathbf{I} - \mathbf{V}_I)^{-1}\mathbf{V}_I(\mathbf{I} - \mathbf{V}_I)^{-1}\mathbf{e}_I}{p'\hat{\sigma}^2} \tag{3.6.5}$$

This result can be better understood by using the spectral decomposition $\mathbf{V}_I = \boldsymbol{\Gamma}\boldsymbol{\Lambda}\boldsymbol{\Gamma}^T$, where $\boldsymbol{\Gamma}$, with columns $\boldsymbol{\gamma}_l$, is an $m \times m$ orthogonal matrix of eigenvectors, and $\boldsymbol{\Lambda}$ is an $m \times m$ diagonal matrix of eigenvalues, $0 \le \lambda_1 \le \ldots \le \lambda_m \le 1$:

$$\begin{aligned}D_I &= \frac{\mathbf{e}_I^T(\boldsymbol{\Gamma}\boldsymbol{\Gamma}^T - \boldsymbol{\Gamma}\boldsymbol{\Lambda}\boldsymbol{\Gamma}^T)^{-1}\boldsymbol{\Gamma}\boldsymbol{\Lambda}\boldsymbol{\Gamma}^T(\boldsymbol{\Gamma}\boldsymbol{\Gamma}^T - \boldsymbol{\Gamma}\boldsymbol{\Lambda}\boldsymbol{\Gamma}^T)^{-1}\mathbf{e}_I}{p'\hat{\sigma}^2} \\ &= \frac{(\boldsymbol{\Gamma}^T\mathbf{e}_I)^T(\mathbf{I} - \boldsymbol{\Lambda})^{-1}\boldsymbol{\Lambda}(\mathbf{I} - \boldsymbol{\Lambda})^{-1}(\boldsymbol{\Gamma}^T\mathbf{e}_I)}{p'\hat{\sigma}^2}\end{aligned} \tag{3.6.6}$$

If $\lambda_m = 1$, the inverse in (3.6.6) does not exist, the data remaining after the cases indexed by I are removed are rank deficient, and a unique estimator $\hat{\boldsymbol{\beta}}_{(\mathrm{I})}$ does not exist. When $\lambda_m = 1$, we set $D_\mathrm{I} = +\infty$. If $\lambda_m < 1$, a scalar version of (3.6.6) is given by

$$D_\mathrm{I} = \frac{1}{p'} \sum_{l=1}^{m} \left[h_l^2 \frac{\lambda_l}{1-\lambda_l} \right] \tag{3.6.7}$$

where, for $l = 1, 2, \ldots, m$

$$h_l^2 = \frac{(\boldsymbol{\gamma}_l^\mathrm{T} \mathbf{e}_\mathrm{I})^2}{\hat{\sigma}^2 (1-\lambda_l)} \tag{3.6.8}$$

Under normality, the h_l^2 are identically distributed. The form (3.6.7) for D_I is directly comparable to D_i, except D_I is given as a sum over m orthogonal directions of squared residuals times fixed components, while for D_i, $m = 1$.

Other norms

The other choices for norms discussed in Section 3.5.1 can also be generalized to the case $m > 1$ with little difficulty. In particular, if a lower-dimensional norm corresponding to $\boldsymbol{\psi} = \mathbf{Z}\boldsymbol{\beta}$ is of interest, then $D_\mathrm{I}^{\psi} \equiv D_\mathrm{I}(\mathbf{M}, c)$, with $\mathbf{M}$, c defined by (3.5.8), provides the appropriate norm. One can show that $qD_\mathrm{I}^{\psi} \leqslant p' D_\mathrm{I}$ for all I and $\boldsymbol{\psi}$, so if D_I is negligible, so is D_I^{ψ}. In the special case where $Z = (\mathbf{0}, \mathbf{I}_q)$, and $(\mathbf{X}_1, \mathbf{X}_2)$ is the conforming partition of $\mathbf{X}$, D_I^{ψ} becomes

$$q\hat{\sigma}^2 D_\mathrm{I}^{\psi} = \mathbf{e}_\mathrm{I}^\mathrm{T} (\mathbf{I} - \mathbf{V}_\mathrm{I})^{-1} (\mathbf{V}_\mathrm{I} - \mathbf{U}_\mathrm{I})(\mathbf{I} - \mathbf{V}_\mathrm{I})^{-1} \mathbf{e}_\mathrm{I} \tag{3.6.9}$$

where $\mathbf{U}_\mathrm{I}$ is the appropriate principal minor of $\mathbf{X}_1 (\mathbf{X}_1^\mathrm{T} \mathbf{X}_1)^{-1} \mathbf{X}_1^\mathrm{T}$.

The internally scaled norm for SIC_I can be obtained by following the derivation in Section 3.5.2. In practice, computation of this norm for $m > 1$ is likely to be impractical because of the need to compute $(\mathbf{I} - \mathbf{V}_\mathrm{I})^{-1}$ (or its eigenvalues) for all possible subsets of size m.

3.6.1 GENERALIZED POTENTIAL

For $m = 1$, potential has been defined as essentially the fixed part of the characterizing measure $D_i(\mathbf{M}, c)$. Since each of the fixed parts of the measures given in Table 3.5.4 is a monotonic function of v_{ii}, these norms provide equivalent information on potential and v_{ii} is a reasonable summary. When $m > 1$, the notion of potential is more elusive since $D_\mathrm{I}(\mathbf{M}, c)$ will not conveniently factor into fixed and random parts. However, useful insights can be obtained from an investigation of D_{I}.

Dividing and multiplying the right side of expression (3.6.7) for D_{I} by Σh_k^2 gives

$$p' D_{\mathrm{I}} = (\Sigma h_k^2) \sum_{l=1}^{m} \left(\frac{h_l^2}{\Sigma h_k^2} \right) \frac{\lambda_l}{1-\lambda_l} \tag{3.6.10}$$

This form can be simplified in two ways. First, by definition

$$\Sigma h_k^2 = \frac{\mathbf{e}_{\mathrm{I}}^{\mathrm{T}} (\mathbf{I} - \mathbf{V}_{\mathrm{I}})^{-1} \mathbf{e}_{\mathrm{I}}}{\hat{\sigma}^2} = r_{\mathrm{I}}^2 \tag{3.6.11}$$

which is the generalization of r_i^2 given at (2.2.19). Next, define $q_l^2 = h_l^2 / \Sigma h_k^2$. Under normality, each q_l^2 follows a Beta distribution with parameters 1/2 and $(m-1)/2$; their joint distribution is Dirichlet. We can therefore write

$$D_{\mathrm{I}} = \frac{r_{\mathrm{I}}^2}{p'} \sum q_l^2 \frac{\lambda_l}{1-\lambda_l} = \frac{r_{\mathrm{I}}^2}{p'} Q_{\mathrm{I}} \tag{3.6.12}$$

where $Q_{\mathrm{I}} = \Sigma q_l^2 \lambda_l / (1 - \lambda_l)$. This form corresponds closely to that for $m = 1$, since D_{I} is factored into r_{I}^2, which measures the degree to which $(\mathbf{Y}_{\mathrm{I}}, \mathbf{X}_{\mathrm{I}})$ is an outlying set, and a potential-like term that has random components that are independent of the parameters in the model.

Several observations concerning Q_{I} can be made by simultaneously considering $\{q_l^2\}$ and the eigenvalues of $\mathbf{V}_{\mathrm{I}}$. First,

$$\frac{\lambda_1}{1-\lambda_1} \le Q_{\mathrm{I}} \le \frac{\lambda_m}{1-\lambda_m} \tag{3.6.13}$$

This interval is nonstochastic. If λ_m is small, the cases indexed by I will have little potential regardless of the observed values of $\{q_l^2\}$. For example, if each v_{ii}, $i \in \mathrm{I}$, is small, it follows that Q_{I} must be small since $\lambda_m \le \mathrm{tr}(\mathbf{V}_{\mathrm{I}}) = \Sigma v_{ii}$. On the other hand, if λ_1 is large, the cases must have high potential. Since $\lambda_1 \le \min(v_{ii})$, a necessary condition for λ_1 to be large is all v_{ii} must be large; that is, each case individually must have high potential.

For example, suppose $m = 2$ and $\mathrm{I} = (i, j)$, with $v = v_{ii} = v_{jj}$ (x_i and x_j lie on the same elliptical contour). If, in addition, $v_{ij} = 0$, then $\mathbf{V}_{\mathrm{I}} = v\mathbf{I}$, $\lambda_1 = \lambda_2 = v$ and $D_{\mathrm{I}} = D_i + D_j$. In this very special example, the potential for this pair is large or small according to the size of v.

If $m > p'$, at most p' of the eigenvalues of $\mathbf{V}_{\mathrm{I}}$ are nonzero since $\mathbf{V}_{\mathrm{I}}$ has rank of at most p'. Hence, for $m \geq p'$, we may write

$$Q_{\mathrm{I}} = \sum_{l=m-p'+1}^{m} q_l^2 \frac{\lambda_l}{1-\lambda_l}, \qquad m \geq p'$$

and the potential interval (3.6.13) has a lower bound of zero.

In situations where λ_1 is small but λ_m is large, Q_{I} depends on the values of $\{q_l^2\}$. When $m = 2$, q_l^2 is distributed as Beta $(\frac{1}{2}, \frac{1}{2})$, which has a U-shaped density with most of its probability massed near 0 and 1. Q_{I} will therefore tend to fall near one of the extremes of (3.6.13), and $\mathbf{e}_{\mathrm{I}}$ will tend to fall along one of the eigenvectors of $\mathbf{V}_{\mathrm{I}}$. When $m > 2$, the density of each q_l^2 is reverse J-shaped, with mode at zero. Thus when $m > p'$, Q_{I} will tend to be small since each q_l^2 will be small on the average.

For any $m \geq 2$ and under a correctly specified linear model, the *expected potential* is

$$\begin{aligned} E(Q_{\mathrm{I}}) &= \sum E(q_l^2)\lambda_l/(1-\lambda_l) = \frac{1}{m}\sum \lambda_l/(1-\lambda_l) \\ &= \frac{1}{m}\operatorname{tr}(\mathbf{V}_{\mathrm{I}}(\mathbf{I}-\mathbf{V}_{\mathrm{I}})^{-1}) \end{aligned} \tag{3.6.14}$$

When the cases indexed by I form an outlying set under the shift model $\mathbf{Y} = \mathbf{X}\boldsymbol{\beta} + \mathbf{D}\boldsymbol{\phi} + \boldsymbol{\varepsilon}$, $\hat{\sigma}^2 r_{\mathrm{I}}^2/\sigma^2$ has a noncentral chi-squared distribution with noncentrality parameter

$$\mathbf{u}^{\mathrm{T}}\mathbf{u} = \boldsymbol{\phi}^{\mathrm{T}}(\mathbf{I}-\mathbf{V}_{\mathrm{I}})\boldsymbol{\phi}/\sigma^2 \tag{3.6.15}$$

One can show that, under this model, the joint distribution of $\{q_l^2\}$ depends on $\mathbf{u} = \sigma^{-1}(\mathbf{I}-\boldsymbol{\Lambda})^{1/2}\boldsymbol{\Gamma}^{\mathrm{T}}\boldsymbol{\phi}$. With $m = 2$, one can show that

$$Eq_1^2 = \frac{1-\exp(-\frac{1}{2}\mathbf{u}^{\mathrm{T}}\mathbf{u})}{\frac{1}{2}\mathbf{u}^{\mathrm{T}}\mathbf{u}}\left[\frac{1}{2} - \frac{u_1^2}{\mathbf{u}^{\mathrm{T}}\mathbf{u}}\right] + \frac{u_1^2}{\mathbf{u}^{\mathrm{T}}\mathbf{u}} \tag{3.6.16}$$

Clearly, outliers can occur in ways which force the potential to be large or small.

EXAMPLE 3.6.1. $m = 2$. For illustration, consider the situation in which $m = 2$ and $v_{ii} = v$ for $i \in \mathrm{I}$. Let ρ denote the correlation between the residuals indexed by I. Then $\lambda_1 = v - (1-v)|\rho|$, $\lambda_2 = v + (1-v)|\rho|$ and the associated eigenvectors are $(1, \operatorname{sign}(\rho))/\sqrt{2}$ and $(1, -\operatorname{sign}(\rho))/\sqrt{2}$, respectively. If $|\rho|$ or $(1-v)$ is small, the potential will be essentially deterministic and equal to $v/(1-v)$, which may be large if the points in question are remote. This situation is illustrated for $p = p' = 2$ by points $\mathbf{x}_1$ and $\mathbf{x}_2$ in Fig. 3.6.2. The points $\mathbf{x}_1$ and $\mathbf{x}_2$ lie along the axes of the ellipse $\mathbf{x}^T(\mathbf{X}^T\mathbf{X})^{-1}\mathbf{x} = v$ and thus $|\rho| = 0$. A configuration for which $|\rho|$ may be large is illustrated by points $\mathbf{x}_1$ and $\mathbf{x}_3 = -\mathbf{x}_1$. For these points, $\rho = v/(1-v)$ and thus $\lambda_1 = 0$ and $\lambda_2 = 2v$, and Q_{I} depends on the orientation of $\mathbf{e}_{\mathrm{I}}$ relative to the eigenvectors $(1, 1)/\sqrt{2}$ and $(1, -1)/\sqrt{2}$. If the elements of $\mathbf{e}_{\mathrm{I}}$ are of opposite sign and approximately equal in absolute value, Q_{I} will be near its maximum, $2v/(1-2v)$. On the other hand, if the elements of $\mathbf{e}_{\mathrm{I}}$ are approximately equal, Q_{I} will be near zero. Similar comments apply to a replicated pair where $\rho = -v/(1-v)$. □

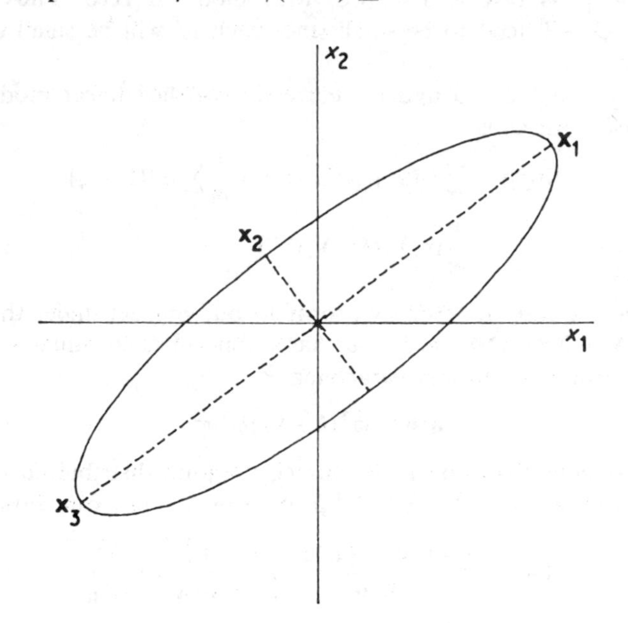

Figure 3.6.2 Contour of constant v_{ii}

Clearly, for potential to be large the maximum eigenvalue of $\mathbf{V}_{\mathrm{I}}$ must be large. As illustrated in Example 3.6.1, this will occur for $m = 2$ if the residual correlation (Appendix A.3) between the two cases is large in absolute value. However, the associated interpretation depends on the sign of this correlation as well as its absolute value. If the correlation is large and negative, then the two cases are probably near each other and may be judged simultaneously. If the correlation is large and positive, the cases will lie on opposite edges of the sampled region and simultaneous judgments of such cases may not be desirable.

A complete characterization of potential relative to D_{I} requires knowing $\lambda_l/(1-\lambda_l)$, $l = 1, 2, \ldots, m$. However D_{I} depends on $\lambda_l/(1-\lambda_l)$ only through Q_{I} which is statistically independent of $\hat{\sigma}^2 r_{\mathrm{I}}^2$. There are a variety of ways to summarize this information on potential. The interval $[\lambda_1/(1-\lambda_1), \lambda_m/(1-\lambda_m)]$, the expected potential (3.6.14) and the *maximum potential* $\lambda_m/(1-\lambda_m)$ (or just λ_m) are reasonable candidates. We believe that the maximum potential is the most desirable single number summary since it characterizes configurations of the rows of $\mathbf{X}$ that can lead to highly influential groups of cases for reasons that are independent of the fit.

Using (3.6.4), the multiple case norm $D_{\mathrm{I}}(\mathbf{M}, c)$ can be written in a form which allows a general definition of maximum potential:

$$D_{\mathrm{I}}(\mathbf{M}, c) = \frac{\mathbf{e}_{\mathrm{I}}^{\mathrm{T}}(\mathbf{I}-\mathbf{V}_{\mathrm{I}})^{-1}\mathbf{X}_{\mathrm{I}}(\mathbf{X}^{\mathrm{T}}\mathbf{X})^{-1}\mathbf{M}(\mathbf{X}^{\mathrm{T}}\mathbf{X})^{-1}\mathbf{X}_{\mathrm{I}}^{\mathrm{T}}(\mathbf{I}-\mathbf{V}_{\mathrm{I}})^{-1}\mathbf{e}_{\mathrm{I}}}{c}$$

$$= r_{\mathrm{I}}^2 \frac{\hat{\sigma}^2}{c} Q_{\mathrm{I}}(\mathbf{M}) \qquad (3.6.17)$$

where

$$Q_{\mathrm{I}}(\mathbf{M}) = \frac{\mathbf{e}_{\mathrm{I}}^{\mathrm{T}}(\mathbf{I}-\mathbf{V}_{\mathrm{I}})^{-1}\mathbf{X}_{\mathrm{I}}(\mathbf{X}^{\mathrm{T}}\mathbf{X})^{-1}\mathbf{M}(\mathbf{X}^{\mathrm{T}}\mathbf{X})^{-1}\mathbf{X}_{\mathrm{I}}^{\mathrm{T}}(\mathbf{I}-\mathbf{V}_{\mathrm{I}})^{-1}\mathbf{e}_{\mathrm{I}}}{\mathbf{e}_{\mathrm{I}}^{\mathrm{T}}(\mathbf{I}-\mathbf{V}_{\mathrm{I}})^{-1}\mathbf{e}_{\mathrm{I}}} \qquad (3.6.18)$$

As in the case when $m = 1$, we consider only two choices for c: Choose $c = k\hat{\sigma}^2$ or $k\hat{\sigma}_{(\mathrm{I})}^2$ where $k > 0$ is a constant that does not depend on $\mathbf{X}$ or $\mathbf{Y}$. With $\mathbf{M}$ fixed and c chosen as above, we define the maximum potential relative to $\mathbf{M}$ as $\max_{\mathbf{e}_{\mathrm{I}}}[Q_{\mathrm{I}}(\mathbf{M})]$. From the definition of $Q_{\mathrm{I}}(\mathbf{M})$, it follows that

$$\max_{\mathbf{e}_{\mathrm{I}}}[Q_{\mathrm{I}}(\mathbf{M})] = \lambda_{\max}[(\mathbf{I}-\mathbf{V}_{\mathrm{I}})^{-1/2}\mathbf{X}_{\mathrm{I}}(\mathbf{X}^{\mathrm{T}}\mathbf{X})^{-1}\mathbf{M}(\mathbf{X}^{\mathrm{T}}\mathbf{X})^{-1} \times \mathbf{X}_{\mathrm{I}}^{\mathrm{T}}(\mathbf{I}-\mathbf{V}_{\mathrm{I}})^{-1/2}] \qquad (3.6.19)$$

where $\lambda_{\max}[\mathbf{A}]$ denotes the maximum eigenvalue of the matrix $\mathbf{A}$. For $\mathbf{M} = \mathbf{X}^{\mathrm{T}}\mathbf{X}$, this reduces to $\lambda_m/(1-\lambda_m)$ as before. For the measure with $\mathbf{M} = \mathbf{X}_{(\mathrm{I})}^{\mathrm{T}}\mathbf{X}_{(\mathrm{I})}$, the maximum potential is λ_m. Thus, the choices $\mathbf{M} = \mathbf{X}^{\mathrm{T}}\mathbf{X}$ and $\mathbf{M} = \mathbf{X}_{(\mathrm{I})}^{\mathrm{T}}\mathbf{X}_{(\mathrm{I})}$ provide essentially the same information about the maximum potential of a particular configuration of the rows of $\mathbf{X}$ to be influential.

Alternative measures of potential

A fixed measure of potential can be defined by appealing to the volume argument analogous to that used for Wilks' statistic (3.5.22). In this formulation, the potential at $\mathbf{X}_I$ is measured by

$$\frac{|\mathbf{X}_{(I)}^T \mathbf{X}_{(I)}|}{|\mathbf{X}^T \mathbf{X}|} = |\mathbf{I} - \mathbf{V}_I| = \Pi (1 - \lambda_i) \tag{3.6.20}$$

High potential or remote sets are indicated by small values of this measure, which is based on the internal dispersion of the $\mathbf{x}_i^T$s in much the same way as the internal norms are based on the dispersion of the sample versions of the influence curve.

The measure (3.6.20) appears in a factorization of a statistic for detecting influential cases by Andrews and Pregibon (1978) (see Chapter 4). It was advocated as a generalized 'leverage' measure by Draper and John (1981).

The type of potential being measured by (3.6.20) seems to differ fundamentally from that measured by the expected or maximum potential. These latter measures judge the potential of a set of cases in the determination of $\hat{\boldsymbol{\beta}}$, while (3.6.20) measures the degree to which $\mathbf{X}_I$ is isolated from the remaining rows of $\mathbf{X}$ in the p'-dimensional space defined by the explanatory variables. As pointed out by Draper and John (1981), cases which have high potential according to (3.6.20) need not have high potential in the estimation of $\boldsymbol{\beta}$. In reference to the second situation in Example 3.6.1, for example, $|\mathbf{I} - \mathbf{V}_I| = (1 - \lambda_1)(1 - \lambda_2) = (1 - 0)(1 - 2v) = 1 - 2v$. If v is large the pair of points will be judged to have high potential according to (3.6.20). However, if $\mathbf{e}_I^T \propto (1, 1)$ the points will have no potential and thus no influence on $\hat{\boldsymbol{\beta}}$.

Although (3.6.20) is not directly relevant to an investigation of the cases that influence $\hat{\boldsymbol{\beta}}$, the information it provides may be useful in other phases of an analysis. If, for example, it were possible to design for the collection of additional data, knowing which of the present points are remote in the factor space would certainly be helpful.

EXAMPLE 3.6.2. ADAPTIVE SCORE DATA NO. 4. To illustrate some of the previous comments on potential we consider two pairs of cases from the data given by Mickey *et al.* (1967). The model is simple linear regression and the $n = 21$ cases are plotted in Fig. 2.2.1. As indicated in the plot, cases 2, 18, and 19 are in question.

Table 3.6.1 gives the case statistics D_i, r_i^2, and v_{ii} for $i = 2, 18$, and 19. Case 19 appears to be an outlier from the assumed model. As shown in Example 2.2.3, the p-value associated with case 19 is between 0.0409 and 0.0425. Although case 19 appears as an outlier, it has relatively little influence. Removal of this observation would move $\hat{\boldsymbol{\beta}}$ to the edge of a

Table 3.6.1 *Selected case statistics, adaptive score data*

Case, i	r_i^2	v_{ii}	D_i
2	0.89	0.155	0.08
18	0.73	0.652	0.68
19	7.97	0.053	0.22

20% confidence ellipse. Case 18, on the other hand, fits the model quite well, but is influential because of the associated high potential $v_{18,18} = 0.652$. Removal of this observation would move $\hat{\boldsymbol{\beta}}$ to the edge of a 48% confidence ellipse. Case 2, as well as the remaining cases in the data, would probably go unnoticed when inspecting individual case statistics.

Consider next the highly influential pair (2, 18), $D_{(2,18)} = 6.37$. Removal of this pair would move $\hat{\boldsymbol{\beta}}$ to the edge of a 99.2% confidence ellipse. However, this does not appear to be an outlying pair since $t^2_{(2,18)} = 2.01$. This pair must, therefore, be influential because of the associated potential. In fact, $Q_\text{I} = 3.50$, which lies near one end of the interval [0.012, 3.85] computed from (3.6.13). Q_I depends on the eigenvalues ($\lambda_1 = 0.012$ and $\lambda_2 = 0.794$) of $\mathbf{V}_{(2,18)}$ and on the chance orientation of $\mathbf{e}_{(2,18)}$ relative to the associated eigenvectors. The observed value of Q_I is large for this pair since $\mathbf{e}^\text{T}_{(2,18)} = (-9.57, -5.54)$ is in the direction of the eigenvector associated with λ_2, $(-9.57, -4.53)$. However, since the lower endpoint of the potential range is small, Q_I for this pair does not necessarily have to be large. If e_2 and e_{18} had been of opposite sign, Q_I might have been small enough to make the pair uninfluential. The fact that e_2 and e_{18} have the same sign and thus lie on the same side of the fitted model could be an indication that the model is incorrect; possibly there are outliers present or a quadratic term is needed.

In contrast to the previous situation, the pair (18, 19) is uninfluential, $D_{(18,19)} = 0.15$, but may be outlying, $t^2_{(18,19)} = 6.30$. Of course, the possibility that this is an outlying pair is due in part to the presence of case 19. The observed $Q_\text{I} = 0.037$ is very near the lower end of the potential range [0.036, 2.025]. This value is small because $\mathbf{e}^\text{T}_{(18,19)} = (-5.54, 30.28)$ lies in the direction of the eigenvector corresponding to λ_1, $(-5.54, 33.77)$.

Four possible summary measures of potential are given in Table 3.6.2 for three pairs of cases. From the information in the first

Table 3.6.2 *Measures of potential, m = 2, adaptive score data*

Pair	$\frac{1}{2}\mathrm{tr}[\mathbf{V}_I(\mathbf{I}-\mathbf{V}_I)^{-1}]$	$\frac{1}{2}\mathrm{tr}[\mathbf{V}_I]$	$\lvert\mathbf{I}-\mathbf{V}_I\rvert$	$\lambda_m/(1-\lambda_m)$
(2, 18)	1.93	0.41	0.20	3.85
(18, 19)	1.03	0.35	0.32	2.02
(11, 18)	1.07	0.37	0.30	2.07

three columns of this table, it may be difficult to form firm judgments about the potential of the new pair (11, 18). □

3.6.2 COMPUTING D_I

One goal in examining subsets of $m > 1$ cases is to find groups of cases that, while not individually influential, are influential *en bloc*. Finding influential subsets which include smaller influential subsets may add little information because the observed influence of the subset will be due, in part, to the influence of the smaller subset. Conversely, finding an uninfluential subset that includes one or more cases that are singly or jointly influential would not decrease the interest in those cases. Thus, good candidates for inclusion in subsets will have small distance values for $m = 1$, but they may well have relatively large values of v_{ii} or r_i^2.

Alternatively, it may be desirable to consider the possibility that the individual cases in an influential subset are related (for example, by time or location). In this situation, good candidates for inclusion in subsets will include influential cases.

EXAMPLE 3.6.3. CLOUD SEEDING NO. 10. The above remarks can be illustrated by reference to Fig. 3.6.3, which contains a semigraphical summary of λ_{max}, t_I^2, and D_I for $m = 2$ in combination with the cube root model for the cloud seeding data. In the display, rows and columns correspond to case indices; thus, for example, the symbol in row 5, column 8 represents the values of the statistics for the pair $I = (5, 8)$. The computed values have been divided into groups so that the more ink used in printing the symbol, the larger the value. The displays illustrate that: (1) subsets with high potential consist of case 2 and any other case, case 18 and any other case and the pair (3, 20); (2) pairs for which t_I^2 is large contain cases 7 or 24; and (3) the influential pairs consist of case 2 and most others, (3, 20) and (7, 18). It is clear that case 2 should be considered as being highly influential, and little is gained by viewing it as one of a pair.

```
Case   λmax                           t²_I                          D_I

  1     #...:..:. ......:#.. ....     ......:... .......... ....     #......... .......:... ....
  2    # ######### ########### ####   . .....:... .......... ....   # ######### ####.##### ####
  3    .# .:..... ......:#.# ..:.     .. ...:... .........# ....    .# ........ ..........# ....
  4    .#. ..:... ......:#.. ....     ... ..:... .......... ....    .#. ..:.... ........... ....
  5    .#:. ..... :.....:#.: ....     .... .:... .......... ....    .#.. ..... ........... ....
  6    :#... :... ..:...:#.. ....     ..... :... .......... ....    .#... .... ........... ....
  7    .#.:.: ... ......:#.. ....     ::::: ::: :::::::::: ::::     .#.:.. ... ........#.. ..:.
  8    .#..... .. ......:#.. ....     ......: .. .......... ...:    .#..... .. ........:.. ....
  9    :#...... . ......:#.. ....     ......:. . .......... ....    .#...... . ........... ....
 10    .#.......  ......:#.. ....     ......:..  .......... ....    .#.......  ........... ....
 11    .#..:.....  .....:#.. ..:.     ......:...  ......... ....    .#.........  .......... ....
 12    .#......... . ....:#.. ....    ......:... . ........ ....    .#......... . ......... ....
 13    .#...:.... .. ...:#.. ....     ......:... .. ....... ....    .#......... .. ........ ....
 14    .#......... ... ..:#.. ....    ......:... ... ...... ....    .#......... ... ....... ....
 15    .#......... .... .:#.. ....    ......:... .... ..... ...:    ............ .... ...... ....
 16    .#......... ..... :#.. ...:    ......:... ..... .... ....    .#......... ..... ..... ...:
 17    :#:::::::: :::::: #:: ::::     ......:... ...... ... ....    :#......... ....... :.. ....
 18    ########### ####### ## ####    ......:... ....... .. ....    .#....#:.. .......: .. ....
 19    .#......... ......:# . ....    ......:... ........ . ....    .#......... ......... . ....
 20    .##.:..... ......:#.  ..:.     ..#...:... .........  ....    .##........ ..........  ....
 21    .#......... ......:#..  .:.    ......:... ..........  ...    .#......... ...........  ...
 22    .#......... ......:#.. . ..    ......:... .......... . ..    .#......... ........... . ..
 23    .#:........ :.....:#.: :. .    ......:... .......... .. .    .#....:... ........... .. .
 24    .#......... .....::#.. ...     ......::.. ...:...... ...     .#......... .....:..... ...
```

Figure 3.6.3 Semigraphical display of λ_{max}, t_I^2, and D_I, cloud seeding data

Among the pairs whose removal does not result in a rank deficient model, the most likely outlying pair is (7, 15) with $t^2_{(7,15)} = 13.82$, the most influential pair is (2, 5) with $D_{(2,5)} = 16.48$, and the pairs with the highest potential are (2, 15) and (2, 5) with λ_{max} equal to 0.9834 and 0.9822, respectively.

The only additional information obtained by an examination of all pairs is for (3, 20). For this pair, $\lambda_{max} = 1$, and its deletion leads to a rank deficient model. These cases require special handling, and, to accommodate them, deletion of a variable (EA) from the model is desirable. □

For $m = 2$ and n not too large, semigraphical displays like Fig. 3.6.3 can be used to present the information about pairs of cases. However, if $m > 2$ or n is large, this summary becomes impractical, and better computational and display methods are needed.

If sufficient computer memory is available to store the residual vector and all of the elements of $\mathbf{V}$, an efficient algorithm for finding multiple case outliers can be based on the Furnival and Wilson (1974) method for subset selection. However, an equivalent algorithm for finding subsets with large D_I is not immediately apparent, since altering a subset by addition, deletion, or substitution of a case can result in substantial changes in the eigenstructure of $\mathbf{V}_I$, and hence in the value of D_I. Even so, complete storage of $\mathbf{V}$ is usually impractical and realistic techniques for finding influential subsets should use only the residuals and the diagonal entries of $\mathbf{V}$. Using only these, upper bounds for D_I can be derived, and only if these are sufficiently large must D_I be computed exactly.

For the first upper bound, since $\lambda_m/(1-\lambda_m)^2 \geq \lambda_l/(1-\lambda_l)^2$, $l = 1, 2, \ldots, m$, D_I can be bounded by

$$D_\mathrm{I} \leq \frac{1}{p'\hat{\sigma}^2} \frac{\lambda_m}{(1-\lambda_m)^2} \sum_{l=1}^{m} (\gamma_l^\mathrm{T} \mathbf{e}_\mathrm{I})^2$$

$$= \frac{1}{p'\hat{\sigma}^2} \frac{\lambda_m}{(1-\lambda_m)} \mathbf{e}_\mathrm{I}^\mathrm{T} \boldsymbol{\Gamma} \boldsymbol{\Gamma}^\mathrm{T} \mathbf{e}_\mathrm{I} \tag{3.6.21}$$

or, since $\boldsymbol{\Gamma}\boldsymbol{\Gamma}^\mathrm{T} = \mathbf{I}$,

$$D_\mathrm{I} \leq \frac{\lambda_m}{(1-\lambda_m)^2} \left(\frac{\sum_{i \in \mathrm{I}} e_i^2}{p'\hat{\sigma}^2} \right) \tag{3.6.22}$$

For this to be useful, λ_m must be replaced by an approximation that can be computed without forming $\mathbf{V}_\mathrm{I}$. Assuming $\mathrm{tr}(\mathbf{V}_\mathrm{I})$ = trace of $\mathbf{V}_\mathrm{I}$ to be less than one, the simplest approximation is $\lambda_m \leq \mathrm{tr}(\mathbf{V}_\mathrm{I})$. Thus,

$$D_\mathrm{I} \leq \frac{\mathrm{tr}(\mathbf{V}_\mathrm{I})}{(1-\mathrm{tr}(\mathbf{V}_\mathrm{I}))^2} \left(\frac{\sum_{i \in \mathrm{I}} e_i^2}{p'\hat{\sigma}^2} \right)$$

or equivalently,

$$D_\mathrm{I} \leq \frac{\sum_{i \in \mathrm{I}} v_{ii}}{\left(1 - \sum_{i \in \mathrm{I}} v_{ii}\right)^2} \frac{\sum_{i \in \mathrm{I}} e_i^2}{p'\hat{\sigma}^2} \tag{3.6.23}$$

The upper bound in (3.6.23) depends only on the single case statistics and provides a potentially different upper bound for each I. For any subset with $\mathrm{tr}(\mathbf{V}_\mathrm{I}) \geq 1$, a better approximation to λ_m is required, which requires forming $\mathbf{V}_\mathrm{I}$. If m is small (2 or 3) exact computation of D_I is probably as efficient as approximating λ_m.

For fixed m, let $T = \max_\mathrm{I}(\Sigma_{i \in \mathrm{I}} v_{ii})$ and $R^2 = \max_\mathrm{I}(\Sigma_{i \in \mathrm{I}} e_i^2)$, where I varies over all subsets of size m under consideration. Two upper bounds for the right side of (3.6.23) are then

$$D_\mathrm{I} \leq \frac{\mathrm{tr}(\mathbf{V}_\mathrm{I})}{(1-\mathrm{tr}(\mathbf{V}_\mathrm{I}))^2} \frac{R^2}{p'\hat{\sigma}^2} \tag{3.6.24}$$

and, if $T < 1$,

$$D_\mathrm{I} \le \frac{T}{(1-T)^2} \frac{\Sigma_{i \in \mathrm{I}} e_i^2}{p' \hat{\sigma}^2} \tag{3.6.25}$$

These last two may be combined to give

$$D_\mathrm{I} \le \frac{T}{(1-T)^2} \frac{R^2}{p' \hat{\sigma}^2} \tag{3.6.26}$$

Clearly, (3.6.23) ≤ (3.6.24) ≤ (3.6.26), and (3.6.23) ≤ (3.6.25) ≤ (3.6.26).

An algorithm for finding all relevant subsets with fixed m can be based on these approximations. First, influential subsets of size smaller than m may be eliminated if desired. Then, the remaining v_{ii} and e_i^2 are ordered, largest to smallest. The four inequalities can then be applied to subsets with $\mathrm{tr}\,(\mathbf{V}_\mathrm{I}) < 1$ in the order (3.6.26), then (3.6.24), or (3.6.25), and finally (3.6.23). Exact computation is required if (3.6.23) is too big. By considering subsets according to the ordered lists of v_{ii} and e_i^2, the subsets that are more likely to be influential are considered first, and once one of the bounds is sufficiently small, no further subsets made up of cases lower in the lists need to be considered. Generally, this method will be useful in data sets with n large relative to p', where $\mathrm{tr}\,(\mathbf{V}_\mathrm{I})$ will usually be less than 1. In smaller data sets, relatively more subsets must be considered. Cook and Weisberg (1980) discuss examples, for $m = 2$ and $m = 3$, and for two data sets, one with $n = 21$, $p = 8$ and the other with $n = 125$, $p = 4$. The results of a simple algorithm are summarized in Table 3.6.3. The number of subsets is less than the total number of possible subsets because cases influential in subsets of a smaller size were not considered as m was increased. While in data set 1 little is

Table 3.6.3 *Computations using bounds. Source: Cook and Weisberg* (1980)

	Data set 1 $n = 21$, $p = 8$	*Data set* 2 $n = 125$, $p = 4$
$m = 2$:		
number of subsets considered	155	7 503
number of applications of inequalities	153	651
number of D_I computed	74	5
$m = 3$:		
number of subsets considered	560	302 621
number of applications of inequalities	560	74 802
number of D_I computed	520	727

gained by use of the inequalities, in data set 2, with large n, substantial decrease in computation is apparent.

EXAMPLE 3.6.4. DRILL DATA. In this example, we consider a data set obtained from an experiment to characterize the performance of a certain type of drill bit over a range of drilling conditions. For each experimental run, the work piece and drill were placed at opposite ends of a lathe and the values of the following design variables were set:

S = speed of rotation of the work piece in surface feet per minute;
F = feed rate in inches per revolution (rate at which the drill passes through the work piece);
D = diameter of the drill bit.

The rate of rotation of the drill bit was held constant throughout the experiment. The response variable Y is the axial load (thrust) on the drill bit during the drilling process.

The experimental runs were originally arranged in a completely randomized composite design, but the experiment was prematurely terminated for reasons that are unimportant in this analysis. The data, as provided by M. R. Delozier of Kennametal, Inc., Latrobe, Pennsylvania, are given in Table 3.6.4. The coarseness of the responses is due to rounding in the measurement technique; the responses for each combination of S, F and D are from replicate runs. The portion of the design that was completed is shown graphically in Fig. 3.6.4. The size of each point is proportional to the number of replicates at that point.

Since the possibility that the response is a nonlinear function of the explanatory variables cannot be discounted, we tentatively adopt the second-order response surface model

$$Y = \beta_0 + \beta_1 S + \beta_2 F + \beta_3 D + \beta_{12} SF + \beta_{13} SD + \beta_{23} FD + \beta_{11} S^2 + \beta_{22} F^2 + \beta_{33} D^2 + \varepsilon \qquad (3.6.27)$$

Figure 3.6.5 gives plots of $L_{\max}(\lambda)$ versus λ for the power family of transformations for the second-order model (3.6.27) and for the first-order subset model $Y = \beta_0 + \beta_1 S + \beta_2 F + \beta_3 D + \varepsilon$. Evidently a transformation can improve the fit of the second-order model, but will not result in a significantly improved fit for the subset model. The magnitude of differences between the ordinates of the two curves shows that including the cross product and quadratic terms does improve the fit regardless of the transformation selected. While the likelihood analysis clearly suggests that some transformation is necessary, it

Table 3.6.4 *Drill data. Source: M. R. Delozier*

Case	*S*	*F*	*D*	*Y*
1	450	0.0060	1.50	430
2	450	0.0060	1.50	368
3	300	0.0045	1.50	306
4	300	0.0045	1.50	306
5	481	0.0050	2.25	894
6	375	0.0045	2.25	813
7	450	0.0060	3.00	969
8	450	0.0060	3.00	969
9	375	0.0065	2.25	976
10	300	0.0045	3.00	727
11	300	0.0045	3.00	606
12	375	0.0050	1.00	276
13	450	0.0045	1.50	338
14	450	0.0045	1.50	399
15	300	0.0060	1.50	368
16	300	0.0060	1.50	368
17	375	0.0050	2.25	894
18	375	0.0050	2.25	732
19	375	0.0050	2.25	813
20	375	0.0050	2.25	894
21	375	0.0050	2.25	732
22	375	0.0050	2.25	813
23	375	0.0050	2.25	813
24	375	0.0050	2.25	894
25	375	0.0050	2.25	813
26	269	0.0050	2.25	813
27	450	0.0045	3.00	727
28	450	0.0045	3.00	485
29	300	0.0060	3.00	969
30	300	0.0060	3.00	847
31	375	0.0050	3.50	1126

provides little help for deciding on a particular choice since the asymptotic 95% confidence interval contains most of the power transformations used in practice. In this analysis we use the logarithmic transformation LY, since it is near the maximum likelihood estimate and has been found to be appropriate in past analyses of similar data. Transformations of the design variables will not be considered in this example.

The mean squares for lack of fit and pure error from the second-order model using LY are 0.0779 and 0.0114, respectively, and the

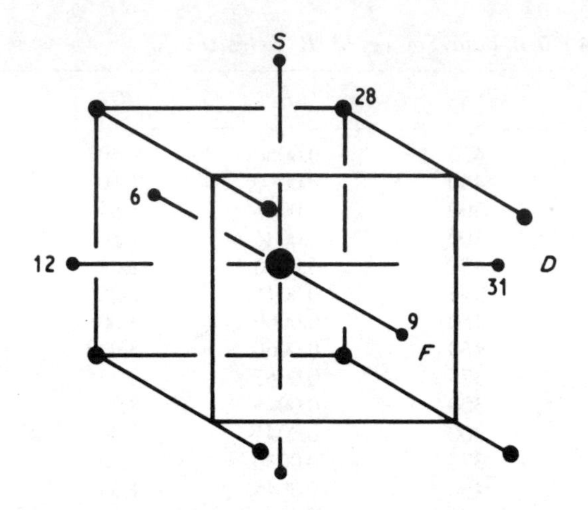

Figure 3.6.4 Design for the drill data; locations of selected cases are indicated

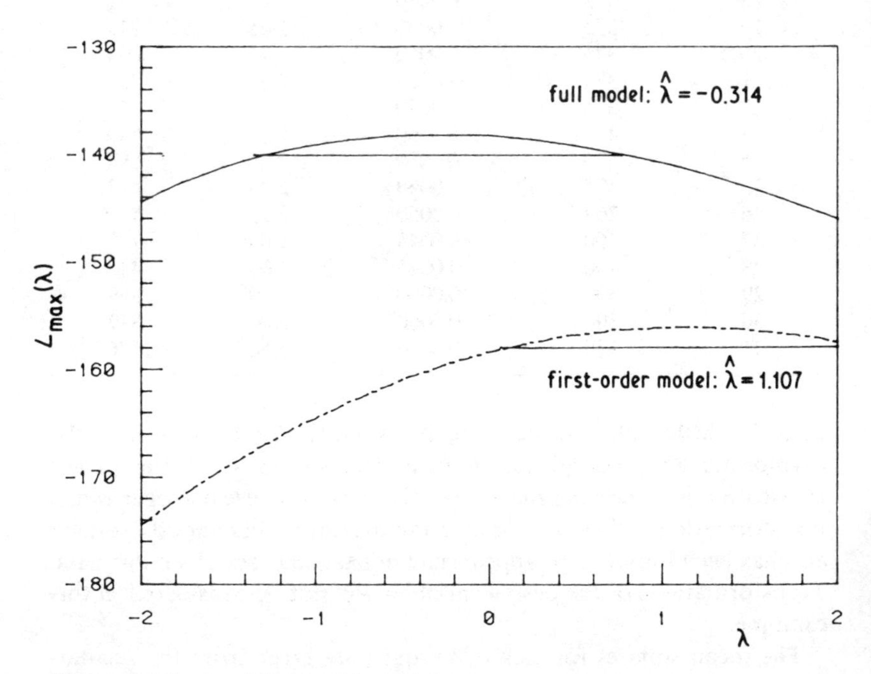

Figure 3.6.5 $L_{\max}(\lambda)$ versus λ, drill data

corresponding F-statistic is equal to 6.8 with 5 and 16 df. Even with the logarithmic transformation, the fit of the model does not seem adequate. Similar results are found for other transformations contained in the 95% confidence interval given in Fig. 3.6.5. Rather than attempting to build a more complicated model, we next consider various diagnostics applied to the second-order model.

The added variable plot of the constructed variable for the power family is given in Fig. 3.6.6. No single case seems to be greatly influencing the transformation, although cases 5, 9, and 31 form a group in the upper-left corner and may be jointly influential. Figure 3.6.7 gives a scatter plot of the Studentized residuals for the data with LY as response versus the fitted values. Aside from showing that cases 5, 9, 28, and 31 have absolute Studentized residuals larger than 2, this plot is of little help. Case 9 has the largest Studentized residual, and $t_9 = 3.36$; the Bonferroni p-value is 0.097. When the mean square for pure error is used to estimate σ^2, $r_9 = 4.26$. With this substitution, r_9 has a nominal $t(16)$-distribution since case 9 is not replicated. The corresponding p-value using the Bonferroni inequality is 0.019.

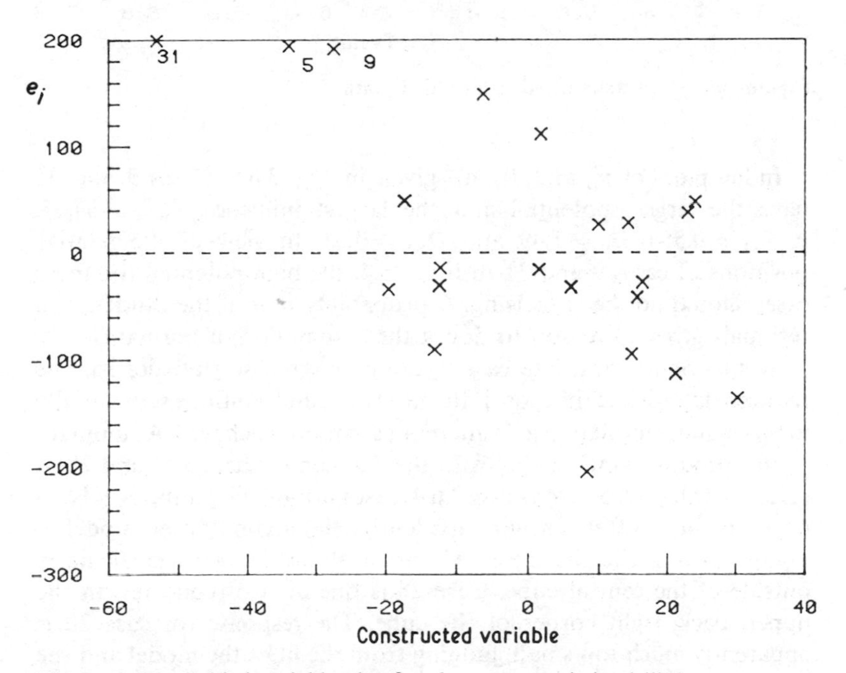

Figure 3.6.6 Added variable plot for the score statistic, drill data

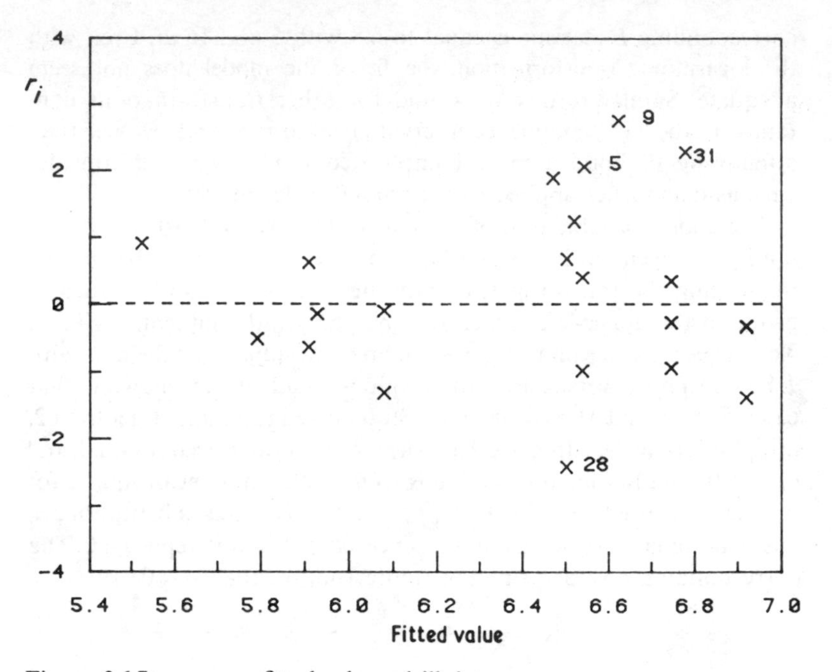

Figure 3.6.7 r_i versus fitted values, drill data

Index plots of v_{ii} and D_i are given in Fig. 3.6.8. Cases 9 and 31 have the largest potential and the largest influence, $v_{9,\,9} = 0.663$, $v_{31,\,31} = 0.550$, $D_9 = 1.49$ and $D_{31} = 0.63$. In view of the relative positions of cases 9 and 31 in Fig. 3.6.4, the high potential for these cases should not be surprising. A probability plot of the Studentized residuals gives no reason to doubt the assumption of normality.

At this point we delete case 9, examine the case statistics for the reduced data, delete the most influential case, and continue sequentially in this manner until the least squares fit seems well behaved. A summary of this process, which ended with the deletion of cases 6, 9, and 28, is given in Table 3.6.5. From Fig. 3.6.4, cases 6 and 9 lie on the F axis on opposite sides of the origin. Evidently, the second-order model is unable to describe the observed thrust along this axis, particularly outside of the central cube. Case 28 is one of two replicates on the upper, back, right corner of the cube. The response for case 28 is apparently much too small, judging from the fit of the model and the response at the second replicate.

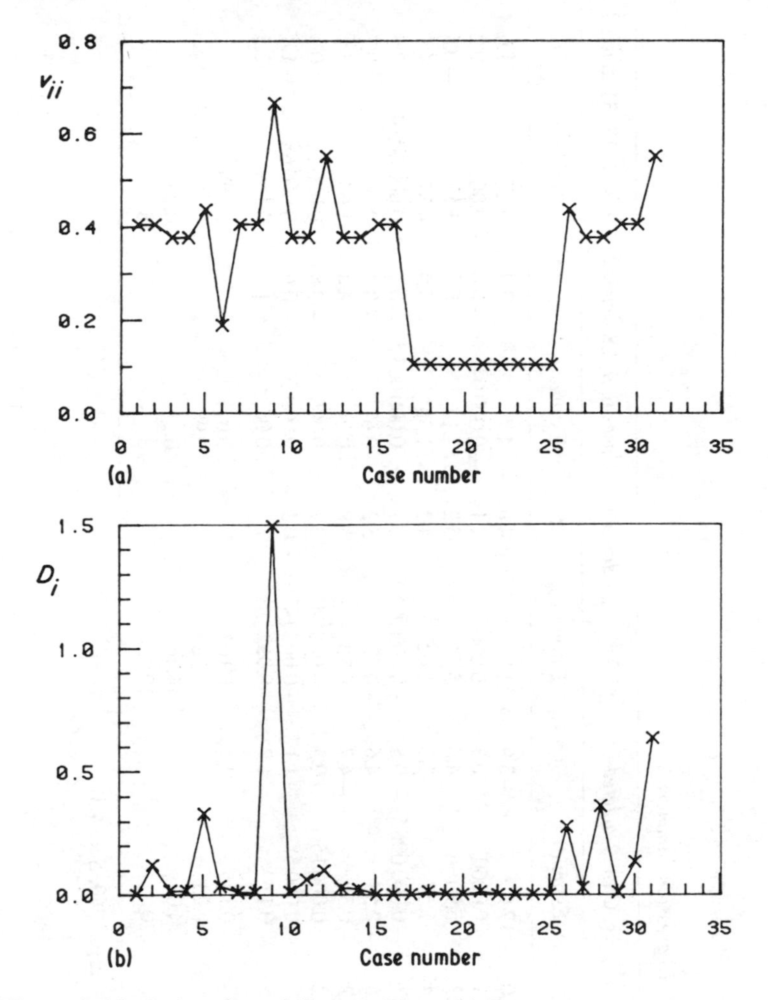

Figure 3.6.8 Index plots, drill data. (a) v_{ii}. (b) D_i

As a check on the above sequential procedure, we computed t_{I}^2 and D_{I} for all possible I with $m = 2$ and 3. Table 3.6.6 gives the four largest values of t_{I}^2 and D_{I} for $m - 2$ and $m = 3$. The most likely outlying triplet contains cases 6, 9, and 28, as identified previously. The agreement between sequential and simultaneous methods cannot, of course, be guaranteed in general. The Bonferroni p-value for $\mathrm{I} = (6, 9, 28)$ is 0.003. The most influential triplet is $\mathrm{I} = (9, 12, 31)$ with $D_{(9,12,31)} = 10.84$.

Table 3.6.5 *Drill data.* (*a*) *Regression summaries*

	All data		*Case 9 deleted*		*Cases 9, 28 deleted*		*Cases 6, 9, 28 deleted*		*Cases 9, 12, 31 deleted*	
	Estimate	*t*	*Estimate*	*t*	*Estimate*	*t*	*Estimate*	*t*	*Estimate*	*t*
Intercept	−6.7	−2.0	−12.6	−3.9	−11.9	−4.6	−14.7	−7.4	5.26	1.0
S	0.017	2.1	0.0060	0.8	0.0066	1.1	−0.000041	−0.0	−0.0015	−0.2
F	2.68	2.3	5.90	4.4	5.54	5.2	7.17	8.4	−1.84	−0.8
D	1.92	3.4	1.57	3.3	1.62	4.2	1.38	5.0	4.87	4.8
S^2	−0.000021	−2.3	−0.0000065	−0.8	−0.0000051	−0.7	0.0000035	0.7	0.0000034	0.5
F^2	−0.27	−2.6	−0.58	−4.6	−0.52	−5.2	−0.67	−8.4	0.16	0.7
D^2	−0.38	−5.2	−0.30	−4.7	−0.30	−5.8	−0.24	−6.4	−1.05	−4.9
SF	0.00015	0.2	0.00015	0.3	−0.00034	−0.7	−0.00028	−0.8	0.00015	0.3
SD	−0.00068	−0.9	−0.00068	−1.1	−0.00014	−0.3	−0.00020	−0.6	−0.00068	−1.4
FD	0.11	1.5	0.11	1.9	−0.062	1.3	0.067	1.9	0.119	2.5
$\hat{\sigma}$	0.165		0.135		0.107		0.077		0.106	
R^2	0.90		0.93		0.96		0.98		0.95	
F^*_{lof}	6.85		4.03		4.46		0.33		0.89	
df	21		20		19		18		18	

* F for lack of fit (see Weisberg, 1980, Sec. 4.3)

Table 3.6.5 *Drill data.* (*b*) *Case statistics*

Statistic	*All data*	*Case 9 deleted*	*Cases 9, 28 deleted*
D_6	0.03	0.32	0.41
D_{28}	0.36	0.48	
D_9	1.49		
$v_{6,6}$	0.187	0.284	0.29
$v_{28,28}$	0.376	0.377	
$v_{9,9}$	0.663		
t_6	1.90(16)*	3.59(16)	4.10(15)
t_{28}	−4.92(15)	−4.69(15)	
t_9	4.26(16)		

* df for nominal t, using available orthogonal pure error.

Table 3.6.6 *Selected case statistics for* $m = 2$ *and* $m = 3$, *drill data*

Subset	D_I	t_I^2
(12, 31)	5.82	16.10
(6, 9)	3.32	15.35
(5, 26)	2.39	14.32
(2, 9)	2.38	10.70
(28, 31)	2.15	21.16
(9, 28)	1.66	15.29
(9, 12, 31)	10.84	11.15
(12, 27, 31)	6.58	14.47
(12, 26, 31)	6.14	10.37
(8, 12, 31)	6.10	10.59
(6, 9, 28)	3.04	27.07
(12, 28, 31)	5.24	26.65
(5, 26, 28)	2.26	26.40

Removal of this triplet will displace $\hat{\boldsymbol{\beta}}$ to the edge of a 99.9997% confidence region. The least squares fit of the second-order model without this influential triplet is summarized in the final columns of Table 3.6.5(a).

Of the five points identified in this analysis, four (6, 9, 12, 31) are single replicates on the D and F axes and two (6, 9) of these four are contained in the outlying triplet. Any analysis of these data will be strongly dependent on the validity of these four cases and, unless the precise form of the model is known, conclusions will be tentative at best. At this point, little can be gained by further analysis of these data, since conclusions must depend so heavily on the four unreplicated points. Useful statements concerning the relationships between the variables will require more experimental runs.

Box and Draper (1975) propose a design criterion that will help avoid the ambiguity inherent in this analysis: To minimize the effects of a small proportion of outlying responses on the fitted values, choose a design to minimize the dispersion of the v_{ii}s, $\Sigma\,(v_{ii}-\bar{v})^2/n$. For fixed n and p', this is equivalent to choosing a design to minimize $\Sigma\,v_{ii}^2$ since $\bar{v}=p'/n$ is fixed. The design points in this example give $\min(v_{ii}) = 0.104$, $\max(v_{ii}) = 0.663$, $\bar{v} = 0.323$ and $\Sigma\,(v_{ii}-\bar{v})^2/n = 0.0247$. One way that this design can be improved is to move 6 of the 9 center points to replicate the previously unreplicated points, giving $\min(v_{ii}) = 0.190$, $\max(v_{ii}) = 0.382$, $\bar{v} = 0.323$ and $\Sigma\,(v_{ii}-\bar{v})^2/n = 0.003$. Generally, it is necessary to replicate the remote points in a composite design to gain some robustness against outliers. □

CHAPTER 4

Alternative approaches to influence

'The path by which we rise to knowledge must be made smooth and beaten in its lower steps, and often ascended and descended, before we can scale our way to any eminence, much less climb to the summit.'

HERSCHEL, *op. cit.*

The diagnostic statistics presented in the last chapter share a common heritage: they all depend on the same perturbation scheme, namely case deletion, and they all use a sample influence curve to monitor changes in the resulting analysis. These methods seem to have found wide acceptance because of their intuitive appeal and computational simplicity. Other approaches to the problem of assessing influence can be developed by altering either the method of perturbation, or by changing the aspect of the analysis that is monitored. In this chapter we look at several methods that do not depend directly on the influence curve, but do use case deletion perturbation schemes. There are both advantages and perils in these other approaches. A principal danger is the possibility of designing a measure that has no firm theoretical basis; a useful measure must refer to some specific part of the analysis and the values of the derived statistics must be monotonic measures of what is meant by influence. The main advantage in other approaches is the possibility of monitoring factors other than changes in the location estimates. The methods based on the sample influence curve, for example, are largely insensitive to changes in estimated scale; other methods can take an alternate view.

We consider three approaches to influence that generally meet the requirements of the last paragraph. The first of these compares the volume of confidence ellipsoids based on full and reduced samples, thereby directly including changes in estimated scale in the measure. The second related measure is due to Andrews and Pregibon (1978) and can be thought of as a general omnibus diagnostic, although it is weakly dependent on the structure of the regression problem. We then turn to a Bayesian predictivist procedure in which predictive distributions of future observations are compared. This method is more comprehensive than the others, combining several aspects of the analysis into a single measure. After a comparison of influence measures, we briefly discuss methods that can be used to calibrate the various influence measures.

4.1 Volume of confidence ellipsoids

One possible measure of the uncertainty in estimating a vector of parameters is the volume of a corresponding confidence ellipsoid (Cook and Weisberg, 1980). This volume is also related to various measures of design optimality with smaller volumes corresponding to more informative designs. A reasonable measure of influence that responds to this uncertainty or information is the change in volume when a subset of cases is removed. Computation of this measure is straightforward, since the volume of an ellipsoid is proportional to the inverse square root of the determinant of the appropriate cross product matrix.

To obtain a general measure, reorder $\mathbf{X}$ so that the last $q \leq p'$ columns of $\mathbf{X}$ correspond to the coefficients of interest and partition $\mathbf{X} = (\mathbf{X}_1, \mathbf{X}_2)$ with $\mathbf{X}_2$ $n \times q$. Similarly, define $\mathbf{C} = (\mathbf{0}, \mathbf{I}_q)$, so $\boldsymbol{\psi} = \mathbf{C}\boldsymbol{\beta}$ is the coefficient vector of interest. A $(1-\alpha) \times 100\%$ confidence ellipsoid for $\boldsymbol{\psi}$ based on $\hat{\boldsymbol{\psi}} = \mathbf{C}\hat{\boldsymbol{\beta}}$ is

$$\begin{aligned}\mathscr{E}(\boldsymbol{\psi}) = \{\boldsymbol{\psi}^* | (\boldsymbol{\psi}^* - \hat{\boldsymbol{\psi}})^{\mathrm{T}}(\mathbf{C}(\mathbf{X}^{\mathrm{T}}\mathbf{X})^{-1}\mathbf{C}^{\mathrm{T}})^{-1}(\boldsymbol{\psi}^* - \hat{\boldsymbol{\psi}}) \\ \leq q\hat{\sigma}^2 F(1-\alpha; q, n-p')\}\end{aligned} \tag{4.1.1}$$

If a subset of m cases indexed by I is deleted, then the corresponding ellipsoid based on $\hat{\boldsymbol{\psi}}_{(\mathrm{I})} = \mathbf{C}\hat{\boldsymbol{\beta}}_{(\mathrm{I})}$ is

$$\begin{aligned}\mathscr{E}_{(\mathrm{I})}(\boldsymbol{\psi}) = \{\boldsymbol{\psi}^* | (\boldsymbol{\psi}^* - \hat{\boldsymbol{\psi}}_{(\mathrm{I})})^{\mathrm{T}}(\mathbf{C}(\mathbf{X}_{(\mathrm{I})}^{\mathrm{T}}\mathbf{X}_{(\mathrm{I})})^{-1}\mathbf{C}^{\mathrm{T}})^{-1}(\boldsymbol{\psi}^* - \hat{\boldsymbol{\psi}}_{(\mathrm{I})}) \\ \leq q\hat{\sigma}_{(\mathrm{I})}^2 F(1-\alpha; q, n-p'-m)\}\end{aligned} \tag{4.1.2}$$

The volumes of the two ellipsoids are

$$Vol(\mathscr{E}(\boldsymbol{\psi})) \propto (q\hat{\sigma}^2 F_q)^{q/2} |\mathbf{C}(\mathbf{X}^{\mathrm{T}}\mathbf{X})^{-1}\mathbf{C}^{\mathrm{T}}|^{1/2} \tag{4.1.3}$$

and

$$Vol(\mathscr{E}_{(\mathrm{I})}(\boldsymbol{\psi})) \propto (q\hat{\sigma}_{(\mathrm{I})}^2 F_q^m)^{q/2} |\mathbf{C}(\mathbf{X}_{(\mathrm{I})}^{\mathrm{T}}\mathbf{X}_{(\mathrm{I})})^{-1}\mathbf{C}^{\mathrm{T}}|^{1/2} \tag{4.1.4}$$

where we adopt the shorthand $F_q = F(1-\alpha; q, n-p')$ and $F_q^m =$

$F(1-\alpha; q, n-p'-m)$. The logarithm of the ratio of (4.1.4) to (4.1.3) is

$$VR_{\mathrm{I}}(\psi) = \log\left[\frac{Vol(\mathscr{E}_{(\mathrm{I})}(\psi))}{Vol(\mathscr{E}(\psi))}\right]$$

$$= \tfrac{1}{2}\log\left[\frac{|\mathbf{C}(\mathbf{X}_{(\mathrm{I})}^{\mathrm{T}}\mathbf{X}_{(\mathrm{I})})^{-1}\mathbf{C}^{\mathrm{T}}|}{|\mathbf{C}(\mathbf{X}^{\mathrm{T}}\mathbf{X})^{-1}\mathbf{C}^{\mathrm{T}}|}\left(\frac{\hat{\sigma}_{(\mathrm{I})}^{2}F_{q}^{m}}{\hat{\sigma}^{2}F_{q}}\right)^{q}\right] \quad (4.1.5)$$

The results in Appendix A.2 can be used to simplify (4.1.5): $\hat{\sigma}_{(\mathrm{I})}^2/\hat{\sigma}^2 = (n-p'-r_{\mathrm{I}}^2)/(n-p'-m)$ and the ratio of determinants can be shown to equal

$$\frac{|\mathbf{I}-\mathbf{U}_{\mathrm{I}}|}{|\mathbf{I}-\mathbf{V}_{\mathrm{I}}|}$$

where $\mathbf{U} = \mathbf{X}_1(\mathbf{X}_1^{\mathrm{T}}\mathbf{X}_1)^{-1}\mathbf{X}_1^{\mathrm{T}}$ is the projection on the columns of $\mathbf{X}$ that are not of direct interest, and $\mathbf{U}_{\mathrm{I}}$ and $\mathbf{V}_{\mathrm{I}}$ are $m \times m$ submatrices of $\mathbf{U}$ and $\mathbf{V}$, respectively. Combining these results into (4.1.5) and simplifying gives

$$VR_{\mathrm{I}}(\psi) = -\tfrac{1}{2}\log|\mathbf{I}-\mathbf{V}_{\mathrm{I}}| + \tfrac{1}{2}\log|\mathbf{I}-\mathbf{U}_{\mathrm{I}}|$$

$$-\frac{q}{2}\log\left(\frac{n-p'-m}{n-p'-r_{\mathrm{I}}^{2}}\frac{F_{q}}{F_{q}^{m}}\right) \quad (4.1.6)$$

For $m = 1$, two choices for q are of general interest. First, if $q = p'$, then $\mathbf{C} = \mathbf{I}$, $|\mathbf{I}-\mathbf{V}_{\mathrm{I}}| = 1 - v_{ii}$, $|\mathbf{I}-\mathbf{U}_{\mathrm{I}}| = 1$ and (4.1.6) becomes

$$VR_i' = \log\left[\frac{Vol(\mathscr{E}_{(i)})}{Vol(\mathscr{E})}\right] = -\tfrac{1}{2}\log(1-v_{ii}) - \frac{p'}{2}\log\left(\frac{n-p'-1}{n-p'-r_i^2}\frac{F_{p'}}{F_{p'}^{1}}\right)$$

(4.1.7)

Apart from the ratio of F-values, this is equivalent to the statistic COVRATIO given in Belsley *et al.* (1980). Alternatively, if the intercept is ignored then $\mathbf{C} = (\mathbf{0}, \mathbf{I}_p)$, $|\mathbf{I}-\mathbf{V}_{\mathrm{I}}| = 1 - v_{ii}$, $|\mathbf{I}-\mathbf{U}_{\mathrm{I}}| = 1 - 1/n$, and (4.1.6) becomes

$$VR_i = \log\left[\frac{Vol(\mathscr{E}_{(i)}(\psi))}{Vol(\mathscr{E}(\psi))}\right] = -\tfrac{1}{2}\log(1-v_{ii}) + \tfrac{1}{2}\log(1-1/n)$$

$$-\frac{p}{2}\log\left(\frac{n-p'-1}{n-p'-r_i^2}\frac{F_p}{F_p^1}\right) \quad (4.1.8)$$

This form is recommended for general use in situations when the origin lies well outside the region of applicability of the model. This will happen often when the explanatory variables are not centered.

The log volume measure can be positive or negative. A negative measure means that deletion of the cases indexed by I decreases volume and hence *increases* precision. This will occur for $m = 1$ if r_i^2 is large but v_{ii} is small. A positive value of this ratio implies a larger volume for the reduced data, and less precision. This will occur in general for $m = 1$ whenever v_{ii} is large. The volume measure seems to balance the effects of the residual and the potential, and these in turn pull the measure in opposite directions.

4.2 The Andrews and Pregibon diagnostic

A distinct alternative method for detecting influential cases in linear regression was suggested by Andrews and Pregibon (1978). Initially, consider the effects of an outlier in $\mathbf{Y}$ and an outlying row of $\mathbf{X}$ separately. First, the deletion of a case corresponding to an outlier in $\mathbf{Y}$ will tend to result in a marked reduction in the residual sum of squares. The residual sum of squares, therefore, is a diagnostic for detecting influential cases arising because of an outlier in $\mathbf{Y}$. Second, as seen in Section 4.1, the influence of a row of $\mathbf{X}$ is at least in part reflected by the change in $|\mathbf{X}^{\mathrm{T}}\mathbf{X}|$ when the row is deleted. If $|\mathbf{X}^{\mathrm{T}}\mathbf{X}|$ changes substantially when $\mathbf{x}_i$ is deleted, then the corresponding case $(y_i, \mathbf{x}_i^{\mathrm{T}})$ will have a large influence on $\hat{\boldsymbol{\beta}}$ or, minimally, $\mathrm{Var}(\hat{\boldsymbol{\beta}})$.

Andrews and Pregibon suggest that these separate diagnostics based on change in the residual sum of squares and $|\mathbf{X}^{\mathrm{T}}\mathbf{X}|$ be combined into a single diagnostic based on the change in $(n-p')\hat{\sigma}^2 \times |\mathbf{X}^{\mathrm{T}}\mathbf{X}|$ resulting from the deletion of one or more cases. Specifically, they suggest the ratio

$$\begin{aligned} R_{\mathrm{I}} &= \frac{(n-p'-m)\hat{\sigma}^2_{(\mathrm{I})}|\mathbf{X}^{\mathrm{T}}_{(\mathrm{I})}\mathbf{X}_{(\mathrm{I})}|}{(n-p')\hat{\sigma}^2|\mathbf{X}^{\mathrm{T}}\mathbf{X}|} \\ &= \frac{(n-p'-m)\hat{\sigma}^2_{(\mathrm{I})}}{(n-p')\hat{\sigma}^2}|\mathbf{I}-\mathbf{V}_{\mathrm{I}}| \\ &= \left(1-\frac{r_{\mathrm{I}}^2}{n-p'}\right)|\mathbf{I}-\mathbf{V}_{\mathrm{I}}| \end{aligned} \tag{4.2.1}$$

as a measure of the collective influence of the cases indexed by I.

A form for R_{I} which allows additional insight into its behavior can be obtained as follows. Let $\mathbf{X}^* = (\mathbf{X}, \mathbf{Y})$, the matrix of explanatory variables augmented with $\mathbf{Y}$. From Appendix A.2,

$$\begin{aligned} |\mathbf{X}^{*\mathrm{T}}\mathbf{X}^*| &= |\mathbf{X}^{\mathrm{T}}\mathbf{X}||\mathbf{Y}^{\mathrm{T}}\mathbf{Y}-\mathbf{Y}^{\mathrm{T}}\mathbf{X}(\mathbf{X}^{\mathrm{T}}\mathbf{X})^{-1}\mathbf{X}^{\mathrm{T}}\mathbf{Y}| \\ &= (n-p')\hat{\sigma}^2|\mathbf{X}^{\mathrm{T}}\mathbf{X}| \end{aligned} \tag{4.2.2}$$

Thus, (4.2.1) can be represented as

$$R_{\mathrm{I}} = \frac{|\mathbf{X}^{*\mathrm{T}}_{(\mathrm{I})}\mathbf{X}^{*}_{(\mathrm{I})}|}{|\mathbf{X}^{*\mathrm{T}}\mathbf{X}^{*}|} \tag{4.2.3}$$

Several immediate observations can be made from this form. First, R_{I} is a unitless measure. Second, $R_{\mathrm{I}}^{-1/2} - 1$ corresponds to the proportional change in the volume of an ellipsoid generated by $\mathbf{X}^{*\mathrm{T}}\mathbf{X}^{*}$ when the cases indexed by I are deleted. Small values of R_{I} correspond to influential cases. Finally, R_{I} is invariant under permutations of the columns of $\mathbf{X}^{*}$ and thus the vector of responses $\mathbf{Y}$ is not given special recognition. For this reason, R_{I} does not make full use of the structure of the regression problem. If there is interest in particular aspects of the problem, then it may be desirable to use other measures that reflect those interests directly. On the other hand, R_{I} may serve effectively as an omnibus measure of influence.

Under normality, $(n-p'-m)\hat{\sigma}^2_{(\mathrm{I})}/(n-p')\hat{\sigma}^2$ follows a Beta distribution with parameters $(n-p'-m)/2$ and $m/2$, so R_{I} is proportional to a Beta random variable and reference values based on moments can be easily calculated.

For comparative purposes, it is convenient to take minus one half the logarithm of R_{I}, which is

$$AP_{\mathrm{I}} \equiv -\tfrac{1}{2}\log(R_{\mathrm{I}}) = -\tfrac{1}{2}\log|\mathbf{I}-\mathbf{V}_{\mathrm{I}}| + \tfrac{1}{2}\log\left(\frac{n-p'}{n-p'-r_{\mathrm{I}}^2}\right) \tag{4.2.4}$$

This statistic will now be large for influential cases, and can be compared to the analogous volume ratio based on a p'-dimensional ellipsoid (4.1.7). The two statistics differ primarily by signs and relative weights of the two terms, and by a factor of $-1/(n-p'-r_i^2)$ in the second logarithm. If $(n-p')$ is large enough to ignore this last factor, these statistics use the same information but combine it differently.

The determination of R_{I} for all subsets of m cases can be a formidable computational task. Andrews and Pregibon (1978) discuss strategies for approaching this problem.

4.3 Predictive influence

In this section, we present a Bayesian method for assessing the influence of cases on the prediction of future observations. The method, developed by Johnson and Geisser (1979, 1980), uses Kullback–Leibler divergences to measure the difference between predictive densities based on full and reduced data sets. The discussion here is restricted to the linear model (2.1.1), although the technique is quite general and applicable in many other situations. We first assume that σ^2 is known and later extend the methodology to the more common situation in which σ^2 is unknown. The former situation is easier to study since the corresponding analytic details are relatively uncomplicated.

4.3.1 KULLBACK–LEIBLER DIVERGENCES AND PREDICTIVE DENSITIES WITH σ^2 KNOWN

Let $\mathbf{Y}$ denote an n-vector of random variables that can be represented by the linear model (2.1.1) and assume that the errors $\boldsymbol{\varepsilon}$ follow an n-dimensional normal distribution with mean $\mathbf{0}$ and covariance $\sigma^2\mathbf{I}$, $N_n(\mathbf{0}, \sigma^2\mathbf{I})$. Given the observed value $\mathbf{y}$ of $\mathbf{Y}$, we suppose that the goal is to predict a q-dimensional vector $\mathbf{Y}_f$ of future observations that are represented by the linear model

$$\mathbf{Y}_f = \mathbf{X}_f\boldsymbol{\beta} + \boldsymbol{\varepsilon}_f$$

where $\boldsymbol{\varepsilon}_f$ is $N_q(\mathbf{0}, \sigma^2\mathbf{I})$, $\mathbf{X}_f$ is a $q \times p'$ known matrix of explanatory variables and $\boldsymbol{\beta}$ is the same as that in (2.1.1).

The predictive density for $\mathbf{Y}_f$ given $\mathbf{y}$, $\mathbf{X}$, $\mathbf{X}_f$, and σ^2, is a standard Bayesian tool for inference about $\mathbf{Y}_f$ (Aitchison and Dunsmore, 1975; Geisser, 1965, 1971). Predictive densities are free of unknown parameters by construction. The mean and median of the predictive density are obvious choices for point predictions while the spread and shape of the predictive density reflect the uncertainty of prediction. To obtain the predictive density, it is first necessary to find the posterior density of the unknown parameter $\boldsymbol{\beta}$.

Let $f(\cdot|\boldsymbol{\mu}, \boldsymbol{\Sigma})$ denote the density for a $N_n(\boldsymbol{\mu}, \boldsymbol{\Sigma})$ random vector. Following Johnson and Geisser, we assume the improper prior $p(\boldsymbol{\beta})\mathrm{d}\boldsymbol{\beta} \propto \mathrm{d}\boldsymbol{\beta}$ for $\boldsymbol{\beta}$. The posterior density $p(\boldsymbol{\beta}|\mathbf{y})$ for $\boldsymbol{\beta}$ given $\mathbf{Y} = \mathbf{y}$ is

$$p(\boldsymbol{\beta}|\mathbf{y}) = \frac{f(\mathbf{y}|\mathbf{X}\boldsymbol{\beta}, \sigma^2\mathbf{I})p(\boldsymbol{\beta})}{\int f(\mathbf{y}|\mathbf{X}\boldsymbol{\beta}, \sigma^2\mathbf{I})p(\boldsymbol{\beta})\mathrm{d}\boldsymbol{\beta}} \tag{4.3.1}$$

The corresponding predictive density for $\mathbf{Y}_f$ given $\mathbf{y}$, $\mathbf{X}$, $\mathbf{X}_f$, and σ^2 is

$$\int f(\mathbf{y}_f|\mathbf{X}_f\boldsymbol{\beta}, \sigma^2\mathbf{I})p(\boldsymbol{\beta}|\mathbf{y})\mathrm{d}\boldsymbol{\beta} = f(\mathbf{y}_f|\mathbf{X}_f\hat{\boldsymbol{\beta}}, \sigma^2(\mathbf{I} + \mathbf{X}_f(\mathbf{X}^{\mathrm{T}}\mathbf{X})^{-1}\mathbf{X}_f^{\mathrm{T}})) \tag{4.3.2}$$

As implied by the notation, this predictive density is $N_q(\mathbf{X}_f\hat{\boldsymbol{\beta}}, \sigma^2[\mathbf{I} + \mathbf{X}_f(\mathbf{X}^{\mathrm{T}}\mathbf{X})^{-1}\mathbf{X}_f^{\mathrm{T}}])$ and is obtained by averaging the sampling density of the future observations with respect to the posterior distribution of

$\boldsymbol{\beta}$. A useful property of the predictive density (4.3.2) is that it will converge almost surely to the sampling density of $\mathbf{Y}_f$ (Johnson and Geisser, 1979).

The influence of a collection of cases I on prediction can be determined by comparing the predictive density based on the full data to the corresponding density obtained after removing the cases in question. From (4.3.2), the predictive density for the reduced data is $N_q(\mathbf{X}_f\hat{\boldsymbol{\beta}}_{(\mathrm{I})}, \sigma^2[\mathbf{I}+\mathbf{X}_f(\mathbf{X}_{(\mathrm{I})}^{\mathrm{T}}\mathbf{X}_{(\mathrm{I})})^{-1}\mathbf{X}_f^{\mathrm{T}}])$. Influence is reflected by changes in both the location and shape of the predictive density. Of course, one way to compare these densities and thus assess influence is to compare the locations and scales separately. This quite naturally leads to developments similar to those in Chapter 3 and Section 4.1.

A comprehensive method for comparing predictive densities can be based on the Kullback–Leibler measure of divergence, defined as follows. Let g_i, $i = 1, 2$, be densities and let E_i be the expectation operator with respect to g_i. The Kullback–Leibler divergence measure $d(g_1, g_2)$ is defined by

$$d(g_1, g_2) = E_1[\log(g_1/g_2)] = \int \log(g_1/g_2)g_1(x)\mathrm{d}x \qquad (4.3.3)$$

This measure will be positive if g_1 and g_2 are different and will equal zero if $g_1 = g_2$. If $f_1 = N_n(\boldsymbol{\mu}_1, \boldsymbol{\Sigma}_1)$ and $f_2 = N_n(\boldsymbol{\mu}_2, \boldsymbol{\Sigma}_2)$, assuming that $\boldsymbol{\Sigma}_1$, $\boldsymbol{\Sigma}_2$ are positive definite, it is not hard to verify that

$$2d(f_1, f_2) = (\boldsymbol{\mu}_1 - \boldsymbol{\mu}_2)^{\mathrm{T}}\boldsymbol{\Sigma}_2^{-1}(\boldsymbol{\mu}_1 - \boldsymbol{\mu}_2) + \log(|\boldsymbol{\Sigma}_2|/|\boldsymbol{\Sigma}_1|) + \mathrm{tr}(\boldsymbol{\Sigma}_1\boldsymbol{\Sigma}_2^{-1}) - n \qquad (4.3.4)$$

The first term on the right of (4.3.4) corresponds to the distance between centers of f_1 and f_2 relative to contours of constant density for f_2. The second term compares the volumes of ellipsoids based on the two distributions and it will be zero only if the volumes are equal. The third term, $\mathrm{tr}(\boldsymbol{\Sigma}_1\boldsymbol{\Sigma}_2^{-1})$, may be conveniently viewed as a 'remainder' that compares the eigenstructure of $\boldsymbol{\Sigma}_1$ to that of $\boldsymbol{\Sigma}_2$. For example, if $\boldsymbol{\Sigma}_1$ and $\boldsymbol{\Sigma}_2$ commute and thus have the same eigenvectors, then $\mathrm{tr}(\boldsymbol{\Sigma}_1\boldsymbol{\Sigma}_2^{-1})$ is simply the sum of the ratios of the eigenvalues of $\boldsymbol{\Sigma}_1$ to the corresponding eigenvalues of $\boldsymbol{\Sigma}_2$.

4.3.2 PREDICTIVE INFLUENCE FUNCTIONS, σ^2 KNOWN

The predictive distributions for the full and reduced data sets are $f = N_q(\mathbf{X}_f\hat{\boldsymbol{\beta}}, \sigma^2[\mathbf{I}+\mathbf{X}_f(\mathbf{X}^{\mathrm{T}}\mathbf{X})^{-1}\mathbf{X}_f^{\mathrm{T}}])$ and $f_{(\mathrm{I})} = N_q(\mathbf{X}_f\hat{\boldsymbol{\beta}}_{(\mathrm{I})}, \sigma^2[\mathbf{I} + \mathbf{X}_f(\mathbf{X}_{(\mathrm{I})}^{\mathrm{T}}\mathbf{X}_{(\mathrm{I})})^{-1}\mathbf{X}_f^{\mathrm{T}}])$, respectively. The Kullback–Leibler divergence measure can be computed in two ways, depending on which of these distributions is associated with f_1 and which with f_2 in (4.3.3). From (4.3.4), we see that distance between centers is computed relative to f_2, which suggests associating f_2 with the full data predictive density. We adopt this idea and, following Johnson and Geisser, we call $d(f_{(I)}, f)$ a predictive influence function (*PIF*).

EXAMPLE 4.3.1. PREDICTIVE INFLUENCE WHEN $q = 1, m = 1, \sigma^2$ KNOWN. Let $\mathbf{x}_f^T = \mathbf{X}_f, v_f = \mathbf{x}_f^T(\mathbf{X}^T\mathbf{X})^{-1}\mathbf{x}_f$ and $v_{if} = \mathbf{x}_f^T(\mathbf{X}^T\mathbf{X})^{-1}\mathbf{x}_i$. The predictive density f based on the full data for a single future prediction at x_f is $N(\mathbf{x}_f^T\hat{\boldsymbol{\beta}}, \sigma^2(1+v_f))$ and the corresponding density $f_{(i)}$ based on the reduced data is $N\{\mathbf{x}_f^T\hat{\boldsymbol{\beta}}_{(i)}, \sigma^2[1+v_f+v_{if}^2/(1-v_{ii})]\}$.

Using (4.3.4) and after a little algebra, the *PIF* $d(f_{(i)}, f; \mathbf{x}_f)$ for a single prediction at $\mathbf{x}_f$ can be written as

$$2d(f_{(i)}, f; \mathbf{x}_f) = D_i(\mathbf{X}^T\mathbf{X}, \sigma^2)\rho_{if}^2 \frac{v_f}{1+v_f} - \log\left[1+\rho_{if}^2\left(\frac{v_{ii}}{1-v_{ii}}\right) \times \left(\frac{v_f}{1+v_f}\right)\right] + \left[1+\rho_{if}^2\left(\frac{v_{ii}}{1-v_{ii}}\right)\left(\frac{v_f}{1+v_f}\right)\right] - 1 \quad (4.3.5)$$

where $\rho_{if}^2 = v_{if}^2/v_{ii}v_f$ is the squared correlation between $\mathbf{x}_i^T\hat{\boldsymbol{\beta}}$ and $\mathbf{x}_f^T\hat{\boldsymbol{\beta}}$. Thus, the behavior of this *PIF* depends on $D_i(\mathbf{X}^T\mathbf{X}, \sigma^2)$, ρ_{if}^2, v_f, and v_{ii}. With σ^2 replaced by $\hat{\sigma}^2$, the first term on the right of (4.3.5) is the same as that obtained from a comparison of point predictions in the frequentist approach discussed in Section 3.5; see (3.5.18) and the subsequent discussion.

The second and third terms on the right of (4.3.5) depend only on $1+\rho_{if}^2 v_{ii} v_f/[(1+v_f)(1-v_{ii})]$, the ratio of the variance associated with $f_{(i)}$ to that of f. Since this ratio is always ≥ 1, the variance of the predictive distribution cannot decrease when a case is deleted and σ^2 is known. The change in variance will tend to be large when v_{ii} is large and $\mathbf{x}_f = \mathbf{x}_i$.□

To use a *PIF*, it is first necessary to specify $\mathbf{X}_f$, the matrix containing the points in the factor space that correspond to future predictions. This is clearly a disadvantage since $\mathbf{X}_f$ will not normally be known during the development of the model. To overcome this problem and thus make the *PIF*s more available for use as routine diagnostics, Johnson and Geisser (1980) suggest using $\mathbf{X}$ in place of $\mathbf{X}_f$. When $\mathbf{X} = \mathbf{X}_f$, we will write d_I for $d(f_{(I)}, f)$ where $f_{(I)} = N_n(\mathbf{X}\hat{\boldsymbol{\beta}}_{(I)}, \sigma^2[\mathbf{I}+\mathbf{X}(\mathbf{X}_{(I)}^T\mathbf{X}_{(I)})^{-1}\mathbf{X}^T])$ and $f = N_n(\mathbf{X}\hat{\boldsymbol{\beta}}, \sigma^2(\mathbf{I}+\mathbf{V}))$.

To obtain a relatively simple form for d_I, we substitute into (4.3.4) term by term. First, the change in centers is

$$
\begin{aligned}
(\mu_1 - \mu_2)^T \Sigma_2^{-1} (\mu_1 - \mu_2) &= (\hat{\beta} - \hat{\beta}_{(I)})^T X^T (I + V)^{-1} X (\hat{\beta} - \hat{\beta}_{(I)}) / \sigma^2 \\
&= (\hat{\beta} - \hat{\beta}_{(I)})^T X^T (I - \tfrac{1}{2} V) X (\hat{\beta} - \hat{\beta}_{(I)}) / \sigma^2 \\
&= D_I(X^T X, 2\sigma^2) \qquad (4.3.6)
\end{aligned}
$$

Thus, the distance between μ_1 and μ_2 is measured by a member of the class of norms of the *SIC*, $D_I(M, c)$, with σ^2 in place of $\hat{\sigma}^2$. Also, this form is closely related to the influence curve for prediction obtained as a result of a frequentist approach.

Next, the change in volume is measured by

$$
\log|\Sigma_2| - \log|\Sigma_1| = \log|I + V| - \log|I + X(X_{(I)}^T X_{(I)})^{-1} X^T|
$$

Since V is a rank p' symmetric, idempotent matrix, the eigenvalues of $I + V$ are 2 with multiplicity p' and 1 with multiplicity $n - p'$ and $|I + V| = 2^{p'}$. Next, using Appendix A.2 to evaluate the partitioned form of $|I + X(X_{(I)}^T X_{(I)})^{-1} X^T|$ that results from the partition $X^T = (X_{(I)}^T, X_I^T)$, it follows that

$$
|I + X(X_{(I)}^T X_{(I)})^{-1} X^T| = 2^{p'} |I + \tfrac{1}{2} V_I (I - V_I)^{-1}|
$$

Combining terms, the change in volume can be obtained from the determinant of a single $m \times m$ matrix,

$$
\log|\Sigma_2| - \log|\Sigma_1| = -\log|I + \tfrac{1}{2} V_I (I - V_I)^{-1}| \qquad (4.3.7)
$$

that depends only on the eigenvalues of V_I.

The final term of d_I is

$$
\begin{aligned}
\operatorname{tr}[\Sigma_1 \Sigma_2^{-1}] &= \operatorname{tr}[(I + X(X_{(I)}^T X_{(I)})^{-1} X^T)(I + V)^{-1}] \\
&= \operatorname{tr}[(I + X(X_{(I)}^T X_{(I)})^{-1} X^T)(I - \tfrac{1}{2} V)] \\
&= n - \frac{p'}{2} + \tfrac{1}{2} \operatorname{tr}[X^T X (X_{(I)}^T X_{(I)})^{-1}] \\
&= n + \tfrac{1}{2} \operatorname{tr}[V_I (I - V_I)^{-1}] \qquad (4.3.8)
\end{aligned}
$$

which again depends on the eigenvalues of V_I. Finally, combining the last three results, d_I can be expressed as

$$
d_I = D_I(X^T X, 4\sigma^2) - \tfrac{1}{2} \log|I + \tfrac{1}{2} V_I (I - V_I)^{-1}| + \tfrac{1}{4} \operatorname{tr}[V_I (I - V_I)^{-1}] \qquad (4.3.9)
$$

The *PIF* d_I depends on only e_I and V_I. The predictive approach, therefore, utilizes the same building blocks as the previous approaches.

The main difference is in how the predictive approach combines this information to produce one overall measure of influence.

The form for d_I given in (4.3.9) is perhaps the most useful for the purposes of computation since all quantities are calculated from the full data. For interpretation, however, the identity

$$\mathbf{X}_I(\mathbf{X}_{(I)}^T\mathbf{X}_{(I)})^{-1}\mathbf{X}_I^T = \mathbf{V}_I(\mathbf{I}-\mathbf{V}_I)^{-1}$$

is useful: $\text{tr}\,[\mathbf{V}_I(\mathbf{I}-\mathbf{V}_I)^{-1}]$ is proportional to the sum of the variances of the estimated values, based on the reduced data, at the cases indexed by I. In addition, under a correct linear model,

$$E[D_I(\mathbf{X}^T\mathbf{X}, 4\sigma^2)] = \tfrac{1}{4}\,\text{tr}\,[\mathbf{V}_I(\mathbf{I}-\mathbf{V}_I)^{-1}]$$

which is the average squared distance between the centers of the ellipsoids associated with the predictive densities based on the full and reduced data and is proportional to the expected potential discussed in Section 3.6.1.

4.3.3 PREDICTIVE INFLUENCE FUNCTIONS, σ^2 UNKNOWN

When σ^2 is unknown, the predictive densities are multivariate Student densities rather than multivariate normal. Let $S_n(\nu, \boldsymbol{\mu}, \boldsymbol{\Sigma})$ denote an n-dimensional Student density with ν degrees of freedom, location parameter $\boldsymbol{\mu}$ and dispersion matrix $\boldsymbol{\Sigma}$. Assuming the joint prior $p(\boldsymbol{\beta}, \sigma^2)\,d\boldsymbol{\beta}\,d\sigma^2 \propto \sigma^{-2}\,d\boldsymbol{\beta}\,d\sigma^2$ and setting $\mathbf{X}_f = \mathbf{X}$, the predictive densities based on the full and reduced data sets are

$$S_n(n-p', \mathbf{X}\hat{\boldsymbol{\beta}}, \hat{\sigma}^2(\mathbf{I}+\mathbf{V}))$$

and

$$S_n(n-p'-m, \mathbf{X}\hat{\boldsymbol{\beta}}_{(I)}, \hat{\sigma}_{(I)}^2(\mathbf{I}+\mathbf{X}(\mathbf{X}_{(I)}^T\mathbf{X}_{(I)})^{-1}\mathbf{X}^T)),$$

respectively. Unfortunately, the *PIF*s based on these densities are complicated and difficult to study. Johnson and Geisser (1980) use normal densities to approximate the predictive Student densities, and then develop the corresponding approximate *PIF*s along the lines indicated above.

For $\nu > 2$, the covariance matrix for a multivariate Student random variable is $[\nu/(\nu-2)]\boldsymbol{\Sigma}$. It is reasonable to use

$$N_n\left(\mathbf{X}\hat{\boldsymbol{\beta}}, \left(\frac{n-p'}{n-p'-2}\right)\hat{\sigma}^2(\mathbf{I}+\mathbf{V})\right)$$

and

$$N_n\left(\mathbf{X}\hat{\boldsymbol{\beta}}_{(\mathrm{I})}, \left(\frac{n-m-p'}{n-m-p'-2}\right)\hat{\sigma}^2_{(\mathrm{I})}(\mathbf{I}+\mathbf{X}(\mathbf{X}^{\mathrm{T}}_{(\mathrm{I})}\mathbf{X}_{(\mathrm{I})})^{-1}\mathbf{X}^{\mathrm{T}})\right)$$

densities to approximate the predictive densities based on the full and reduced data, respectively. The approximate *PIF* $\tilde{d}_{\mathrm{I}}$ can now be developed by following the steps in the σ^2 known case. The terms that measure the change in volume and eigenstructure depend on the ratio $\hat{\sigma}^2_{(\mathrm{I})}/\hat{\sigma}^2$ and thus on r_{I}^2. The approximate *PIF* may be written as

$$\tilde{d}_{\mathrm{I}} = \left(\frac{n-p'-2}{n-p'}\right)D_{\mathrm{I}}(\mathbf{X}^{\mathrm{T}}\mathbf{X}, 4\hat{\sigma}^2) + \tfrac{1}{4}k_{\mathrm{I}}\,\mathrm{tr}\,[\mathbf{V}_{\mathrm{I}}(\mathbf{I}-\mathbf{V}_{\mathrm{I}})^{-1}]$$

$$-\tfrac{1}{2}\log|\mathbf{I}+\tfrac{1}{2}\mathbf{V}_{\mathrm{I}}(\mathbf{I}-\mathbf{V}_{\mathrm{I}})^{-1}| + \frac{n}{2}[k_{\mathrm{I}} - \log(k_{\mathrm{I}}) - 1] \quad (4.3.10)$$

where

$$k_{\mathrm{I}} = \left(\frac{n-p'-2}{n-m-p'-2}\right)\left(1-\frac{r_{\mathrm{I}}^2}{n-p'}\right)$$

Apart from constants, the difference between $\tilde{d}_{\mathrm{I}}$ and d_{I} is in the presence of k_{I} in the former measure. Since k_{I} is a decreasing function of r_{I}^2, it will be small when the cases indexed by I do not conform to the assumed model.

The special case $m = 1$ is informative,

$$\tilde{d}_i = \frac{p'(n-p'-2)}{4(n-p')}D_i + \frac{k_i}{4}\left(\frac{v_{ii}}{1-v_{ii}}\right) - \tfrac{1}{2}\log\left(1+\tfrac{1}{2}\frac{v_{ii}}{1-v_{ii}}\right)$$

$$+\frac{n}{2}(k_i - \log(k_i) - 1) \quad (4.3.11)$$

Thus, $\tilde{d}_i$ depends only on n, p', r_i^2, and v_{ii}, and is a monotonically increasing function of v_{ii} when n, p', and r_i^2 are fixed. With n, p', v_{ii} fixed, $\tilde{d}_i$ is a convex function of r_i^2, and, if v_{ii} is small, the minimum of $\tilde{d}_i$ can occur with $r_i^2 > 0$. As a practical matter, the fact that $\tilde{d}_i$ is not always monotonic in r_i^2 may not be important, since the minimum will occur for a very small value of r_i.

If the Kullback–Leibler divergence is computed with the roles of the full and reduced densities interchanged, the resulting measure is somewhat more complicated. In particular, the part of the measure that compares centers uses a metric that is different for each choice of i, and thus is not directly comparable from case to case; see Johnson and Geisser (1979) for further details.

An alternative to choosing $\mathbf{X}_f = \mathbf{X}$

In Example 4.3.1, we discussed the *PIF* for a single prediction at $\mathbf{x}_f$ when $m = 1$ and σ^2 is assumed known. Here, we discuss the corresponding results for σ^2 unknown. Notation, unless otherwise defined, follows that in Example 4.3.1. The predictive densities for the full and reduced data sets are Student densities which may be approximated by normal densities as before. Let $\tilde{d}_i(\mathbf{x}_f)$ be the approximate *PIF* obtained using the normal approximation.

Using (4.3.4) and the result of Example 4.3.1, it can be verified that

$$2\tilde{d}_i(\mathbf{x}_f) = \frac{p'(n-p'-2)}{n-p'} D_i \rho_{if}^2 \frac{v_f}{1+v_f} + k_i \rho_{if}^2 \frac{v_{ii}}{1-v_{ii}} \frac{v_f}{1+v_f} - \log\left[1 + \rho_{if}^2 \frac{v_{ii}}{1-v_{ii}} \frac{v_f}{1+v_f}\right] + k_i - \log(k_i) - 1 \qquad (4.3.12)$$

where

$$k_i = \left(\frac{n-p'-2}{n-p'-3}\right)\left(1 - \frac{r_i^2}{n-p'}\right)$$

as before. The difference between (4.3.12) and the analogous expression in Example 4.3.1 is in the presence of k_i.

As indicated previously, the usefulness of $\tilde{d}_i(\mathbf{x}_f)$ as a routine diagnostic is limited because of the requirement that $\mathbf{x}_f$ be specified *a priori*. Indeed, this limitation was the motivation behind Johnson and Geisser's suggestion to use $\mathbf{X}_f = \mathbf{X}$ for routine checking. A potential problem with this approach, however, is that $\tilde{d}_i(\mathbf{x}_f)$ may be large for some points $\mathbf{x}_f$ that are not adequately reflected by the diagnostic resulting from setting $\mathbf{X}_f = \mathbf{X}$. This can be overcome by using $\tilde{d}_i^* = \max[d_i(\mathbf{x}_f)]$ with the maximum taken over all possible values of $\mathbf{x}_f$, so that, for each i, the *PIF* is evaluated at the point $\mathbf{x}_f^*$ where the influence is maximized. This is the same as one of the approaches used in the discussion of the frequentist approach to prediction given in Section 3.5. If $\tilde{d}_i^*$ is small then it can safely be concluded that the i-th case is uninfluential for any single prediction. The same conclusion does not necessarily follow when $\tilde{d}_i$ is small, since there may exist points for which $\tilde{d}_i(\mathbf{x}_f)$ is relatively large. If $\tilde{d}_i^*$ is large then predictions around $\mathbf{x}_f^*$ will be seriously influenced by the i-th case. Further investigation may be necessary to determine the stability of predictions in other regions.

It is easily verified that $\tilde{d}_i(\mathbf{x}_f)$ is monotonically increasing in $\rho_{if}^2[v_f/(1+v_f)]$ and that it depends on $\mathbf{x}_f$ only through this term.

Thus, to maximize $\tilde{d}_i(\mathbf{x}_f)$ by choice of $\mathbf{x}_f$ it is sufficient to maximize $\rho_{if}^2[v_f/(1+v_f)]$. From Appendix A.3 it follows that

$$\max_{\mathbf{x}_f}\left(\rho_{if}^2\frac{v_f}{1+v_f}\right)=\frac{1}{v_{ii}}\left[v_{ii}-\frac{1}{n+1}\right] \tag{4.3.13}$$

Substitution into (4.3.12) yields

$$2\tilde{d}_i^* = \left(\frac{n-p'-2}{n-p'}\right)r_i^2\frac{[v_{ii}-1/(n+1)]}{1-v_{ii}}+k_i\frac{[v_{ii}-1/(n+1)]}{1-v_{ii}}$$
$$-\log\left\{1+\frac{[v_{ii}-1/(n+1)]}{1-v_{ii}}\right\}+k_i-\log(k_i)-1 \tag{4.3.14}$$

The first term, which measures location differences, is proportional to (3.5.21), the analogous measure from the frequentist approach. The remaining terms are similar to those in $\tilde{d}_i$, but are adjusted to give differential weights to the various components. For example, each of the final three terms of $\tilde{d}_i^*$ is $1/n$ times the corresponding term in $\tilde{d}_i$. Each of the remaining terms in $\tilde{d}_i^*$ can be obtained from the corresponding term in $\tilde{d}_i$ by replacing $\frac{1}{2}[v_{ii}/(1-v_{ii})]$ with $[v_{ii}-1/(n+1)]/(1-v_{ii})$. These relationships suggest that $\tilde{d}_i$ may be relatively more sensitive to the removal of cases with large values of r_i^2 while $\tilde{d}_i^*$ will be more sensitive to cases with large v_{ii}.

4.4 A comparison of influence measures

Thus far, we have considered no less than four distinct types of diagnostic statistics to assess influence, each with many variations. A comparison of the various measures can be useful. As representatives of the normed influence curves, we will consider for $m=1$, $D_i = D_i(\mathbf{X}^T\mathbf{X}, p'\hat{\sigma}^2)$ and $D_i' = D_i(\mathbf{X}^T\mathbf{X}, p'\hat{\sigma}_{(i)}^2)$. To represent the volume ratios, we use VR_i' defined by (4.1.7) and the logarithm of the Andrews–Pregibon measure AP_i defined by (4.2.4). Finally, two measures based on the Bayesian predictivist approach, $\tilde{d}_i$ defined by (4.3.11) and $\tilde{d}_i^*$ defined by (4.3.14), will be compared. The major omissions from this list are the measures that require specification of a set of combinations of coefficients of interest for study and the internally scaled measures. These latter measures may have different behavior than the overall measures, depending on the structure of a specific problem.

When cases are considered one at a time, all of these influence measures are functions of r_i^2, v_{ii}, and the constants, n, p', and, for the volume ratio, a ratio of percentage points of F. Thus, all the statistics use the same building blocks but combine the information differently. The behavior of these statistics can be studied by comparing them for various combinations of n, p', v_{ii}, and r_i^2. Figure 4.4.1 contains plots of all six measures versus v_{ii} for $n = 50$, $p' = 5$, and a different value of r_i^2 in each plot, $r_i^2 = 0$, 1, 4, and 9, respectively. Since the statistics have different calibrations, we compare the qualitative shapes of the curves rather than their values.

When $r_i^2 = 0$, $\hat{\boldsymbol{\beta}} - \hat{\boldsymbol{\beta}}_{(i)} = 0$, and both D_i and D_i' are exactly 0 for all values of v_{ii}. The other measures do not have this property, and all become larger for v_{ii} large. The Andrews–Pregibon measure and the volume measures behave like a constant times $\log(1 - v_{ii})$, while the predictive measures respond only to much larger values of v_{ii}. For $r_i^2 = 1$, the two distance measures D_i and D_i' are identical and require moderately large values of v_{ii} to exhibit influence. The volume and Andrews–Pregibon measures are not sensitive to the increase from $r_i^2 = 0$ to $r_i^2 = 1$ and exhibit essentially the same behavior as in

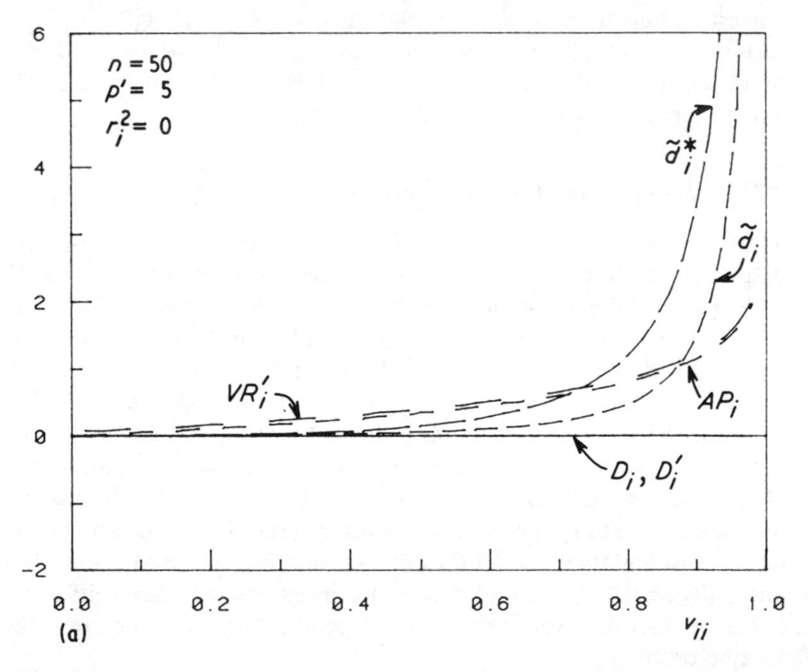

Fig. 4.4.1

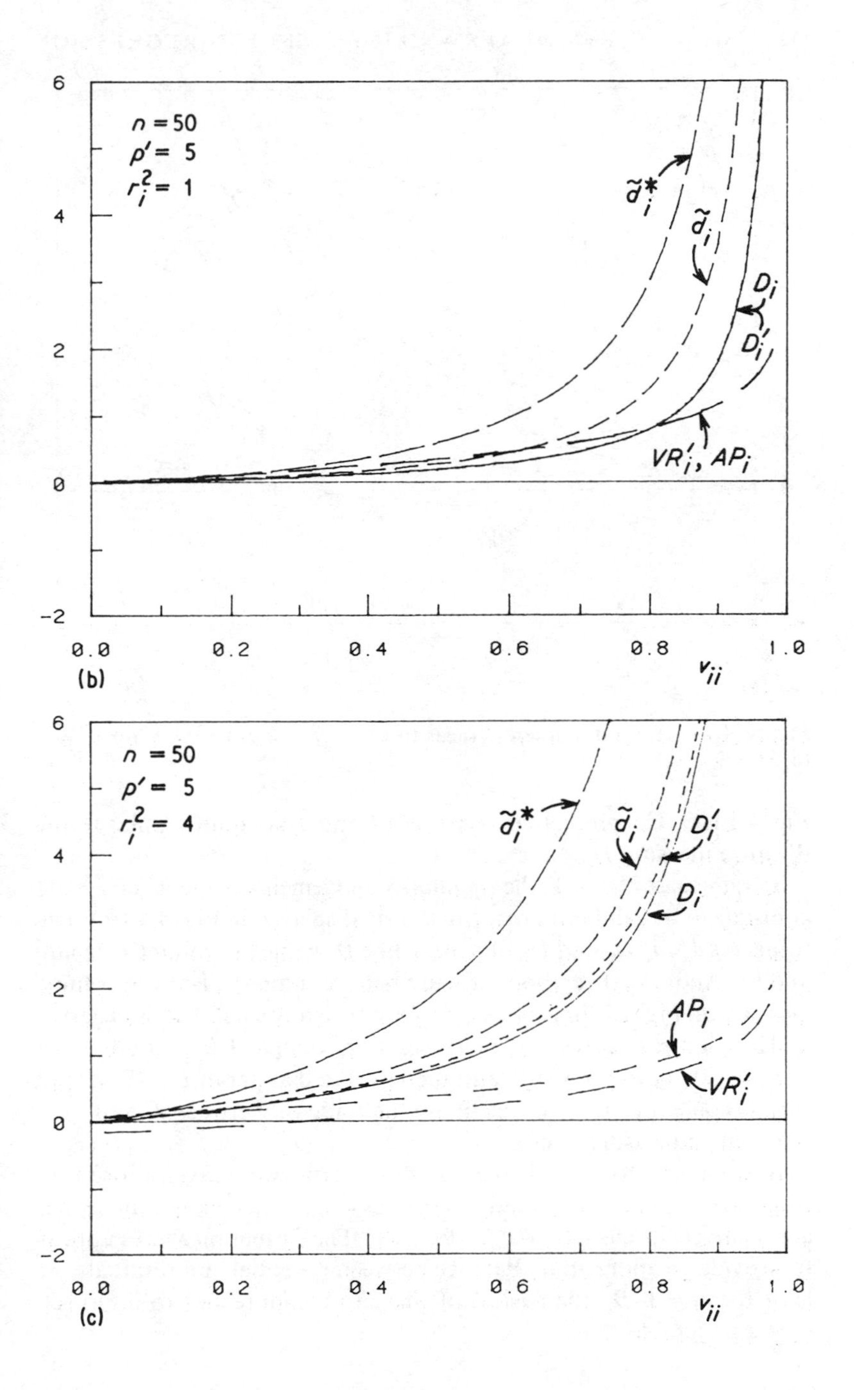

n = 50
ρ′ = 5
r_i^2 = 1
d̃_i*
d̃_i
D_i
D_i′
VR_i′, AP_i
0.0
0.2
0.4
0.6
0.8
1.0
v_ii
(b)
n = 50
ρ′ = 5
r_i^2 = 4
d̃_i*
d̃_i
D_i′
D_i
AP_i
VR_i′
0.0
0.2
0.4
0.6
0.8
1.0
v_ii
(c)

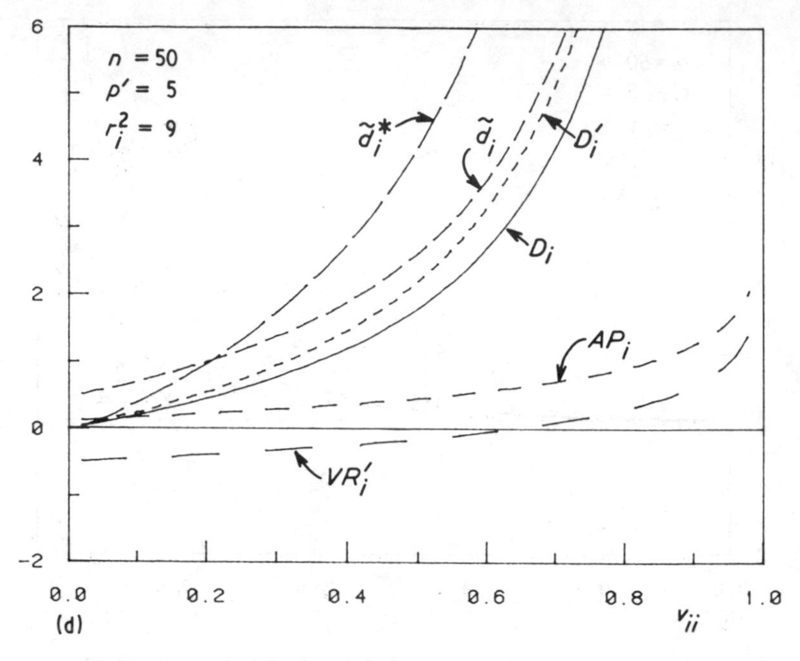

Figure 4.4.1 Several influence measures. (a) $r_i^2 = 0$ (b) $r_i^2 = 1$ (c) $r_i^2 = 4$ (d) $r_i^2 = 9$

Fig. 4.4.1(a). The predictive measures $\tilde{d}_i^*$ and $\tilde{d}_i$ are quite similar to the distance measure D_i.

As r_i increases from 1, the qualitative judgements made when $r_i^2 = 1$ continue to be valid but are more clearly displayed in Fig. 4.4.1(c). The measures $\tilde{d}_i^*$, $\tilde{d}_i$, D_i, and D_i' all behave like D_i, while the volume measure and the Andrews–Pregibon measure behave similarly. For the volume measures in Fig. 4.4.1 (c) and (d), if v_{ii} is sufficiently small VR_i' is negative, and it becomes positive as v_{ii} increases. For example, VR_i' is about -0.5 at $v_{ii} = 0.02$ in Fig. 4.4.1(d), and increases to 0 at about $v_{ii} = 0.65$ and then becomes positive. In this figure the trade-off between r_i^2 and v_{ii} in the volume measure is clear.

In summary, for $m = 1$, the measures form two classes: those that respond to r_i^2 and v_{ii} essentially as D_i does (D_i, D_i', $\tilde{d}_i$, $\tilde{d}_i^*$), and those that are relatively insensitive to v_{ii} (AP_i, VR_i'). The former measures appear to provide an appropriate balance between potential and residuals. At least for $m = 1$, D_i, the easiest of these to compute and to interpret, seems preferable.

For $m > 1$, the comparison between the influence measures is much more complicated, but some general comments are in order. First, the volume measures VR'_I and AP_I depend only on $(n-p'-r_I^2)/(n-p')$, the eigenvalues of $\mathbf{V}_I$, and constants. If $n-p'$ is large, these measures are relatively insensitive to r_I^2. The measures D_I, D'_I, $\tilde{d}_I$, and $\tilde{d}_I^*$ all have a lead term like D_I and hence they behave similarly. These depend not only on r_I^2 and the eigenvalues of $\mathbf{V}_I$, but also on the orientation of the vector $\mathbf{e}_I$ of residuals in an appropriate geometry. Thus, two groups of cases with identical r_I^2 and eigenvalues of $\mathbf{V}_I$ need not have the same influence as measured by D_I. For these measures, then, the notion of an influential subset is more complicated, and the discussion of potential in Section 3.6.1 is relevant.

Draper and John (1981) conducted a detailed examination of the relative merits of AP_I and D_I. In addition to showing that AP_I may isolate cases that are not outliers or influential for parameter estimation, they show by example that the reverse may also happen: The Andrews–Pregibon statistic cannot be guaranteed to locate outliers or cases that are influential for $\hat{\boldsymbol{\beta}}$. They recommend the study of r_I^2 (essentially their Q_k), $|\mathbf{I}-\mathbf{V}_I|$, and D_I. In the larger class of statistics discussed here, it is clear that their advice is sound, although their choice of potential measures $|\mathbf{I}-\mathbf{V}_I|$ may be replaced by one of those discussed in Section 3.6.1.

EXAMPLE 4.4.1. CLOUD SEEDING NO. 11. Table 4.4.1 lists several of the influence statistics discussed in this chapter for the cloud seeding data; see also Table 3.5.5. The subset ψ is chosen as in Example 3.5.3. The important observation from this table is that the ordering of cases on influence is different for the various statistics. Computation of them all can lead to confusing conclusions. A more reasonable approach is to adopt one of the measures – possibly D_i – as the standard and use additional measures as called for by specific concerns.□

Calibration

The various influence measures discussed in this and the previous chapter each provide a way of ordering individual or groups of cases based on their impact on a selected characteristic of the analysis. Experience with a given measure will provide additional insight that can be useful for an understanding of the importance of its magnitude. Beyond this, however, there are only a few methods of calibration

Table 4.4.1 *More influence measures, cloud seeding data*

Case	D_i	VR_i	$VR_i(\psi)$	R_i	$\tilde{d}_i$	$\tilde{d}_i^*$
1	0.030	0.343	0.184	0.439	0.196	0.367
2	4.555	0.782	0.392	0.023	18.579	36.696
3	0.062	0.359	0.166	0.363	0.310	0.633
4	0.037	0.144	0.060	0.639	0.104	0.187
5	0.030	0.250	0.095	0.568	0.125	0.216
6	0.014	0.414	0.206	0.371	0.234	0.448
7	0.729	−1.158	−0.571	0.222	3.499	3.151
8	0.078	−0.261	−0.124	0.600	0.305	0.302
9	0.000	0.330	0.188	0.588	0.099	0.106
10	0.069	0.071	0.043	0.575	0.176	0.327
11	0.014	0.251	0.123	0.650	0.087	0.113
12	0.007	0.187	0.096	0.807	0.048	0.037
13	0.001	0.319	0.166	0.612	0.093	0.092
14	0.000	0.252	0.125	0.828	0.063	0.017
15	0.126	−0.182	−0.080	0.522	0.390	0.538
16	0.000	0.317	0.186	0.626	0.090	0.082
17	0.318	0.318	0.231	0.225	1.039	2.166
18	0.320	0.669	0.485	0.090	2.493	5.239
19	0.000	0.280	0.153	0.739	0.072	0.036
20	0.062	0.359	0.211	0.363	0.310	0.633
21	0.004	0.257	0.123	0.745	0.068	0.045
22	0.027	0.180	0.098	0.657	0.089	0.148
23	0.149	0.202	0.133	0.378	0.434	0.886
24	0.315	−0.384	−0.188	0.370	1.057	1.380

available. As mentioned previously, many of the measures of the form $D_I(\mathbf{M}, c)$ can be monotonically transformed to a more familiar scale that does not depend on n and p' by comparing $D_I(\mathbf{M}, c)$ to the percentage points of the appropriate F-distribution. For example, the knowledge that the removal of case 1 would move the least squares estimate of $\boldsymbol{\beta}$ to the edge of a 95% confidence region while the removal of case 2 would move the same estimate to the edge of a 5% region is surely more useful than just knowing that case 1 is more influential than case 2. In addition, half-normal plots with a simulated envelope (see Sections 2.3.4 and 3.5.3) can be used in combination with any of the influence measures to help avoid problems of overinterpretation. These techniques are intended as aids to interpretation and not as foundations for accept–reject rules or p-values.

Dempster and Gasko-Green (1981) suggest methods for sequentially removing individual cases and determining conditional p-values that can be used to formulate stopping criteria. Their methods are based on the repeated application of a selection rule to determine the most discrepant case at each stage. The class of available selection rules is large and includes many of the influence measures discussed previously. Belsley *et al.* (1980) discuss other methods such as the use of gaps for determining the cases that require further attention.

CHAPTER 5

Assessment of influence in other problems

'In the study of nature, we must not, therefore, be scrupulous as to *how* we reach to a knowledge of such general facts: provided only we verify them carefully when once detected, we must be content to seize them wherever they are to be found.'

HERSCHEL, *op. cit.*

Most of the methods for the analysis of influence presented in earlier chapters depend on the elegance of the linear least squares regression problem. The use of the sample influence curve to measure influence is aided by the algebraic updating formulae in Appendix A.2 that allow computations to be done from full sample statistics; interpretation of normed influence measures is made clear by appeal to elliptical confidence regions that characterize linear least squares regression.

As mentioned in Section 3.4, the extension of the sample versions of the influence curve to other problems is conceptually straightforward. As a practical matter, however, the use of these ideas can be expensive since exact updating formulae are generally lacking. To compute the sample influence curve for a parameter $\boldsymbol{\theta}$, for example, values of $\hat{\boldsymbol{\theta}}_{(i)}$, $i = 1, 2, \ldots, n$, in addition to the complete data estimate are needed, and each of these may require iteration. In addition, the definition of a residual and the choice of a norm can be troubling. Norms of the sample influence curve based on elliptical contours will not always be appropriate.

In this chapter we discuss ways in which diagnostics for linear least squares regression might be extended to more complex situations. In the next section we present a general definition of residuals and suggest an extended version of the v_{ii}s. A general approach to influence, including likelihood-based measures, is discussed in Section 5.2. A relatively inexpensive approximation of the sample influence curve is suggested and this in turn leads to the problem of judging the accuracy of the approximations. Sections 5.3–5.5 contain discussions of non-linear least squares, logistic regression and robust regression, respectively. We comment briefly on several other problems in Section 5.6.

The general purpose of this chapter is to suggest ideas rather than specific diagnostics. Except perhaps for logistic regression, the specific methods presented have not been studied in great detail and more work is required before definite recommendations can be given.

5.1 A general definition of residuals

Cox and Snell (1968) define residuals for a fairly general class of models and suggest a method for determining their first two moments. This in turn leads to a generalization of some of the diagnostics for linear least squares regression to more complex models.

Assume that the i-th response y_i is a known function g_i of an unknown parameter vector $\boldsymbol{\theta}$ and an unobservable error ε_i,

$$y_i = g_i(\boldsymbol{\theta}, \varepsilon_i), \qquad i = 1, 2, \ldots, n \tag{5.1.1}$$

The errors ε_i are assumed to be continuous, independent and identically distributed with a completely known distribution, so location and scale parameters are not distinguished. This formulation excludes some standard models such as time series and components of variance problems where the response may depend on the errors in a more complicated way.

Assuming a unique solution for ε_i, (5.1.1) may be re-expressed in the form

$$\varepsilon_i = h_i(y_i, \boldsymbol{\theta}), \qquad i = 1, 2, \ldots, n \tag{5.1.2}$$

Cox and Snell define the i-th residual $\hat{\varepsilon}_i$ by

$$\hat{\varepsilon}_i = h_i(y_i, \hat{\boldsymbol{\theta}}), \qquad i = 1, 2, \ldots, n \tag{5.1.3}$$

where $\hat{\boldsymbol{\theta}}$ is the maximum likelihood estimate of $\boldsymbol{\theta}$. We call $\hat{\varepsilon}_i$ a *maximum likelihood residual* (Cox and Snell call it a crude residual).

Suppose, for example, that (5.1.1) is the usual linear model written as $y_i = \mathbf{x}_i^{\mathrm{T}}\boldsymbol{\beta} + \sigma\varepsilon_i$, where the ε_is are independent, identically distributed normal random variables with $E(\varepsilon_i) = 0$ and var $(\varepsilon_i) = 1$. If

$$\boldsymbol{\theta}^{\mathrm{T}} = (\boldsymbol{\beta}^{\mathrm{T}}, \sigma),$$

then

$$\begin{aligned}\hat{\varepsilon}_i &= (y_i - \mathbf{x}_i^{\mathrm{T}}\hat{\boldsymbol{\beta}})/\hat{\sigma} \\ &= e_i/(\textstyle\sum e_j^2/n)^{1/2}\end{aligned}$$

and $\hat{\varepsilon}_i$ is a standardized version of the ordinary residual e_i. (In this chapter, $\hat{\sigma}^2$ is the maximum likelihood estimator of σ^2.)

In general, moments of the maximum likelihood residuals cannot be obtained explicitly. Useful approximations to $E(\hat{\varepsilon}_i)$ and $E(\hat{\varepsilon}_i^2)$ can, however, be obtained from a quadratic expansion of (5.1.3) about $\boldsymbol{\theta}$,

$$\hat{\varepsilon}_i \cong \varepsilon_i + (\hat{\boldsymbol{\theta}} - \boldsymbol{\theta})^{\mathrm{T}}\dot{\mathbf{H}}_i(\boldsymbol{\theta}) + \tfrac{1}{2}(\hat{\boldsymbol{\theta}} - \boldsymbol{\theta})^{\mathrm{T}}\ddot{\mathbf{H}}_i(\boldsymbol{\theta})(\hat{\boldsymbol{\theta}} - \boldsymbol{\theta}) \tag{5.1.4}$$

where $\dot{\mathbf{H}}_i(\boldsymbol{\theta})$ is a q-vector with elements $\partial h_i(y_i, \boldsymbol{\theta})/\partial\theta_j$, and $\ddot{\mathbf{H}}_i(\boldsymbol{\theta})$ is a $q \times q$ matrix with elements $\partial^2 h_i(y_i, \boldsymbol{\theta})/\partial\theta_j\partial\theta_k$, both evaluated at $\boldsymbol{\theta}$. Expressions for $E\hat{\varepsilon}_i$, $\operatorname{var}(\hat{\varepsilon}_i)$, and $\operatorname{cov}(\hat{\varepsilon}_i, \hat{\varepsilon}_j)$ in terms of $\dot{\mathbf{H}}_i(\boldsymbol{\theta})$, $\ddot{\mathbf{H}}_i(\boldsymbol{\theta})$, the expected information matrix, the score vector and the second-order bias of the ML estimator $\hat{\boldsymbol{\theta}}$ are given by Cox and Snell; see also Cox and Hinkley (1974, Chapter 9). These expressions take the form

$$\begin{aligned} E(\hat{\varepsilon}_i) &= E(\varepsilon_i) + a_i \\ \operatorname{var}(\hat{\varepsilon}_i) &= \operatorname{var}(\varepsilon_i) - c_{ii} \\ \operatorname{cov}(\hat{\varepsilon}_i, \hat{\varepsilon}_j) &= c_{ij} \end{aligned} \tag{5.1.5}$$

In all but the simplest situations, the determination of the a_is and c_{ij}s will require a considerable amount of tedious algebra.

In the usual linear regression model, the expressions in (5.1.5) are exact. One finds that $E(\varepsilon_i) = a_i = 0$, $\operatorname{var}(\varepsilon_i) = 1$, $c_{ii} = n(v_{ii} - p'/n)/(n - p')$ and $c_{ij} = -nv_{ij}/(n - p')$, $i \neq j$. Generally, we expect that the c_{ij}s can be viewed as extensions of the v_{ij}s and used as diagnostics in an analogous manner.

Using (5.1.5) a Studentized version ε_i^* of the ML residuals can be defined so that $E(\varepsilon_i^*) = E(\varepsilon_i)$ and $\operatorname{var}(\varepsilon_i^*) = \operatorname{var}(\varepsilon_i)$ to order $1/n$. The motivation for this is analogous to that for the r_is in linear regression: The ε_i^*s provide a better reflection of the ε_is and plots can be interpreted without the complications caused by nonconstant means and variances.

EXAMPLE 5.1.1. LEUKEMIA DATA NO. 1. Leukemia is a type of cancer characterized by an excess of white blood cells. At diagnosis, the count of white blood cells provides a useful measure of the patient's initial condition, more severe conditions being reflected by higher counts. Feigl and Zelen (1965) discuss the use of the white blood cell count as an explanatory variable in models to predict survival time after diagnosis.

Feigl and Zelen (1965) report the survival times in weeks and the white blood cell counts for a sample of 33 patients who died of acute leukemia. In addition, each patient was classified as AG positive or AG

Table 5.1.1 *Leukemia data, y = survival time in weeks, WBC = white blood cell count, and related statistics for* 17 *patients diagnosed as AG positive. Source: Feigl and Zelen* (1965)

Case	*WBC*	y_i	a_i	c_{ii}	$\hat{\varepsilon}_i$	ε_i^*	LD_i	$\hat{v}_{ii}$	$\hat{r}_i$	$D_i^1(\mathbf{X}^T\hat{\mathbf{W}}\mathbf{X}, p')$
1	2 300	65	−0.013	0.315	0.56	0.49	0.03	0.085	−0.61	0.02
2	750	156	−0.088	0.770	0.79	0.70	0.11	0.247	−0.28	0.01
3	4 300	100	0.013	0.160	1.17	1.12	0.03	0.112	0.17	0.00
4	2 600	134	−0.007	0.279	1.23	1.20	0.07	0.170	0.23	0.01
5	6 000	16	0.022	0.106	0.22	0.19	0.02	0.017	−1.68	0.02
6	10 500	108	0.029	0.061	1.94	1.91	0.08	0.116	0.72	0.03
7	10 000	121	0.029	0.063	2.13	2.11	0.11	0.129	0.83	0.05
8	17 000	4	0.028	0.068	0.09	0.07	0.07	0.006	−3.02	0.03
9	5 400	39	0.019	0.121	0.51	0.46	0.01	0.042	−0.70	0.01
10	7 000	143	0.025	0.088	2.12	2.11	0.16	0.147	0.83	0.06
11	9 400	56	0.028	0.066	0.96	0.90	0.00	0.059	−0.05	0.00
12	32 000	26	0.015	0.140	0.80	0.74	0.03	0.070	−0.23	0.00
13	35 000	22	0.013	0.156	0.71	0.65	0.04	0.066	−0.36	0.00
14	100 000	1	−0.037	0.453	0.05	0.03	0.35	0.011	−4.12	0.09
15	100 000	1	−0.037	0.453	0.05	0.03	0.35	0.011	−4.12	0.09
16	52 000	5	−0.002	0.245	0.19	0.15	0.18	0.024	−1.85	0.04
17	100 000	65	−0.037	0.453	3.47	4.18	9.89	0.689	2.38	6.26

negative, indicating the presence or absence of a certain morphologic characteristic in the white cells. The data for the 17 patients classified *AG* positive are given in Table 5.1.1; data for the *AG* negatives are given in a later example. To develop a prediction equation based on the *AG* positive cases, we use a model mentioned by Fiegl and Zelen and expanded upon by Cox and Snell (1968),

$$y_i = [\theta_1 \exp(\theta_2 x_i)]\varepsilon_i, \qquad i = 1, 2, \ldots, n \tag{5.1.6}$$

where y_i is the survival time for the i-th patient, $\varepsilon_i, \ldots, \varepsilon_n$ are independent, standard exponential random variables, and if x_i' is the (base 10) logarithm of the i-th white blood cell count, $x_i = x_i' - \bar{x}'$.

The log likelihood $L(\theta_1, \theta_2)$ is easily found to be

$$L(\theta_1, \theta_2) = -n \log(\theta_1) - \sum y_i \exp(-\theta_2 x_i)/\theta_1 \tag{5.1.7}$$

and the expected information matrix is

$$\begin{pmatrix} n/\theta_1^2 & 0 \\ 0 & \sum x_i^2 \end{pmatrix}$$

Selected contours of constant $L(\theta_1, \theta_2)$ are plotted in Fig. 5.1.1 (the points plotted in this figure will be discussed later). The maximum likelihood estimates $\hat{\theta}_1 = 51.109$ and $\hat{\theta}_2 = -1.110$ were determined using Newton's method.

The ML residuals defined by

$$\hat{\varepsilon}_i = h_i(y_i, \hat{\theta}) = y_i \exp(-\hat{\theta}_2 x_i)/\hat{\theta}_1 \tag{5.1.8}$$

are given in Table 5.1.1. Case 17 has the largest residual, $\hat{\varepsilon}_{17} = 3.47$. If the $\hat{\varepsilon}_i$ are treated as a sample from a standard exponential distribution, the residual for case 17 is not large, since the probability that the largest order statistic exceeds 3.47 is 0.42. Of course, the ML residuals do not have constant expectation or variance, and it is possible that a Studentized version would be more revealing.

For the ML residuals defined by (5.1.8) Cox and Snell provide the approximate moments,

$$\begin{aligned} E(\hat{\varepsilon}_i) &= 1 + \frac{1}{2n} + \tfrac{1}{2}\left(x_i \sum x_j^3 - x_i^2 \sum x_j^2\right)/\left(\sum x_j^2\right)^2 \\ &= 1 + a_i \end{aligned} \tag{5.1.9}$$

and

$$\begin{aligned} \operatorname{var}(\hat{\varepsilon}_i) &= 1 - \frac{1}{n} + \left(x_i \sum x_j^3 - 3x_i^2 \sum x_j^2\right)/\left(\sum x_j^2\right)^2 \\ &= 1 - c_{ii} \end{aligned} \tag{5.1.10}$$

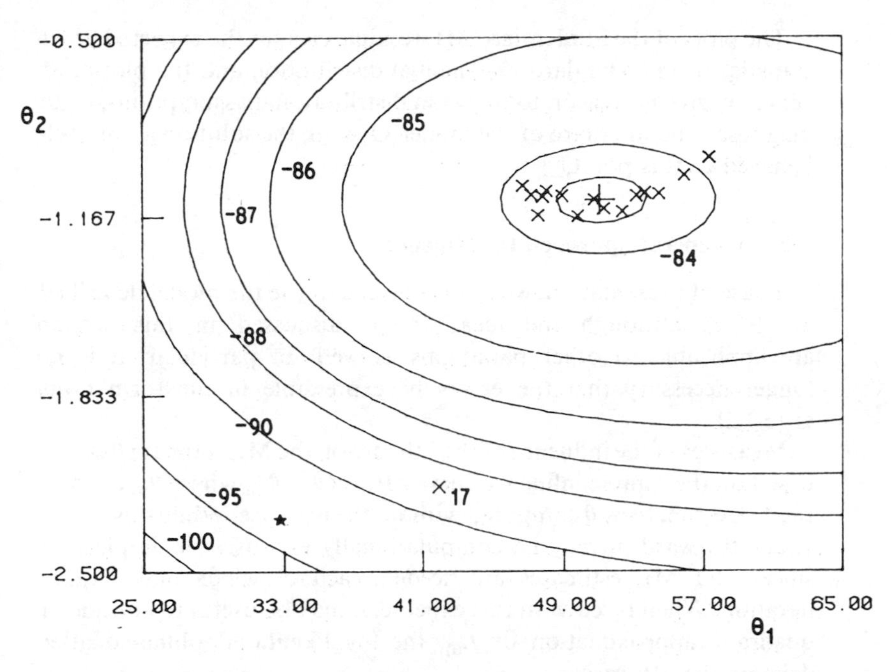

Figure 5.1.1 Likelihood contours for leukemia data. '+' indicates $\hat{\theta}^{\mathrm{T}} = (51.109, -1.110)$ with $L(\hat{\theta}) = -83.88$. '×' indicate $\hat{\theta}_{(i)}$. '*' indicates $\hat{\theta}^{1}_{(17)}$

With the summations fixed, both $E(\hat{\varepsilon}_i)$ and $\mathrm{var}(\hat{\varepsilon}_i)$ are quadratic functions of x_i with maxima occurring at the values of x_i that are closest to $\Sigma x_j^3/2\Sigma x_j^2$ and $\Sigma x_j^3/6\Sigma x_j^2$, respectively. The values of a_i and c_{ii} are given in Table 5.1.1. As expected, the values of c_{ii} are largest at the extremes with $c_{22} = 0.77$ the maximum. In analogy with linear regression, case 2 may have a substantial influence on the ML estimates.

Cox and Snell construct a Studentized version ε_i^* of the ML residuals by using (5.1.9) and (5.1.10) in combination with the transformations $\varepsilon_i^* = \{\hat{\varepsilon}_i/(1-l_i)\}^{1+k_i}$. Assuming that ε_i^* has a standard exponential distribution, it can be sho'vn that appropriate transformations are given by $l_i = -0.21\,c_{ii} - 1.43a_i$ and $k_i = \frac{1}{2}(2a_i + c_{ii})$, $i = 1, 2, \ldots, n$. The values of ε_i^* are also given in Table 5.1.1. The largest difference between $\hat{\varepsilon}_i$ and ε_i^* occurs at case 17, $\varepsilon_{17}^* = 4.18$. The chance that 4.18 would be exceeded in a sample of 17 from a standard exponential distribution is 0.23 so that there is still no reason to suspect case 17 as an outlier.

The plots of the Studentized ML residuals versus the expected order statistics from a standard exponential distribution and the plot of ε_i^* versus x_i give no reason to question distributional assumptions, or to diagnose general failure of the model. Overall, the solution seems well behaved to this point.□

5.2 A general approach to influence

For ease of presentation, we shall continue to use the model described at (5.1.1), although the ideas to be discussed in this section are applicable to other paradigms as well. In particular, it is no longer necessary that the errors be expressible in the form given at (5.1.2).

Measures of the influence of the i-th case on the ML estimate $\hat{\boldsymbol{\theta}}$ can be based on the sample influence curve $SIC_i \propto \hat{\boldsymbol{\theta}} - \hat{\boldsymbol{\theta}}_{(i)}$, where $\hat{\boldsymbol{\theta}}_{(i)}$ denotes the ML estimate of $\boldsymbol{\theta}$ computed without the i-th case. While this idea is straightforward, it may be computationally expensive to implement since $n+1$ ML estimates are needed, each of which may require iteration. When faced with this expense, it may be useful to consider a quadratic approximation of $L_{(i)}$, the log likelihood obtained after deleting the i-th case:

$$L_{(i)}(\boldsymbol{\theta}) \simeq L_{(i)}(\hat{\boldsymbol{\theta}}) + (\boldsymbol{\theta} - \hat{\boldsymbol{\theta}})^{\mathrm{T}}\dot{\mathbf{L}}_{(i)}(\hat{\boldsymbol{\theta}}) + \tfrac{1}{2}(\boldsymbol{\theta} - \hat{\boldsymbol{\theta}})^{\mathrm{T}}\ddot{\mathbf{L}}_{(i)}(\hat{\boldsymbol{\theta}})\,(\boldsymbol{\theta} - \hat{\boldsymbol{\theta}}) \tag{5.2.1}$$

where $\dot{\mathbf{L}}_{(i)}(\hat{\boldsymbol{\theta}})$ is the gradient vector with j-th element $\partial L_{(i)}(\boldsymbol{\theta})/\partial\theta_j$ evaluated at $\boldsymbol{\theta} = \hat{\boldsymbol{\theta}}$ and $\ddot{\mathbf{L}}_{(i)}(\hat{\boldsymbol{\theta}})$ has (j, k)-th element $\partial^2 L_{(i)}(\boldsymbol{\theta})/\partial\theta_j\, \partial\theta_k$, evaluated at $\boldsymbol{\theta} = \hat{\boldsymbol{\theta}}$. If $-\ddot{\mathbf{L}}_{(i)}(\hat{\boldsymbol{\theta}})$ is positive definite, the quadratic approximation is maximized at

$$\hat{\boldsymbol{\theta}}_{(i)}^1 = \hat{\boldsymbol{\theta}} - (\ddot{\mathbf{L}}_{(i)}(\hat{\boldsymbol{\theta}}))^{-1}\dot{\mathbf{L}}_{(i)}(\hat{\boldsymbol{\theta}}) \tag{5.2.2}$$

We refer to $\hat{\boldsymbol{\theta}}_{(i)}^1$ as a one-step approximation to $\hat{\boldsymbol{\theta}}_{(i)}$, since it is the same as would be obtained by a single step of Newton's method using $\hat{\boldsymbol{\theta}}$ as starting values to maximize $L_{(i)}(\boldsymbol{\theta})$ (see Kennedy and Gentle, 1980, Chapter 10).

If $\hat{\boldsymbol{\theta}}_{(i)}$ is not too different from $\hat{\boldsymbol{\theta}}$, and $L_{(i)}(\boldsymbol{\theta})$ is locally quadratic, the one-step estimator should be close to the fully iterated value. For cases that are influential, $\hat{\boldsymbol{\theta}} - \hat{\boldsymbol{\theta}}_{(i)}$ is 'large', the accuracy of the one-step estimator is likely to be lower, but an accurate approximation to $\hat{\boldsymbol{\theta}}_{(i)}$ will not be needed as long as $\hat{\boldsymbol{\theta}} - \hat{\boldsymbol{\theta}}_{(i)}^1$ is sufficiently 'large' to draw our attention for further consideration.

In the linear least squares problem, elliptical norms of the sample influence curve provide a sufficiently rich class of metrics for ordering cases on influence. In more general problems, this class can be overly restrictive, especially if elliptical confidence contours are not appropriate. If we let $t(\boldsymbol{\theta})$ be a function of the q-vector $\boldsymbol{\theta}$, then a general measure can be viewed as any function $m(t(\hat{\boldsymbol{\theta}}), t(\hat{\boldsymbol{\theta}}_{(i)}))$ that maps into the positive real line. Most of the alternative methods for assessing influence given in Chapter 4, for example, can be expressed as members of this general class. However, since $m(t(\hat{\boldsymbol{\theta}}), t(\hat{\boldsymbol{\theta}}_{(i)}))$ is not in general a function of the sample influence curve, the theoretical foundations for influence measures derived from the influence curve may be lacking. Before any alternative measure is to be adopted, its logical foundation must be carefully studied.

An important example of the general measure is derived from the use of contours of the log likelihood function to order cases based on influence. Let $L(\boldsymbol{\theta})$ be the log likelihood based on the complete data. We define a *likelihood distance* LD_i as

$$LD_i = 2[L(\hat{\boldsymbol{\theta}}) - L(\hat{\boldsymbol{\theta}}_{(i)})] \tag{5.2.3}$$

or, using the one-step estimator,

$$LD_i^1 = 2[L(\hat{\boldsymbol{\theta}}) - L(\hat{\boldsymbol{\theta}}_{(i)}^1)] \tag{5.2.4}$$

This is easily seen to be in the general class with $t(\boldsymbol{\theta}) = L(\boldsymbol{\theta})$, and LD_i is not necessarily a function of just the sample influence curve for $\boldsymbol{\theta}$.

The measures LD_i and LD_i^1 may also be interpreted in terms of the asymptotic confidence region (see Cox and Hinkley, 1974, Chapter 9)

$$\{\boldsymbol{\theta}\colon 2[L(\hat{\boldsymbol{\theta}}) - L(\boldsymbol{\theta})] \le \chi^2(\alpha; q)\}$$

where $\chi^2(\alpha; q)$ is the upper α point of the chi-squared distribution with q df, and q is the dimension of $\boldsymbol{\theta}$. LD_i can therefore be calibrated by comparison to the $\chi^2(q)$ distribution.

If the log likelihood contours are approximately elliptical, LD_i can be usefully approximated by Taylor expansion of $L(\hat{\boldsymbol{\theta}}_{(i)})$ around $\hat{\boldsymbol{\theta}}$,

$$L(\hat{\boldsymbol{\theta}}_{(i)}) \cong L(\hat{\boldsymbol{\theta}}) + (\hat{\boldsymbol{\theta}}_{(i)} - \hat{\boldsymbol{\theta}})^{\mathrm{T}}\dot{\mathbf{L}}(\hat{\boldsymbol{\theta}}) + \tfrac{1}{2}(\hat{\boldsymbol{\theta}}_{(i)} - \hat{\boldsymbol{\theta}})^{\mathrm{T}}(\ddot{\mathbf{L}}(\hat{\boldsymbol{\theta}}))\,(\hat{\boldsymbol{\theta}}_{(i)} - \hat{\boldsymbol{\theta}})$$

and, since $\dot{\mathbf{L}}(\hat{\boldsymbol{\theta}}) = 0$,

$$LD_i \cong (\hat{\boldsymbol{\theta}}_{(i)} - \hat{\boldsymbol{\theta}})^{\mathrm{T}}(-\ddot{\mathbf{L}}(\hat{\boldsymbol{\theta}}))\,(\hat{\boldsymbol{\theta}}_{(i)} - \hat{\boldsymbol{\theta}}) \tag{5.2.5}$$

A different approximation can be obtained by replacing the observed information $-\ddot{\mathbf{L}}(\hat{\boldsymbol{\theta}})$ in (5.2.5) by the expected information matrix, evaluated at $\hat{\boldsymbol{\theta}}$. Either of these approximations, however, can be seriously misleading if contours of $L(\boldsymbol{\theta})$ are markedly nonelliptical.

The likelihood distance can be easily modified to accommodate situations in which a subset $\boldsymbol{\theta}_1$ of $\boldsymbol{\theta}$ is of special interest. Let $\boldsymbol{\theta}^{\mathrm{T}} = (\boldsymbol{\theta}_1^{\mathrm{T}}, \boldsymbol{\theta}_2^{\mathrm{T}})$ and $\hat{\boldsymbol{\theta}}_{(i)}^{\mathrm{T}} = (\hat{\boldsymbol{\theta}}_{1(i)}^{\mathrm{T}}, \hat{\boldsymbol{\theta}}_{2(i)}^{\mathrm{T}})$. An asymptotic confidence region for $\boldsymbol{\theta}_1$ is given by

$$\{\boldsymbol{\theta}_1 : 2[L(\hat{\boldsymbol{\theta}}) - L(\boldsymbol{\theta}_1, \boldsymbol{\theta}_2(\boldsymbol{\theta}_1))] \leq \chi^2(\alpha, q_1)\}$$

where q_1 is the dimension of $\boldsymbol{\theta}_1$ and

$$L(\boldsymbol{\theta}_1, \boldsymbol{\theta}_2(\boldsymbol{\theta}_1)) = \max_{\boldsymbol{\theta}_2} [L(\boldsymbol{\theta}_1, \boldsymbol{\theta}_2)]$$

denotes the log likelihood maximized over the parameter space for $\boldsymbol{\theta}_2$ with $\boldsymbol{\theta}_1$ fixed (Cox and Hinkley, 1974, Chapter 9). The asymptotic confidence region measure of the displacement of $\hat{\boldsymbol{\theta}}_1$ when the i-th case is deleted is now

$$\begin{aligned} LD_i(\boldsymbol{\theta}_1 | \boldsymbol{\theta}_2) &= 2[L(\hat{\boldsymbol{\theta}}) - L(\hat{\boldsymbol{\theta}}_{1(i)}, \boldsymbol{\theta}_2(\hat{\boldsymbol{\theta}}_{1(i)}))] \\ &= 2\{L(\hat{\boldsymbol{\theta}}) - \max_{\boldsymbol{\theta}_2} [L(\hat{\boldsymbol{\theta}}_{1(i)}, \boldsymbol{\theta}_2)]\} \end{aligned} \tag{5.2.6}$$

with a similar measure obtained if one-step estimators replace fully iterated ones. This measure is compared to the $\chi^2(q_1)$ distribution for calibration.

As an illustration, consider again the usual linear model $\mathbf{Y} = \mathbf{X}\boldsymbol{\beta} + \sigma\boldsymbol{\varepsilon}$, with the ε_i assumed independent, identically distributed $N(0, 1)$. If $\sigma = \sigma_0$ is known, it is easy to verify that

$$LD_i(\boldsymbol{\beta}) = D_i(\mathbf{X}^{\mathrm{T}}\mathbf{X}, \sigma_0^2) \tag{5.2.7}$$

If σ^2 is unknown but $\boldsymbol{\beta}$ is of special interest, (5.2.6) provides the desired measure with $\boldsymbol{\theta}_1 = \boldsymbol{\beta}$, $\boldsymbol{\theta}_2 = \sigma^2$. One finds

$$L(\hat{\boldsymbol{\theta}}) = L(\hat{\boldsymbol{\beta}}, \hat{\sigma}^2) = -\frac{n}{2}\log(2\pi\hat{\sigma}^2) - \frac{n}{2}$$

and

$$L(\boldsymbol{\beta}, \sigma^2(\boldsymbol{\beta})) = -\frac{n}{2}\log[2\pi\sigma^2(\boldsymbol{\beta})] - \frac{n}{2}$$

where $\sigma^2(\boldsymbol{\beta}) = \sum(y_j - \mathbf{x}_j^{\mathrm{T}}\boldsymbol{\beta})^2/n$. Setting $\boldsymbol{\beta} = \hat{\boldsymbol{\beta}}_{(i)}$,

$$\begin{aligned} LD_i(\boldsymbol{\beta}|\sigma^2) &= n\log[(\sigma^2(\hat{\boldsymbol{\beta}}_{(i)}))/\hat{\sigma}^2] \\ &= n\log\left[\frac{p'}{n-p'}D_i + 1\right] \end{aligned} \tag{5.2.8}$$

Since $LD_i(\boldsymbol{\beta}|\sigma^2)$ is a monotonically increasing function of D_i, it is equivalent to D_i. Finally, the likelihood distance for $(\boldsymbol{\beta}, \sigma^2)$ is found to be

$$LD_i(\boldsymbol{\beta}, \sigma^2) = n \log(\hat{\sigma}^2_{(i)}/\hat{\sigma}^2) + (y_i - \mathbf{x}_i^{\mathrm{T}}\hat{\boldsymbol{\beta}}_{(i)})^2/\hat{\sigma}^2_{(i)} - 1$$

$$= n \log\left[\left(\frac{n}{n-1}\right)\frac{n-p'-1}{t_i^2+n-p'-1}\right] + \frac{t_i^2(n-1)}{(1-v_{ii})(n-p'-1)} - 1 \tag{5.2.9}$$

Interestingly, this expression is guaranteed to be monotonically increasing in t_i^2 only if the model contains a constant. For fixed n, p', and v_{ii}, $LD_i(\boldsymbol{\beta}, \sigma)$ is minimized at $t_i^2 = (n-p'-1)(1-nv_{ii})/(n-1)$ which may be positive if regression is through the origin.

EXAMPLE 5.2.1. LEUKEMIA DATA NO. 2. The individual points plotted in Fig. 5.1.1 represent $\boldsymbol{\theta}_{(i)}^{\mathrm{T}} = (\hat{\theta}_{1(i)}, \hat{\theta}_{2(i)})$, $i = 1, 2, \ldots, 17$, for the leukemia data discussed in Example 5.1.1. Only case 17 deviates far from the full sample ML estimate: $\hat{\boldsymbol{\theta}}_{(17)}^{\mathrm{T}} = (41.920, -2.184)$, while the full sample ML estimates are $\hat{\theta}_1 = 51.109$ and $\hat{\theta}_2 = -1.110$. The likelihood distance measure for case 17 is $LD_{17} = 9.89$. Comparing this value to the percentage points of a $\chi^2(2)$ distribution indicates that the removal of case 17 will displace $\hat{\boldsymbol{\theta}}$ to the edge of a 99% asymptotic confidence region.

The values of LD_i, $i = 1, 2, \ldots, 17$, are given in Table 5.1.1. The second largest value of LD_i, $LD_{14} = LD_{15} = 0.35$, indicates minimal movement so that case 17 is the only individually influential case.

Recall from Example 5.1.1 that case 17 has the largest ML residual, but there was insufficient evidence to reject it as an outlier. The influence of case 17 seems to be due to its large ML residual in combination with the relatively large value of c_{ii}. An inspection of the original data reveals that case 17 corresponds to a patient with a very large white blood cell count (100 000) who survived for a relatively long time. Feigl and Zelen (1965) mention that high white blood cell counts are unreliable so a measurement error in x_{17} may be contributing to the influence of case 17. In any event, conclusions based on such data should be viewed skeptically.

In the preceding discussion, the fully iterated estimates $\hat{\boldsymbol{\theta}}_{(i)}$ were used, but the one-step estimates computed from (5.2.2) would have served as well. When superimposed on Fig. 5.1.1, the one-step estimates for $i = 1, 2, \ldots, 16$ are nearly indistinguishable from the cloud of points around the maximum of the log likelihood. The only noticeable disagreement occurs at case 17. The one-step estimate $\hat{\boldsymbol{\theta}}^1_{(17)}$ falls at the point indicated by a 'star' in the lower left-hand corner of Fig. 5.1.1. Since $\hat{\boldsymbol{\theta}}^1_{(17)}$ is farther from $\hat{\boldsymbol{\theta}}$ than $\hat{\boldsymbol{\theta}}_{(17)}$, LD^1_{17} would still be large.

Finally, the quadratic approximation to LD_i given at (5.2.5) would probably work well in this example since the log likelihood contours are nearly elliptical; the approximating elliptical contours and the one-step estimators were given in Fig. 3.5.1. However, the elliptical approximation is not always applicable since even in this example it is possible to transform the parameters to get clearly nonelliptical contours for the log likelihood function.□

5.3 Nonlinear least squares

The nonlinear regression model is given by

$$y_j = f(\mathbf{x}_j, \boldsymbol{\theta}) + \sigma\varepsilon_j, \qquad j = 1, 2, \ldots, n \tag{5.3.1}$$

where $f(\mathbf{x}_j, \boldsymbol{\theta})$ is a scalar-valued function that is nonlinear in the q-vector of unknown parameters $\boldsymbol{\theta}$, and the ε_j are independent and identically distributed $N(0, 1)$. For this problem, the maximum likelihood estimate $\hat{\boldsymbol{\theta}}$ of $\boldsymbol{\theta}$ can be obtained by minimizing the residual sum of squares,

$$G(\boldsymbol{\theta}) = \sum_{j=1}^{n} (y_j - f(\mathbf{x}_j, \boldsymbol{\theta}))^2 \tag{5.3.2}$$

The problem of determining $\hat{\boldsymbol{\theta}}$ can be treated as a special case of the general unconstrained maximization problem, although special methods that use the fact that $G(\boldsymbol{\theta})$ is quadratic are often appropriate; see Kennedy and Gentle (1980, Section 10.3).

The problem of assessing influence in the nonlinear least squares problem can be approached using the general methods outlined earlier in Sections 5.1 and 5.2. In particular, one-step estimators $\hat{\boldsymbol{\theta}}^1_{(i)}$ of the vectors $\hat{\boldsymbol{\theta}}_{(i)}$ that minimize the objective functions

$$G_{(i)}(\boldsymbol{\theta}) = \sum_{j \neq i} (y_j - f(\mathbf{x}_j, \boldsymbol{\theta}))^2, \qquad i = 1, 2, \ldots, n \tag{5.3.3}$$

can be found by application of the result given by Equation (5.2.2). However, particularly interesting results can be obtained if we allow a further approximation. We suppose that, in a neighborhood about $\hat{\boldsymbol{\theta}}$, $f(\mathbf{x}_j, \boldsymbol{\theta})$ is approximately linear,

$$f(\mathbf{x}_j, \boldsymbol{\theta}) \cong f(\mathbf{x}_j, \hat{\boldsymbol{\theta}}) + \hat{\mathbf{z}}_j^{\mathrm{T}}(\boldsymbol{\theta} - \hat{\boldsymbol{\theta}}) \tag{5.3.4}$$

where $\hat{\mathbf{z}}_j^{\mathrm{T}}$ is the j-th row of the $n \times q$ Jacobian matrix $\hat{\mathbf{Z}}$,

$$\hat{\mathbf{z}}_j^{\mathrm{T}} = \left[\frac{\partial f(\mathbf{x}_j, \boldsymbol{\theta})}{\partial \theta_1}, \ldots, \frac{\partial f(\mathbf{x}_j, \boldsymbol{\theta})}{\partial \theta_q}\right]_{\boldsymbol{\theta} = \hat{\boldsymbol{\theta}}} \tag{5.3.5}$$

If the approximation (5.3.4) is substituted into $G_{(i)}(\boldsymbol{\theta})$ defined by (5.3.3), the resulting objective function is minimized at

$$\hat{\boldsymbol{\theta}}_{(i)}^1 = \hat{\boldsymbol{\theta}} + (\hat{\mathbf{Z}}_{(i)}^{\mathrm{T}} \hat{\mathbf{Z}}_{(i)})^{-1} \hat{\mathbf{Z}}_{(i)}^{\mathrm{T}} \mathbf{e}_{(i)}$$

where $\mathbf{e}$ is the n-vector with elements $e_j = y_j - f(\mathbf{x}_j, \hat{\boldsymbol{\theta}})$. This form corresponds to that obtained by using a single step of the Gauss–Newton method (see Kennedy and Gentle, 1980, Chapter 10). The last equation is simplified, with the aid of Appendix A.2 and the fact that $\hat{\mathbf{Z}}^{\mathrm{T}}\mathbf{e} = \mathbf{0}$ to give a more usual form. Defining $\hat{v}_{ii} = \hat{\mathbf{z}}_i^{\mathrm{T}} (\hat{\mathbf{Z}}^{\mathrm{T}}\hat{\mathbf{Z}})^{-1} \hat{\mathbf{z}}_i$, we find (Fox, Hinkley and Larntz, 1980)

$$\hat{\boldsymbol{\theta}}_{(i)}^1 = \hat{\boldsymbol{\theta}} - \frac{(\hat{\mathbf{Z}}^{\mathrm{T}}\hat{\mathbf{Z}})^{-1} \hat{\mathbf{z}}_i e_i}{1 - \hat{v}_{ii}} \tag{5.3.6}$$

When this particular algorithm is used to produce the one-step estimators, the nonlinear least squares problem is essentially replaced by a linear one, with the role of $\mathbf{X}$ taken by $\hat{\mathbf{Z}}$. Most of the diagnostics and residual analyses for linear least squares may be expected to apply at least approximately in nonlinear least squares. In particular, an approximate Studentized residual is

$$\hat{r}_i = \frac{e_i}{\hat{\sigma}\sqrt{(1 - \hat{v}_{ii})}} \tag{5.3.7}$$

where $\hat{\sigma}^2 = G(\hat{\boldsymbol{\theta}})/n$. An elliptical norm of the sample influence curve is

$$D_i(\hat{\mathbf{Z}}^{\mathrm{T}}\hat{\mathbf{Z}}, q\hat{\sigma}^2) = (\hat{\boldsymbol{\theta}} - \hat{\boldsymbol{\theta}}_{(i)})^{\mathrm{T}} (\hat{\mathbf{Z}}^{\mathrm{T}}\hat{\mathbf{Z}}) (\hat{\boldsymbol{\theta}} - \hat{\boldsymbol{\theta}}_{(i)})/q\hat{\sigma}^2$$

When $\hat{\boldsymbol{\theta}}_{(i)}$ is replaced by the one-step approximation $\hat{\boldsymbol{\theta}}_{(i)}^1$, this norm becomes

$$D_i^1(\hat{\mathbf{Z}}^{\mathrm{T}}\hat{\mathbf{Z}}, q\hat{\sigma}^2) = \frac{\hat{r}_i^2}{q} \frac{\hat{v}_{ii}}{1 - \hat{v}_{ii}} \tag{5.3.8}$$

In this and the following two sections, we continue to use $D_i(\cdot, \cdot)$ to denote an elliptical norm. The parameter under consideration should be clear from context. One step versions will be denoted by $D_i^1(\cdot, \cdot)$.

The use of elliptical norms for influence, whether based on one-step or fully iterated estimates, may be inappropriate for some nonlinear problems if $G(\boldsymbol{\theta})$ has markedly nonelliptical contours. In many problems, elliptical confidence regions can be badly biased (Beale, 1960), and the bias may depend on the parameterization chosen for the model (Bates and Watts, 1980, 1981; Hamilton, Watts and Bates, 1982). The problem of choosing a parameterization can have important effects on the analysis of influence.

Alternative norms for $\hat{\boldsymbol{\theta}} - \hat{\boldsymbol{\theta}}^1_{(i)}$ or $\hat{\boldsymbol{\theta}} - \hat{\boldsymbol{\theta}}_{(i)}$ that are less dependent on the shape of contours of $G(\boldsymbol{\theta})$ can be suggested, but these will require considerably more computation. The first of these norms is derived from the form for D_i given by (3.5.6) as a norm of the change in the vector of fitted values. For the nonlinear regression problem, this becomes

$$FD_i^1 = \frac{1}{q\hat{\sigma}^2} \sum_{j=1}^{n} (f(\mathbf{x}_j, \hat{\boldsymbol{\theta}}) - f(\mathbf{x}_j, \hat{\boldsymbol{\theta}}^1_{(i)}))^2 \tag{5.3.9}$$

When $f(\mathbf{x}_j, \boldsymbol{\theta})$ is exactly linear in $\boldsymbol{\theta}$, D_i and FD_i^1 are proportional; otherwise, they may be quite different. When the parameterization of the model is at issue, FD_i^1 may be the preferred statistic since it depends on the parameterization only through approximation of $\hat{\boldsymbol{\theta}}_{(i)}$. If $\hat{\boldsymbol{\theta}}_{(i)}$ is used in place of $\hat{\boldsymbol{\theta}}^1_{(i)}$, FD_i is invariant under choice of parameterization.

Finally, we consider measures derived directly from log likelihood displacement. With reference to the $(q+1)$-dimensional contours for $(\boldsymbol{\theta}, \sigma^2)$, the measure is

$$\begin{aligned} LD_i^1(\boldsymbol{\theta}, \sigma^2) &= 2[L(\hat{\boldsymbol{\theta}}, \hat{\sigma}^2) - L(\hat{\boldsymbol{\theta}}^1_{(i)}, (\hat{\sigma}^1_{(i)})^2)] \\ &= n \log\left[\left(\frac{n}{n-1}\right)\frac{G_{(i)}(\hat{\boldsymbol{\theta}}^1_{(i)})}{G(\hat{\boldsymbol{\theta}})}\right] - n + (n-1)\frac{G(\hat{\boldsymbol{\theta}}^1_{(i)})}{G_{(i)}(\hat{\boldsymbol{\theta}}^1_{(i)})} \end{aligned} \tag{5.3.10}$$

where $G_{(i)}(\boldsymbol{\theta}) = G(\boldsymbol{\theta}) - (y_i - f(x_i, \boldsymbol{\theta}))^2$. When $\boldsymbol{\theta}$ alone is considered, the resulting measure from (5.2.6) is

$$LD_i^1(\boldsymbol{\theta}|\sigma^2) = n \log[G(\hat{\boldsymbol{\theta}}^1_{(i)})/G(\hat{\boldsymbol{\theta}})] \tag{5.3.11}$$

As with FD_i^1, computation of either likelihood norm requires a pass through the data for each i, so these will be useful generally only if n is not too large.

EXAMPLE 5.3.1. DUNCAN'S DATA. Duncan (1978) discusses a set of artificial data with $n = 24$, and for which the appropriate model is

$$y_j = \frac{\theta_1}{\theta_1 - \theta_2}[\exp(-\theta_2 x_j) - \exp(-\theta_1 x_j)] + \sigma\varepsilon_j$$

The data are given in Columns 2 and 3 of Table 5.3.1. The remaining columns of the table give e_i, $\hat{r}_i$, $\hat{v}_{ii}$, $D_i^1(\hat{\mathbf{Z}}^T\hat{\mathbf{Z}}, 2\hat{\sigma}^2)$, FD_i^1, and the two likelihood distances computed from one-step estimates given the maximum likelihood estimate $\hat{\boldsymbol{\theta}}^T = (0.1989, 0.4454)$. From these statistics, case 9 appears as a candidate for a possible outlier, and it is clearly influential in this problem by any of the measures, assuming that the one-step approximation is adequate. To explore the adequacy of the approximation, we have computed the fully iterated estimators $\hat{\boldsymbol{\theta}}_{(i)}$ for each i, using the modified Gauss–Newton algorithms with $\hat{\boldsymbol{\theta}}$ as the starting value. No more than three iterations were required to obtain about four-digit accuracy on $\hat{\boldsymbol{\theta}}_{(i)}$. The correspondence between $\hat{\boldsymbol{\theta}}_{(i)}$ and $\hat{\boldsymbol{\theta}}_{(i)}^1$ was very good, with the largest deviation occurring for case 9. Figure 5.3.1 is a contour plot of $G(\boldsymbol{\theta})$ for this problem, with the fully iterated $\hat{\boldsymbol{\theta}}_{(i)}$ added to the plot. In addition $\hat{\boldsymbol{\theta}}_{(9)}^1$ is indicated.

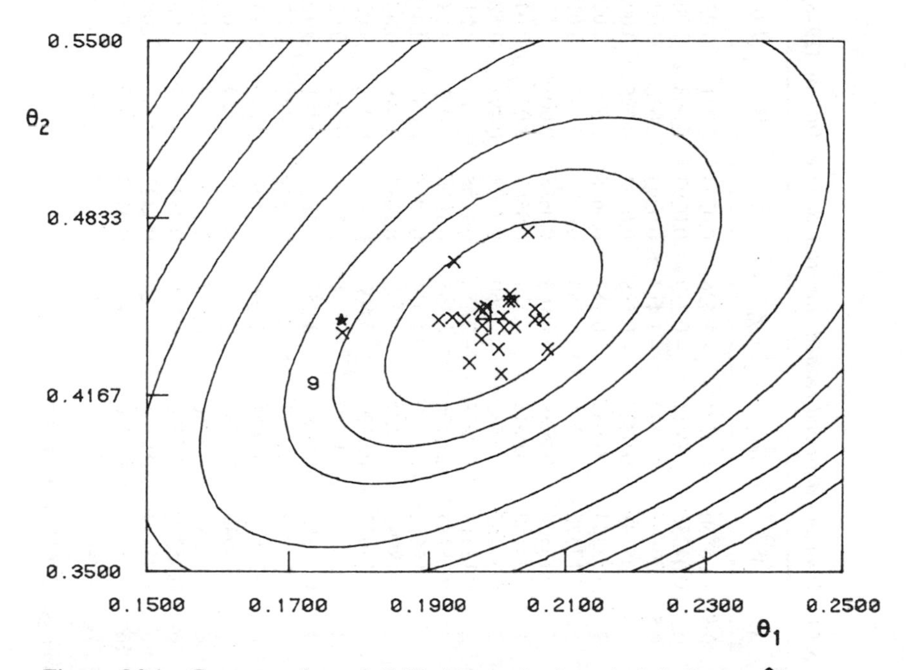

Figure 5.3.1 Contour plot of $G(\boldsymbol{\theta})$, Duncan's data. '+' indicates $\hat{\boldsymbol{\theta}}^T = (0.1989, 0.4454)$, where $G(\hat{\boldsymbol{\theta}}) = 0.070279$. The points plotted are $\hat{\boldsymbol{\theta}}_{(i)}$. The point at the '∗' is $\hat{\boldsymbol{\theta}}_{(9)}^1$

Table 5.3.1 *Duncan's data and related statistics. Source: Duncan* (1978)

Case	x_i	y_i	e_i	$\hat{r}_i$	$\hat{v}_{ii}$	D_i^1	FD_i^1	$LD_i^1(\boldsymbol{\theta}, \sigma^2)$	$LD_i^1(\boldsymbol{\theta}\|\sigma^2)$	$\hat{\theta}^1_{1(i)}$	$\hat{\theta}^1_{2(i)}$
1	0.5	0.00530	−0.0794	−1.426	0.0293	0.031	0.030	0.104	0.064	0.2051	0.4514
2	0.5	0.04356	−0.0411	−0.739	0.0293	0.008	0.008	0.021	0.017	0.2021	0.4485
3	0.5	0.00603	−0.0787	−1.413	0.0293	0.030	0.029	0.100	0.062	0.2051	0.4513
4	0.5	0.05198	−0.0327	−0.588	0.0293	0.005	0.005	0.019	0.011	0.2015	0.4479
5	1	0.15303	0.0085	0.157	0.0688	0.001	0.001	0.022	0.002	0.1978	0.4446
6	1	0.17526	0.0308	0.564	0.0688	0.012	0.012	0.034	0.025	0.1951	0.4427
7	1	0.15337	0.0089	0.163	0.0688	0.001	0.001	0.022	0.002	0.1978	0.4446
8	1	0.20580	0.0613	1.124	0.0688	0.047	0.048	0.107	0.102	0.1913	0.4399
9	2	0.36962	0.1586	2.960	0.1008	0.491	0.523	4.854	1.103	0.1773	0.4446
10	2	0.18513	−0.0258	−0.482	0.1008	0.013	0.013	0.039	0.028	0.2024	0.4455
11	2	0.25143	0.0405	0.755	0.1008	0.032	0.032	0.072	0.070	0.1934	0.4452
12	2	0.15610*	−0.0549	−1.024	0.1008	0.059	0.058	0.125	0.124	0.2064	0.4457
13	4	0.18093	−0.0474	−0.888	0.1089	0.048	0.050	0.108	0.108	0.2008	0.4333
14	4	0.19627	−0.0320	−0.600	0.1089	0.022	0.023	0.056	0.049	0.2002	0.4372
15	4	0.26221	0.0339	0.636	0.1089	0.025	0.024	0.058	0.052	0.1975	0.4540
16	4	0.15962	−0.0687	−1.287	0.1089	0.101	0.107	0.256	0.231	0.2016	0.4279
17	8	0.11619	−0.0253	−0.486	0.1529	0.021	0.022	0.057	0.047	0.1960	0.4340
18	8	0.20856	0.0671	1.290	0.1529	0.150	0.138	0.319	0.291	0.2065	0.4758
19	8	0.18540	0.0439	0.844	0.1529	0.064	0.061	0.129	0.129	0.2039	0.4653
20	8	0.09583	−0.0456	−0.877	0.1529	0.069	0.074	0.156	0.156	0.1937	0.4247
21	16	0.05278	0.0200	0.360	0.0393	0.003	0.003	0.021	0.005	0.2005	0.4490
22	16	0.01473	−0.0181	−0.327	0.0393	0.002	0.002	0.021	0.005	0.1975	0.4421
23	16	0.05738	0.0246	0.443	0.0393	0.004	0.004	0.021	0.008	0.2008	0.4498
24	16	0.02519	−0.0076	−0.138	0.0393	0.000	0.000	0.021	0.001	0.1983	0.4440

* Given as 0.25610 by Duncan

At least for this one problem, we have found that the one-step influence measures provide the same qualitative information as the fully iterated ones and an influential case is clearly identified. In some problems where the $G(\boldsymbol{\theta})$ surface is less well behaved, we should expect that the one-step procedures will not work as well. Further research and experience with these methods is required.□

5.4 Logistic regression and generalized linear models

Although the logistic regression model does not fall in the general framework for residuals given in Section 5.1, the results for the assessment of influence given in Section 5.2 can be applied. We first consider influence assessment, and then present several alternatives for defining residuals.

Consider a sample $\mathbf{Y}^{\mathrm{T}} = (y_1, y_2, \ldots, y_n)$ of independent random variables such that y_j is binomially distributed $B(n_j, p_j)$ with n_j known and p_j unknown. The logistic regression model specifies the relationship

$$\eta_j = \operatorname{logit}(p_j) = \log[p_j/(1-p_j)] = \mathbf{x}_j^{\mathrm{T}}\boldsymbol{\beta}, \qquad j = 1, 2, \ldots, n, \qquad (5.4.1)$$

where $\mathbf{x}_1, \mathbf{x}_2, \ldots, \mathbf{x}_n$ are p'-vectors of explanatory variables and $\boldsymbol{\beta}$ is an unknown parameter vector. In such models, estimation of $\boldsymbol{\beta}$ is typically a major concern.

The log likelihood for $\boldsymbol{\eta} = \mathbf{X}\boldsymbol{\beta}$ is

$$L(\boldsymbol{\eta}) = L(\mathbf{X}\boldsymbol{\beta}) = \sum_{j=1}^{n} [y_j \mathbf{x}_j^{\mathrm{T}}\boldsymbol{\beta} - a_j(\mathbf{x}_j^{\mathrm{T}}\boldsymbol{\beta}) + b_j(y_j)] \qquad (5.4.2)$$

where $a_j(z) = n_j \log[1+\exp(z)]$ and $b_j(z) = \log\binom{n_j}{z}$. The maximum likelihood estimate $\hat{\boldsymbol{\beta}}$ of $\boldsymbol{\beta}$ is often found using Newton's method.

Once $\hat{\boldsymbol{\beta}}$ is obtained, a one-step estimator $\hat{\boldsymbol{\beta}}_{(i)}^1$ of $\hat{\boldsymbol{\beta}}_{(i)}$ can be found using the general results of Section 5.2. Following Pregibon (1981), but using different notation, define $\hat{p}_j = \exp(\mathbf{x}_j^{\mathrm{T}}\hat{\boldsymbol{\beta}})/[1+\exp(\mathbf{x}_j^{\mathrm{T}}\hat{\boldsymbol{\beta}})]$ and let $\hat{\mathbf{W}}$ be an $n \times n$ diagonal matrix with j-th diagonal $n_j\hat{p}_j(1-\hat{p}_j)$. Also, let $\hat{\mathbf{s}}$ be an n-vector with j-th element $\hat{s}_j = y_j - n_j\hat{p}_j$. One can show that

$$\dot{\mathbf{L}}_{(i)}(\hat{\boldsymbol{\eta}}) = \mathbf{X}_{(i)}^{\mathrm{T}}\hat{\mathbf{s}}_{(i)}; \qquad \ddot{\mathbf{L}}_{(i)}(\hat{\boldsymbol{\eta}}) = -(\mathbf{X}_{(i)}^{\mathrm{T}}\hat{\mathbf{W}}_{(i)}\mathbf{X}_{(i)}) \qquad (5.4.3)$$

so that, using (5.2.2) and Appendix A.2,

$$\begin{aligned}\hat{\boldsymbol{\beta}}_{(i)}^1 &= \hat{\boldsymbol{\beta}} + (\mathbf{X}_{(i)}^{\mathrm{T}}\hat{\mathbf{W}}_{(i)}\mathbf{X}_{(i)})^{-1}\mathbf{X}^{\mathrm{T}}{}_{(i)}\hat{\mathbf{s}}_{(i)} \\ &= \hat{\boldsymbol{\beta}} - \frac{(\mathbf{X}^{\mathrm{T}}\hat{\mathbf{W}}\mathbf{X})^{-1}\mathbf{x}_i\hat{s}_i}{1-\hat{v}_{ii}}\end{aligned} \qquad (5.4.4)$$

where $\hat{v}_{ii}$ is the i-th diagonal element of $\hat{\mathbf{V}} = \hat{\mathbf{W}}^{1/2}\mathbf{X}(\mathbf{X}^{\mathrm{T}}\hat{\mathbf{W}}\mathbf{X})^{-1}\mathbf{X}^{\mathrm{T}}\hat{\mathbf{W}}^{1/2}$. Pregibon (1981) discusses the accuracy of this one-step approximation and concludes that componentwise the approximation tends to underestimate the fully iterated value, but that this may be unimportant for identifying influential cases.

Measures for the differences $\hat{\boldsymbol{\beta}} - \hat{\boldsymbol{\beta}}_{(i)}$ or $\hat{\boldsymbol{\beta}} - \hat{\boldsymbol{\beta}}^{1}_{(i)}$ can be derived using elliptical approximations, likelihood displacement, or changes in fitted-value vectors as discussed in the last two sections. Following Pregibon (1981), however, we will consider only the first of these,

$$D_i^1(\mathbf{X}^{\mathrm{T}}\hat{\mathbf{W}}\mathbf{X}, p') = \frac{1}{p'}\left[\frac{\hat{s}_i^2}{n_i\hat{p}_i(1-\hat{p}_i)}\right]\frac{\hat{v}_{ii}}{(1-\hat{v}_{ii})^2} \tag{5.4.5}$$

to characterize influence for logistic regression (Pregibon's measure c_i^1 differs from (5.4.5) only by the factor p' in the denominator). Comparison of (5.4.5) to D_i suggests that $\hat{s}_i^2/[n_i\hat{p}_i(1-\hat{p}_i)(1-\hat{v}_{ii})]$ and $\hat{v}_{ii}$ may be interpreted and used in the same way as r_i^2 and v_{ii} in linear regression.

Residuals for logistic regression can be defined in many ways. Equation (5.4.5) and the analogy with linear least squares suggests the quantities

$$\chi_i = \hat{s}_i/[n_i\hat{p}_i(1-\hat{p}_i)]^{1/2} \tag{5.4.6}$$

Landwehr, Pregibon and Shoemaker (1980) and Pregibon (1981) use an alternative set of residuals based on individual components of the log likelihood ratio or deviance statistic, $dev = -2[L(\mathbf{X}\hat{\boldsymbol{\beta}}) - L(\text{logit}\,(y_i/n_i))]$, where $L\,(\text{logit}\,(y_i/n_i))$ is the log likelihood obtained when each η_i is estimated by $\text{logit}\,(y_i/n_i)$. The deviance has an asymptotic $\chi^2(n-p')$ distribution. Components of deviance are defined as

$$dev_i = \pm\sqrt{2}[l_i(\text{logit}\,(y_i/n_i)) - l_i(\mathbf{x}_i^{\mathrm{T}}\hat{\boldsymbol{\beta}})]^{1/2} \tag{5.4.7}$$

where $l_i(\eta) = y_i\eta - a_i(\eta) + b_i(y_i)$ is the log likelihood based on the i-th case only, and the plus sign is used if $\text{logit}(y_i/n) > \mathbf{x}_i^{\mathrm{T}}\hat{\boldsymbol{\beta}}$ and the minus sign is used otherwise. Clearly, $dev = \Sigma\, dev_i^2$. Landwehr *et al.* (1980) advocate the use of dev_i in graphical procedures.

Finally, Cox and Snell (1968) suggest a somewhat more complicated set of residuals based on a transformation to normality proposed by Blom (1954). Let

$$\phi(u) = \int_0^u t^{-1/3}(1-t)^{-1/3}\,\mathrm{d}t, \qquad 0 \le u \le 1$$

The quantity $\phi(u)/\phi(1)$ is the incomplete beta function $I_u(2/3, 2/3)$. The

i-th residual is then

$$\sqrt{n_i}\frac{[\phi(y_i/m_i)-\phi\{\hat{p}_i-\frac{1}{6}(1-2\hat{p}_i)/n_i\}]}{\hat{p}_i^{1/6}(1-\hat{p}_i)^{1/6}}$$

Cox and Snell state that this set of residuals has essentially normal behavior, even for n_i as small as 5 and $p_i = 0.04$. Estimates of the variances of these residuals are given by Cox and Snell.

EXAMPLE 5.4.1. LEUKEMIA DATA NO. 3. The data for all 33 patients are given in Table 5.4.1 in a form appropriate for fitting logistic

Table 5.4.1 *Leukemia data for logistic regression. Source: Feigl and Zelen* (1965)

Case	*WBC*	*AG*	*y*	*n*
1	2 300	1	1	1
2	750	1	1	1
3	4 300	1	1	1
4	2 600	1	1	1
5	6 000	1	0	1
6	10 500	1	1	1
7	10 000	1	1	1
8	17 000	1	0	1
9	5 400	1	0	1
10	7 000	1	1	1
11	9 400	1	1	1
12	32 000	1	0	1
13	35 000	1	0	1
14	52 000	1	0	1
15	100 000	1	1	3
16	4 400	0	1	1
17	3 000	0	1	1
18	4 000	0	0	1
19	1 500	0	0	1
20	9 000	0	0	1
21	5 300	0	0	1
22	10 000	0	0	1
23	19 000	0	0	1
24	27 000	0	0	1
25	28 000	0	0	1
26	31 000	0	0	1
27	26 000	0	0	1
28	21 000	0	0	1
29	79 000	0	0	1
30	100 000	0	0	2

models. We take the response y to be the number of patients surviving at least 52 weeks for each combination of WBC = white blood cell count, and $AG = 1$ for AG positive patients and $AG = 0$ for AG negative patients. The five patients with $WBC = 100\,000$ are collapsed into two groups, one (case 15) consisting of the three AG positives (with one survivor) and one (case 30) consisting of two AG negatives (with no survivors).

The usual summary statistics obtained from fitting the model

$$\text{logit}(p_j) = \beta_0 + \beta_1 WBC + \beta_2 AG \tag{5.4.8}$$

are given in Table 5.4.2(a). The deviance has the value 27.24 with 27 df. There is no indication from this summary that the model is grossly inadequate.

Table 5.4.2 *Logistic regression summaries, Leukemia data*

	(a) Full data		*(b) One case removed*	
	Estimate	*Asymp. s.e.*	*Estimate*	*Asymp. s.e.*
Intercept	-1.307	0.814	0.212	1.083
WBC	-0.318×10^{-4}	0.186×10^{-4}	-0.235×10^{-3}	0.315×10^{-3}
AG	2.261	0.952	2.558	1.234
df	27		27	
Deviance	27.24		19.11	

Index plots of the χ_i, the diagonal elements of $\hat{\mathbf{V}}$ and $D_i^1(\mathbf{X}^T\hat{\mathbf{W}}\mathbf{X}, p')$ are given in Figs. 5.4.1–5.4.3, respectively. Clearly, case 15 is unusual and may be seriously influencing the fit. From Fig. 5.4.1, χ_{15} is not unusually large and thus the influence of case 15 is apparently due to its relative position in the factor space. Case 15 consists of the results for $n_{15} = 3$ AG positive patients with $WBC = 100\,000$. The fact that one of these patients survived for a relatively long time is surely contributing to the influence of this case.

To understand the role of case 15, we could refit the model after removing either all three patients in case 15 or just the suspect patient (patient 17 in Table 5.1.1). For these data, both alternatives lead to essentially the same revised fit. Table 5.4.2(b) summarizes the fitted model after the removal of patient 17 or, equivalently, modifying case 15 by setting $y_{15} = 0$ and $n_{15} = 2$. The summaries for the full and

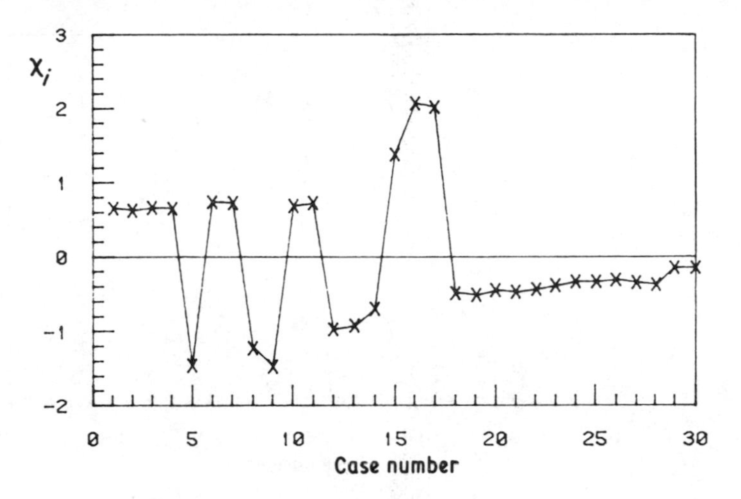

Figure 5.4.1 χ_i versus case number, leukemia data

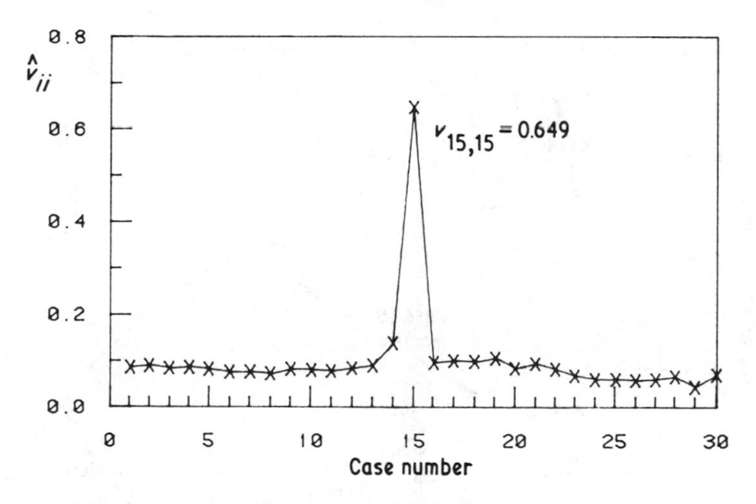

Figure 5.4.2 $\hat{v}_{ii}$ versus case number, leukemia data

reduced data in Table 5.4.2. are clearly quite different. This difference is further illustrated in Fig. 5.4.4 which gives plots of the fitted survival probabilities versus *WBC* and *AG* for the full and reduced data. Surprisingly, the removal of patient 17 increases the estimated probability of survival for patients with small values of *WBC*. The influence of case 15 is certainly overwhelming.

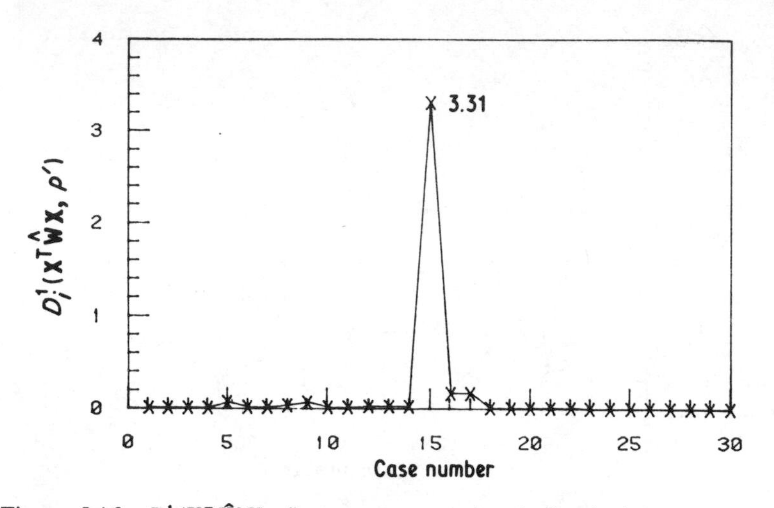

Figure 5.4.3 $D_i^1(\mathbf{X}^{\mathrm{T}}\hat{\mathbf{W}}\mathbf{X}, p')$ versus case number, leukemia data

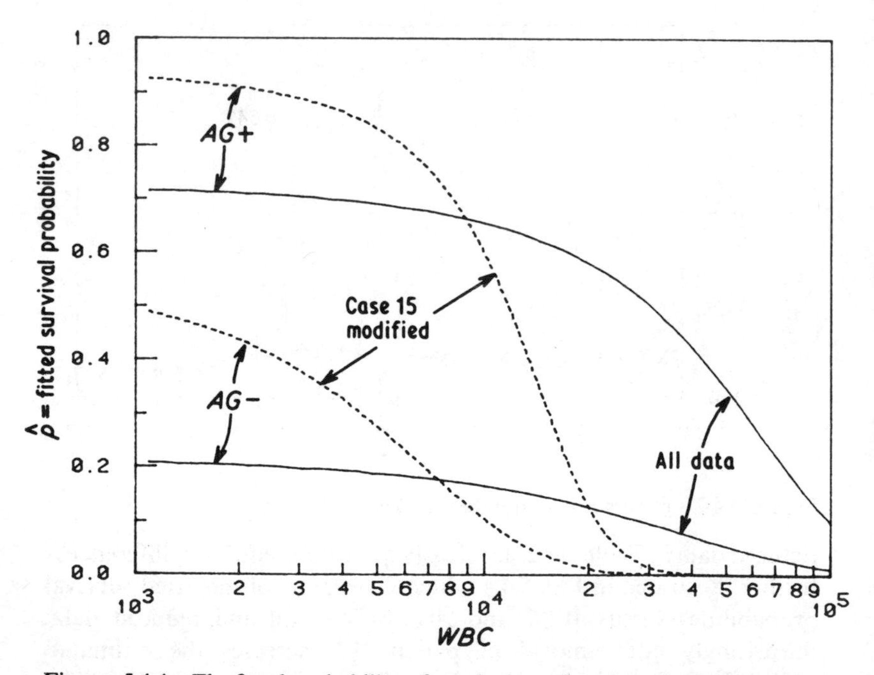

Figure 5.4.4 The fitted probability of survival as a function of *AG* and *WBC*, leukemia data

The influence of patient 17 is of course dependent on the assumed form of the model. One reasonable alternative to model (5.4.8) is obtained by adding the interaction variable $WBC \times AG$ to allow for the possibility that the slopes for the AG positive and AG negative groups may differ. For these data, however, the addition of the interaction term does not lead to a significantly improved fit. For the full data the asymptotic t-value for $WBC \times AG$ is 0.88 and case 15 is still the only influential case. After the removal of case 15, the t-value is 0.38 and no single case is seriously influential.

As another alternative, we could transform WBC via a log transformation, as was done in Example 5.1.1. When this alternative is pursued, the importance of case 15 is lessened. For example, $D_{15}^{1}(\mathbf{X}^{\mathrm{T}}\hat{\mathbf{W}}\mathbf{X}, p') = 0.47$, and the fitted models with and without patient 17 are not as different, as illustrated in Fig. 5.4.5. This reiterates the lesson that the influence of a case can be changed by transformation.□

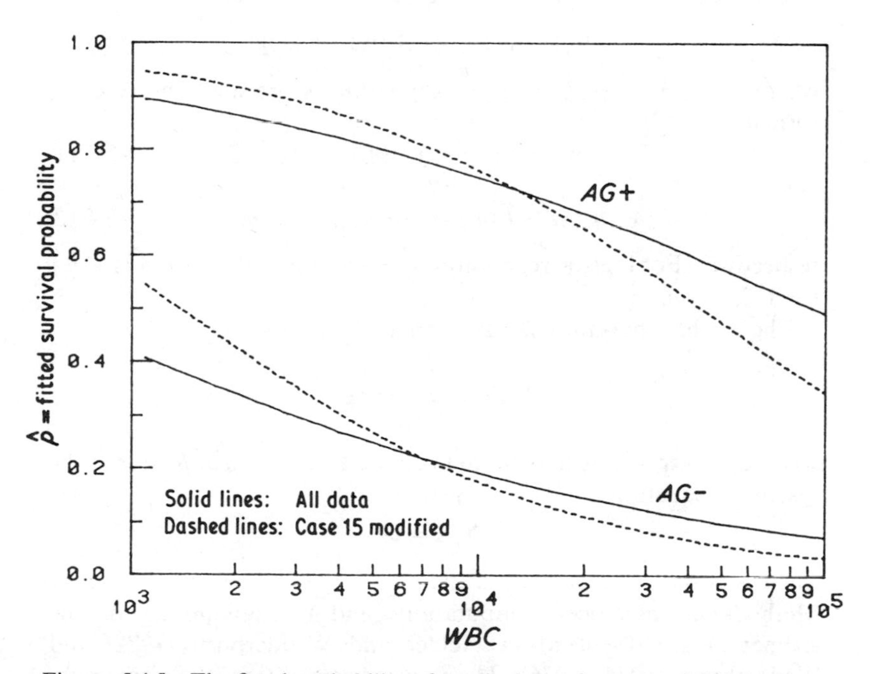

Figure 5.4.5 The fitted probability of survival as a function of WBC, using $\log(WBC)$ as a predictor, leukemia data

Logistic regression is one member of the class of generalized linear models described by Nelder and Wedderburn (1972); see also Wedderburn (1974, 1976). With the appropriate modifications, many of the results for the logistic model can be applied in the larger class.

Let $Y_1, Y_2, \ldots, Y_n$ denote n independent random variables such that Y_i has density

$$f_i(y; \alpha_i) = \exp[y\alpha_i - a_i(\alpha_i) + b_i(y)] \tag{5.4.9}$$

Further, assume that a one-to-one function k can be specified such that

$$\alpha_i = k(\mathbf{x}_i^{\mathrm{T}}\boldsymbol{\beta}), \qquad i = 1, 2, \ldots, n$$

where $\mathbf{x}_i$ is a p'-vector of observable variables, and $\boldsymbol{\beta}$ is an unobservable parameter vector. For logistic regression, $\alpha_i = \log[p_i/(1-p_i)]$, $k(z) = z$, and the other quantities are defined after (5.4.2). The function k is called a *link function* since it provides the link between the parameters α_i and the linear regression function. It is often useful to formulate the link function in terms of $E(y_i)$.

The log likelihood for $\eta_i = \mathbf{x}_i^{\mathrm{T}}\boldsymbol{\beta}$ based on the i-th case only is simply

$$l_i(\eta_i) = y_i k(\eta_i) - A_i(\eta_i) + b_i(y_i) \tag{5.4.10}$$

where $A_i(\eta_i) = a_i(k(\eta_i))$. The corresponding score and observed information are

$$s_i(\eta_i) = \dot{l}_i(\eta_i) = y_i\dot{k}(\eta_i) - \dot{A}_i(\eta_i) \tag{5.4.11}$$

and

$$w_i(\eta_i) = -\ddot{l}_i(\eta_i) = -y_i\ddot{k}(\eta_i) + \ddot{A}_i(\eta_i) \tag{5.4.12}$$

respectively. For logistic regression, $s_i = y_i - n_i p_i$ and $w_i = n_i p_i(1-p_i)$, $i = 1, 2, \ldots, n$.

The log likelihood for $\boldsymbol{\beta}$ based on all n cases is

$$L(\boldsymbol{\beta}) = \sum_{i=1}^{n} l_i(\mathbf{x}_i^{\mathrm{T}}\boldsymbol{\beta})$$

and the corresponding maximum likelihood estimate $\hat{\boldsymbol{\beta}}$ of $\boldsymbol{\beta}$ satisfies the system of equations

$$\mathbf{X}^{\mathrm{T}}\hat{\mathbf{s}} = \mathbf{0}$$

where $\hat{\mathbf{s}}$ is the n-vector with elements $\hat{s}_i = s_i(\mathbf{x}_i^{\mathrm{T}}\hat{\boldsymbol{\beta}})$ defined at (5.4.11). Methods of inference, computations and the uniqueness of the estimators are discussed in Nelder and Wedderburn (1972) and Wedderburn (1974, 1976). Here, we assume that the maximum likelihood estimate is unique.

In general, the diagnostic methods developed for logistic regression also can be used for generalized linear models characterized by (5.4.9) and (5.4.10). In particular, (5.4.4) and (5.4.5) apply with $\hat{\mathbf{W}} = \text{diag}[w_i(\mathbf{x}_i^T\hat{\boldsymbol{\beta}})]$, and $\hat{\mathbf{s}} = (s_i(\mathbf{x}_i^T\hat{\boldsymbol{\beta}}))$, where w_i and s_i are defined at (5.4.12) and (5.4.11), respectively. One possible general extension of the Studentized residual r_i^2 suggested by this procedure is $\hat{r}_i^2 = \hat{s}_i^2/\hat{w}_i(1-\hat{v}_{ii})$.

The $\hat{r}_i^2$ arise also in connection with an extension of the normal theory mean shift outlier model, as outlined in Section 2.2.2, when applied to generalized linear models. One way to describe the possibility that the i-th case is an outlier is to let

$$\eta_j = \mathbf{x}_j^T\boldsymbol{\beta} + \phi d_j \qquad j = 1, 2, \ldots, n \tag{5.4.13}$$

where $d_j = 1$ if $j = i$ and 0 otherwise. This form might be appropriate for the leukemia data, for example, because high white blood cell counts are unreliable. It is easily verified that the maximum likelihood estimator of $\boldsymbol{\beta}$ under (5.4.13) is equal to $\hat{\boldsymbol{\beta}}_{(i)}$, the maximum likelihood estimator of $\boldsymbol{\beta}$ obtained from the original model after deletion of the i-th case. The maximum likelihood estimator of ϕ will satisfy $s_i(\mathbf{x}_i^T\hat{\boldsymbol{\beta}}_{(i)} + \phi) = 0$.

In general, $\hat{r}_i^2$ is a modified version of the score test statistic (Cox and Hinkley, 1974, p. 324) for the hypothesis $\phi = 0$ obtained by substituting the observed information matrix for the expected information matrix. For models with $\ddot{k}(\eta_i) \equiv 0$, the observed and expected information matrices are the same, and $\hat{r}_i^2$ is the score test statistic. This happens, for example, in logistic regression.

EXAMPLE 5.4.2. LEUKEMIA DATA NO. 4. The log likelihood based on the i-th case for the AG positive cases in the leukemia data can be written as

$$l_i(\eta_i) = -y_i\exp[-\log(\theta_1) - \theta_2 x_i] - [\log(\theta_1) + \theta_2 x_i]$$

which is of the form given at (5.4.10) with $k(\eta_i) = -\exp(-\eta_i)$, $A_i(\eta_i) = \eta_i$, $b_i(y_i) = 0$, and $\eta_i = \mathbf{x}_i^T\boldsymbol{\beta}$ with $\boldsymbol{\beta}^T = [\log(\theta_1), \theta_2]$. The corresponding score and observed information are

$$s_i = y_i\exp(-\eta_i) - 1$$

and

$$w_i = y_i\exp(-\eta_i)$$

Thus, $\hat{\mathbf{W}} = \text{diag}[y_i\exp(-\mathbf{x}_i^T\hat{\boldsymbol{\beta}})]$, and $\hat{s}_i = y_i\exp(-\mathbf{x}_i^T\hat{\boldsymbol{\beta}}) - 1$. Values of $D_i^1(\mathbf{X}^T\hat{\mathbf{W}}\mathbf{X}, p')$, $\hat{v}_{ii}$ and $\hat{r}_i$ are given in the last three columns of

Table 5.1.1 for the *AG* positive cases. The information contained in these three statistics is similar to that given by the ε_i^* and c_{ii} obtained in Section 5.1, and the fully iterated influence measure LD_i obtained in Section 5.2 (values for these statistics are also given in Table 5.1.1). The two distance measures LD_i and $D_i^1(\mathbf{X}^T\hat{\mathbf{W}}\mathbf{X}, p')$ show very good agreement, both clearly identifying case 17 as influential. Similarly, the ε_i^* and the $\hat{r}_i$ are closely related, with large ε_i^* corresponding to large $\hat{r}_i$ and small ε_i^* corresponding to large negative $\hat{r}_i$. The agreement between the c_{ii} and the $\hat{v}_{ii}$ is not as strong as between the other statistics. Thus, the c_{ii} and the $\hat{v}_{ii}$ do not appear to contain the same information.□

5.5 Robust regression

In the usual linear regression model $\mathbf{Y} = \mathbf{X}\boldsymbol{\beta} + \boldsymbol{\varepsilon}$, a robust estimate $\tilde{\boldsymbol{\beta}}$ of $\boldsymbol{\beta}$ is obtained by minimizing

$$\sum_{i=1}^{n} \rho[(y_i - \mathbf{x}_i^T\boldsymbol{\beta})/\tilde{\sigma}] \tag{5.5.1}$$

with respect to $\boldsymbol{\beta}$, where ρ is a suitably selected loss function and $\tilde{\sigma}$ is a robust scale estimate that may be determined previously or simultaneously to achieve scale invariance. Estimators that minimize (5.5.1) are called *M-estimators*, a shorthand for maximum likelihood type estimators. For a discussion of robust regression methods see, for example, Huber (1977, 1981) and Hogg (1979).

Robust regression is designed to reduce or bound the influence of outlying responses that often occur when sampling from a symmetric long-tailed distribution. A number of authors, including Huber (1977), caution that robust regression may be ineffective in the presence of remote points in the factor space. Robust estimates can be as sensitive as least squares estimates to such points and it is for this reason that measures of case influence are needed in robust regression.

Many of the methods discussed in Chapter 3 for measuring the influence of the i-th case can be applied to robust regression without change. For example, let all cases have error variance σ^2 except for case i which has var $(\varepsilon_i) = \sigma^2/w_i$, $w_i > 0$, and let $\mathbf{W} = \text{diag}(w_j)$, $w_j = 1$ for all $j \neq i$. Then the influence of case i can be assessed by applying (5.5.1) to the transformed model

$$\mathbf{W}^{1/2}\mathbf{Y} = \mathbf{W}^{1/2}\mathbf{X}\boldsymbol{\beta} + \mathbf{W}^{1/2}\boldsymbol{\varepsilon} \tag{5.5.2}$$

and monitoring the behavior of the corresponding robust estimates $\tilde{\boldsymbol{\beta}}(w_i)$ as w_i is varied.

EXAMPLE 5.5.1. CLOUD SEEDING NO. 12. In this example, we illustrate the preceding remarks by using the loss function (Huber, 1964, 1973),

$$\rho(z) = \begin{cases} \dfrac{z^2}{2}, & |z| \le c \\ c|z| - \dfrac{c^2}{2}, & |z| > c \end{cases} \tag{5.5.3}$$

with $c = 1.345$ to fit model (2.4.23) modified according to (5.5.2) with $i = 2$, to the cloud seeding data. The robust estimates $\tilde{\boldsymbol{\beta}}(w_2)$, $0 < w_2 \le 1$, were obtained via an iterative algorithm based on Newton's method as described in Huber (1977, p. 38) and Holland and Welsch (1977). The value of w_2 was stepped from $w_2 = 1$ to 0; at each step the last value of $(\tilde{\beta}, \tilde{\sigma})$ was used as starting value.

Figurc 5.5.1 contains a plot of the $\tilde{\beta}_{14}(w_2)$ component of $\tilde{\boldsymbol{\beta}}(w_2)$ against w_2. The diagonal line is added for reference; the approximate

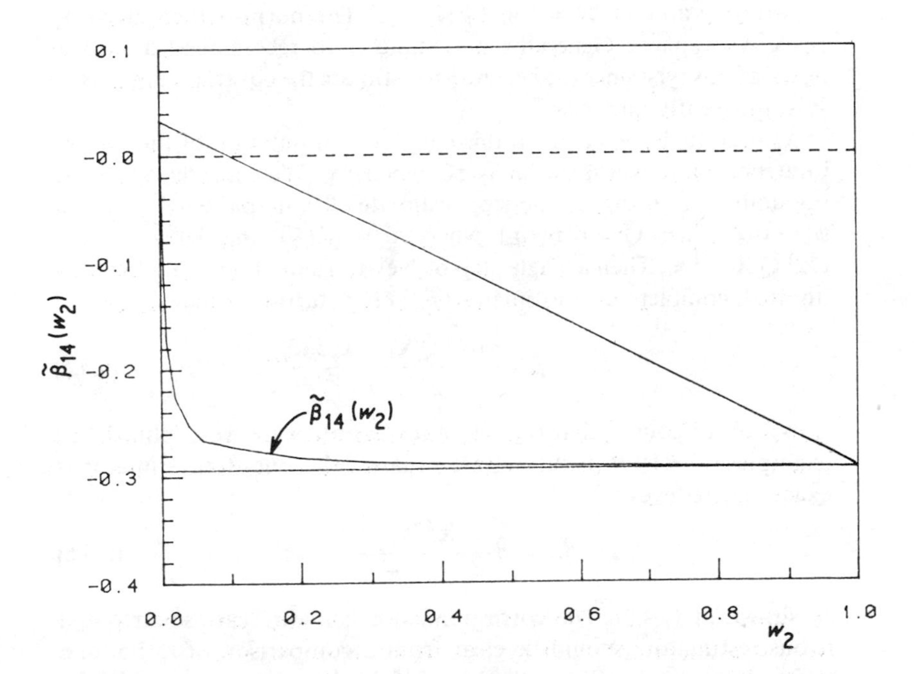

Figure 5.5.1 $\tilde{\beta}_{14}(w_2)$ versus w_2, cloud seeding data. Standard error at $w_2 = 1$ is approximately 0.07

standard error at $w_2 = 1$ is about 0.07. Clearly, $\tilde{\beta}_{14}(w_2)$ is insensitive to perturbations near $w_2 = 1$, but is highly sensitive to perturbations as $w_2 \to 0$. This plot should be compared to the corresponding plot for least squares estimation given in Fig. 3.4.1. In this example the behavior of the least squares and robust estimators are remarkably similar. □

The sample influence curve (3.4.6) for robust regression is

$$SIC_i = (n-1)(\tilde{\boldsymbol{\beta}} - \tilde{\boldsymbol{\beta}}_{(i)}) \tag{5.5.4}$$

where $\tilde{\boldsymbol{\beta}}_{(i)} = \tilde{\boldsymbol{\beta}}(w_i \to 0)$ is the robust estimate of $\boldsymbol{\beta}$ computed without the i-th case. Various useful norms of the sample influence curve can be obtained by following the rationale used in the linear least squares problem. One possible norm of the sample influence curve is $D_i(\mathbf{X}^T\mathbf{X}, p'k)$ where k is a scalar chosen so that $k(\mathbf{X}^T\mathbf{X})^{-1}$ corresponds to an estimate of the asymptotic covariance matrix of $\tilde{\boldsymbol{\beta}}$ (see Hogg, 1979; Huber, 1981, Chapter 7).

A second norm can be based on the iteratively reweighted least squares approach to computation of $\tilde{\boldsymbol{\beta}}$. Let $\tilde{e}_i = (y_i - \mathbf{x}_i^T\tilde{\boldsymbol{\beta}})/\tilde{\sigma}$, $\psi(z) = \mathrm{d}\rho(z)/\mathrm{d}z$ and let $\tilde{\mathbf{W}} = \mathrm{diag}\{\psi(\tilde{e}_i)/\tilde{e}_i\}$. The norm is then given by $D_i(\mathbf{X}^T\tilde{\mathbf{W}}\mathbf{X}, p'\tilde{\sigma}^2)$. Generally, it is difficult to recommend a specific norm of this type since the best way to estimate the covariance matrix of $\tilde{\boldsymbol{\beta}}$ is apparently unknown.

As indicated previously in this chapter, computation of the sample influence curve will most likely be expensive. We consider again the possibility of using a one-step estimate $\tilde{\boldsymbol{\beta}}_{(i)}^1$ in place of $\tilde{\boldsymbol{\beta}}_{(i)}$. Let $\psi' = \mathrm{d}\psi(z)/\mathrm{d}z$, $\tilde{\mathbf{Q}} = \mathrm{diag}(\tilde{q}_i)$ where $\tilde{q}_i = \psi'(\tilde{e}_i)$ and let $\tilde{v}_{ii} = \tilde{q}_i\mathbf{x}_i^T(\mathbf{X}^T\tilde{\mathbf{Q}}\mathbf{X})^{-1}\mathbf{x}_i$. Then a single step of Newton's method using the fully iterated, complete data estimates $(\tilde{\boldsymbol{\beta}}, \tilde{\sigma})$ for starting values gives

$$\tilde{\boldsymbol{\beta}}_{(i)}^1 = \tilde{\boldsymbol{\beta}} - \frac{\tilde{\sigma}(\mathbf{X}^T\tilde{\mathbf{Q}}\mathbf{X})^{-1}\mathbf{x}_i\psi(\tilde{e}_i)}{1-\tilde{v}_{ii}} \tag{5.5.5}$$

provided, of course, that the relevant quantities are well defined (for example, $\tilde{v}_{ii} \neq 1$). For linear least squares, the one-step estimator is exact and reduces to

$$\hat{\boldsymbol{\beta}}_{(i)} = \hat{\boldsymbol{\beta}} - \frac{(\mathbf{X}^T\mathbf{X})^{-1}\mathbf{x}_i e_i}{1-v_{ii}} \tag{5.5.6}$$

as shown at (3.4.6). The correspondence between least squares and robust estimators should be clear from a comparison of (5.5.5) and (5.5.6). In particular, the residuals e_i in (5.5.6) have been replaced by the Winsorized residuals $\tilde{\sigma}\psi(\tilde{e}_i)$ and v_{ii} has been replaced by $\tilde{v}_{ii}$.

A precise characterization of the accuracy of this one-step approximation (5.5.5) is unavailable, but the following observations may help. First, for estimators with redescending ψ-functions, such as Andrews' (1974) sine estimator, $(\mathbf{X}_{(i)}^{\mathrm{T}} \tilde{\mathbf{Q}}_{(i)} \mathbf{X}_{(i)})$ need not be positive definite and the one-step estimator $\tilde{\boldsymbol{\beta}}_{(i)}^{1}$ cannot be guaranteed to decrease the objective function. The one-step estimators can be expected to be more satisfactory for monotone ψ-functions.

Second, it is not difficult to verify that if ψ is piecewise linear (for example, the ψ function corresponding to (5.5.3)) and if the classification of $\tilde{e}_j$ according to the pieces of ψ is the same as the classification of the one-step residuals $\tilde{e}_j^1 = (y_j - \mathbf{x}_j^{\mathrm{T}} \tilde{\boldsymbol{\beta}}_{(i)}^1)/\tilde{\sigma}$ for all $j \neq i$, then $\tilde{\boldsymbol{\beta}}_{(i)}^1 = \tilde{\boldsymbol{\beta}}_{(i)}$. More generally, the accuracy of the one-step approximation seems to depend on the differences $|\tilde{e}_j - \tilde{e}_j^1|$.

Assuming a sufficiently accurate one-step approximation, the effects of remote points in the factor space on robust estimators can be illustrated by using Huber's loss function:

$$\tilde{q}_i = \begin{cases} 1, & |\tilde{e}_i| \leq c \\ 0, & \text{otherwise} \end{cases}$$

$$\tilde{v}_{ii} = \begin{cases} \mathbf{x}_i^{\mathrm{T}} (\mathbf{X}^{\mathrm{T}} \tilde{\mathbf{Q}} \mathbf{X})^{-1} \mathbf{x}_i, & |\tilde{e}_i| \leq c \\ 0, & \text{otherwise} \end{cases}$$

and

$$\tilde{\boldsymbol{\beta}} - \tilde{\boldsymbol{\beta}}_{(i)}^1 = \begin{cases} -\tilde{\sigma}(\mathbf{X}^{\mathrm{T}} \tilde{\mathbf{Q}} \mathbf{X})^{-1} \mathbf{x}_i c, & \tilde{e}_i < -c \\ \dfrac{\tilde{\sigma}(\mathbf{X}^{\mathrm{T}} \tilde{\mathbf{Q}} \mathbf{X})^{-1} \mathbf{x}_i \tilde{e}_i}{1 - \tilde{v}_{ii}}, & -c \leq \tilde{e}_i \leq c \\ \tilde{\sigma}(\mathbf{X}^{\mathrm{T}} \tilde{\mathbf{Q}} \mathbf{X})^{-1} \mathbf{x}_i c, & \tilde{e}_i > c \end{cases}$$

If $-c \leq \tilde{e}_i \leq c$, the influence of the i-th case can be greater than that for least squares since $v_{ii} \leq \tilde{v}_{ii}$, $i = 1, 2, \ldots, n$. Similarly when $|\tilde{e}_i| > c$ the influence of the i-th case will generally be less than that for least squares. Consider, for example, the situation in which $\tilde{q}_i = 0$ but $\tilde{q}_j = 1$ for all $j \neq i$. If $\tilde{e}_i > c$, then

$$(n-1)[\tilde{\boldsymbol{\beta}} - \tilde{\boldsymbol{\beta}}_{(i)}^1] = (n-1)\frac{(\mathbf{X}^{\mathrm{T}}\mathbf{X})^{-1}\mathbf{x}_i}{1 - v_{ii}}(\tilde{\sigma} c)$$

which is the SIC_i for least squares reduced by the factor $c/\tilde{e}_i$. Also, for $j \neq i$

$$\tilde{\boldsymbol{\beta}} - \tilde{\boldsymbol{\beta}}_{(j)}^1 = \tilde{\sigma}(\mathbf{X}^{\mathrm{T}}\mathbf{X})^{-1} \tilde{e}_j \left[\frac{(1 - v_{ii})\mathbf{x}_j + v_{ij}\mathbf{x}_i}{(1 - v_{ii})(1 - v_{jj}) - v_{ij}^2}\right]$$

EXAMPLE 5.5.2. CLOUD SEEDING NO. 13. To illustrate the use of the one-step approximation, we use the cloud seeding data. The full sample estimate for Huber's method was obtained with the least squares estimate as starting values, with Huber's (1977) proposal 2 as the iterative computing method, and using $\tilde{\sigma}$ = median absolute residual/0.6745 to estimate scale. Fifteen iterations were required to get a solution such that the maximum proportional change in any coefficient from the last iteration was less than 0.01. One-step measures $D_i^1(\mathbf{X}^T\tilde{\mathbf{W}}\mathbf{X}, p'\tilde{\sigma}^2)$ and 'fully iterated' $D_i(\mathbf{X}^T\tilde{\mathbf{W}}\mathbf{X}, p'\tilde{\sigma}^2)$ based on 10 iterations were then computed.

Table 5.5.1 lists the two measures for the five cases with the largest values of $D_i(\mathbf{X}^T\tilde{\mathbf{W}}\mathbf{X}, p'\tilde{\sigma}^2)$. With the exception of the clearly influential case 2, agreement between the two measures is adequate, and even case 2 is clearly identified by the one-step measure. Overall, 9 of 24 cases are underestimated using the one-step estimate, but none seriously.

Table 5.5.1 *Five largest influence measures, cloud seeding data*

Case	$D_i^1(\mathbf{X}^T\tilde{\mathbf{W}}\mathbf{X}, p'\tilde{\sigma}^2)$	$D_i(\mathbf{X}^T\tilde{\mathbf{W}}\mathbf{X}, p'\tilde{\sigma}^2)$
2	20.21	9.03
4	0.42	0.49
7	0.34	0.38
17	0.92	0.85
24	0.51	0.37

This analysis has been repeated for Andrews' (1974) sine estimator and for several other data sets. While the results for the sine estimate applied to these data generally agree with the results for the Huber estimate, in other problems we have found the agreement to be much worse. More work is needed to understand the one-step distance measures and their usefulness when applied to the robust estimators. □

5.6 Other problems

In this section we give brief accounts of some of the other problems for which influence has been studied, including the correlation coefficient, discriminant analysis, and linear regression with incomplete data.

5.6.1 CORRELATION COEFFICIENT

Suppose that X_1, X_2 are normal random variables with means μ_1, μ_2, variances σ_1^2, σ_2^2, and covariance σ_{12}. The influence curve for the correlation coefficient $\rho = \sigma_{12}/\sigma_1\sigma_2$ was given by Mallows (1975) as

$$IC(x_1, x_2) = \tilde{x}_1\tilde{x}_2 - \frac{\rho}{2}(\tilde{x}_1^2 + \tilde{x}_2^2) \tag{5.6.1}$$

where $\tilde{x}_j = (x_j - \mu_j)/\sigma_j$, $j = 1, 2$. The empirical influence curve for a sample (x_{1i}, x_{2i}), $i = 1, 2, \ldots, n$, is obtained by substituting the corresponding sample cdf $\hat{F}$ in place of F,

$$EIC(x_1, x_2) = \hat{x}_1\hat{x}_2 - \frac{\hat{\rho}}{2}(\hat{x}_1^2 + \hat{x}_2^2) \tag{5.6.2}$$

where $\hat{x}_j = (x_j - \bar{x}_j)/s_j$, $\bar{x}_j = \Sigma_i x_{ji}/n$ and $s_j^2 = \Sigma_i(x_{ji} - \bar{x}_j)^2/n$, $j = 1, 2$, and $\hat{\rho}$ is the usual estimator of ρ. The sample influence curve is given by

$$SIC_i = (n-1)(\hat{\rho} - \hat{\rho}_{(i)}) \tag{5.6.3}$$

Both sample versions of the influence curve for ρ can provide useful information on the effects of single cases in determining $\hat{\rho}$. In small samples, however, where efficient calculation and methods of display are not a serious issue, the *SIC* seems preferable, as it has a straightforward interpretation and is perhaps the most directly relevant. Devlin *et al.* (1975) suggest the *SIC* for use in detecting outliers that substantially affect $\hat{\rho}$.

For the usual estimator $\hat{\rho}$, the *SIC* can be studied by expressing $\hat{\rho}_{(i)}$ as a function of $\hat{\rho}$ and other full sample statistics. One finds

$$\hat{\rho}_{(i)} = \frac{\hat{\rho} - r_{1i}r_{2i}/n}{[(1 - r_{1i}^2/n)(1 - r_{2i}^2/n)]^{1/2}} \tag{5.6.4}$$

where $r_{ji} = (x_{ji} - \bar{x}_i)/[s_j(1 - 1/n)^{1/2}]$ is the i-th Studentized residual, $i = 1, 2, \ldots, n$, for the j-th marginal sample, $j = 1, 2$. Since r_{ji}^2 is a monotonic function of the normal theory test that the i-th case is a marginal outlier under the mean-shift model, the denominator of (5.6.4) will be small if either x_{1i} or x_{2i} appears to be an outlier when judged against the respective marginal samples. A marginal outlier will have a substantial influence on $\hat{\rho}$.

The numerator in (5.6.4) measures the joint effect of (x_{1i}, x_{2i}) and depends on the location of (x_{1i}, x_{2i}) relative to $(\bar{x}_1, \bar{x}_2)$. If, for example, $\hat{\rho} > 0$ and $x_{1i} < \bar{x}_1$ and $x_{2i} > \bar{x}_2$ (or $x_{i1} > \bar{x}_1$ and $x_{i2} < \bar{x}_2$) then $\hat{\rho} - \hat{\rho}_{(i)} < 0$.

In large samples, an approximation to the sample influence curve may be sufficient. As a first-order approximation, expand the denominator of (5.6.4) in $(r_{1i}^2/n, r_{2i}^2/n)$ in a linear Taylor expansion about (0, 0). For large n, (5.6.3) becomes

$$SIC_i \cong r_{1i}r_{2i} - \frac{\hat{\rho}}{2}(r_{1i}^2 + r_{2i}^2) \tag{5.6.5}$$

which is essentially the empirical influence curve evaluated at (x_{1i}, x_{2i}). Devlin *et al.* (1975) suggest a graphical technique based on this approximation. Let

$$u_1 = \tfrac{1}{2}\left[\frac{\hat{x}_1 + \hat{x}_2}{(1+\hat{\rho})^{1/2}} + \frac{\hat{x}_1 - \hat{x}_2}{(1-\hat{\rho})^{1/2}}\right]$$

and

$$u_2 = \tfrac{1}{2}\left[\frac{\hat{x}_1 + \hat{x}_2}{(1+\hat{\rho})^{1/2}} - \frac{\hat{x}_1 - \hat{x}_2}{(1-\hat{\rho})^{1/2}}\right]$$

so that

$$EIC(x_1, x_2) = (1 - \hat{\rho}^2)u_1 u_2 \tag{5.6.6}$$

The advantage of this form is that the contours of constant influence are hyperbolae. Devlin *et al.* (1975) suggest superimposing selected contours of the *EIC* on scatter plots of (u_1, u_2) and, then reading the approximate influence directly from the plot.

As seen previously, procedures based on the *EIC* should be reasonable approximations to the *SIC* as long as n is large and the r_{ji}^2 are small to moderate. If r_{ji}^2 is large (the case is well removed from the centroid), Devlin *et al.* (1975) suggested that the *EIC* will usually underestimate the *SIC*. For this reason, their graphical procedure is perhaps best used as an initial screen. If a case is found to be influential it may be necessary to conduct a more precise investigation using the *SIC*.

5.6.2 DISCRIMINANT ANALYSIS

Campbell (1978) has considered the use of the influence curve as an aid in detecting outliers in two population normal discriminant analysis. He derives the theoretical and sample influence curves for the usual summary statistics, namely the Mahalanobis D^2, the vector of discriminant means, and the vector of discriminant coefficients (see Lachenbruch, 1975, for precise definitions). Assuming a perturbation in the first population, the influence curve for D^2 evaluated at a point $\mathbf{x}$ is a quadratic function of the difference between the discriminant scores at $\mathbf{x}$ and at the mean of the first population. The sample influence curve, in which estimates replace parameters, corresponds to a v_{ii}-like measure, since there is no component of the influence curve to correspond to a residual. Thus, influential cases for D^2 are those that are more distant (in an appropriate metric) from the mean of the other population. Campbell also discusses a function of the influence curve for D^2 and the other statistics that can be useful in graphical methods for the study of outlying cases.

5.6.3 LINEAR REGRESSION WITH INCOMPLETE DATA

Suppose we wish to fit a linear model of the type discussed earlier in this monograph but values for some of the variables are not observed. We call this a regression problem with incomplete data. Many writers have addressed the problem of estimation of parameters with incomplete data, often assuming that the unobserved data are 'missing at random' (Rubin, 1976), and that the observed data follow a multivariate normal distribution (see, for example, Little, 1979). Computational methods to find the maximum likelihood estimates of parameters of the conditional distribution of the response, given the predictors, have been given by Orchard and Woodbury (1972), Dempster, Laird and Rubin (1977) and Hocking and Marx (1979), among others.

In all of this literature, little or no attention has been paid to the problem of analyzing residuals and assessing influence. Shih (1981) has made first steps in this direction. He defines residuals by essentially using the general approach of Cox and Snell outlined in Section 5.1. If the EM-algorithm of Dempster *et al.* is used for the computations a very elegant result is obtained. At convergence of the algorithm, fill-in values for unobserved values are estimated, and the residuals can then be computed in the usual way based on the filled-in data. Studentized residuals, however, are not as easy to obtain, as the likelihood function, which is needed for the methods of Cox and Snell, is relatively complicated. Shih has also considered the use of one-step estimators, also using the EM algorithm, of the sample influence curve.

Generally, the maximum likelihood residuals seem to be superior to the competitors, such as the residuals computed only from the fully observed cases. However, much more experience with these residuals is required for them to be well understood.

For influence analysis, one can show that incomplete cases will generally not be influential. In addition, the extent to which one-step approximations are useful seems to depend on the covariance structure of the data, and the pattern of the incomplete data.

Appendix

A.1 Weighted least squares

The weighted least squares model is given by

$$\mathbf{Y} = \mathbf{X}\boldsymbol{\beta} + \boldsymbol{\varepsilon} \tag{A.1.1}$$

where all quantities are as defined near (2.1.1), except that $\mathrm{Var}(\boldsymbol{\varepsilon}) = \sigma^2\mathbf{W}^{-1}$ and $\mathbf{W}$ is a known $n \times n$ diagonal matrix with $w_{ii} > 0$. The w_{ii} are often called *case weights.* Although weighted least squares estimators can be computed directly, it is usual to transform to an unweighted least squares problem, and solve this simpler version. Multiplying both sides of (A.1.1) on the left by $\mathbf{W}^{1/2}$,

$$\mathbf{W}^{1/2}\mathbf{Y} = \mathbf{W}^{1/2}\mathbf{X}\boldsymbol{\beta} + \mathbf{W}^{1/2}\boldsymbol{\varepsilon}$$

or, if $\mathbf{Y}^* = \mathbf{W}^{1/2}\mathbf{Y}$, $\mathbf{X}^* = \mathbf{W}^{1/2}\mathbf{X}$, and $\boldsymbol{\varepsilon}^* = \mathrm{W}^{1/2}\boldsymbol{\varepsilon}$,

$$\mathbf{Y}^* = \mathbf{X}^*\boldsymbol{\beta} + \boldsymbol{\varepsilon}^* \tag{A.1.2}$$

Since $\mathrm{Var}(\boldsymbol{\varepsilon}^*) = \sigma^2\mathbf{I}$, it follows immediately that $\hat{\boldsymbol{\beta}} = (\mathbf{X}^{*\mathrm{T}}\mathbf{X}^*)^{-1}\mathbf{X}^{*\mathrm{T}}\mathbf{Y}^*$. Computationally, then, $\hat{\boldsymbol{\beta}}$ can be obtained by multiplying y_i and each element of $\mathbf{x}_i$, including the constant, by $w_{ii}^{1/2}$, and solving the resulting unweighted least squares problem. Using this transformation, the residuals are $\mathbf{e}^* = \mathbf{Y}^* - \mathbf{X}^*\hat{\boldsymbol{\beta}} = \mathbf{W}^{1/2}(\mathbf{Y} - \mathbf{X}\hat{\boldsymbol{\beta}})$, while the correct residuals for the model (A.1.1) are $\mathbf{e} = \mathbf{Y} - \mathbf{X}\hat{\boldsymbol{\beta}}$. The elements of $\mathbf{e}^*$ are sometimes called *weighted residuals,* and of course $\mathbf{e} = \mathbf{W}^{-1/2}\mathbf{e}^*$. Studentized residuals are identical under either formulation. Distance measures are also the same under both formulations, provided, of course, that the correct norm is used. For model (A.1.1), the appropriate norm is $D_i(\mathbf{X}^{\mathrm{T}}\mathbf{W}\mathbf{X}, p'\hat{\sigma}^2)$, which is equivalent to $D_i(\mathbf{X}^{*\mathrm{T}}\mathbf{X}^*, p'\hat{\sigma}^2)$, the correct norm for (A.1.2).

A.2 Updating formulae

Let $\mathbf{A}$ be a $p' \times p'$ rank p' symmetric matrix, and suppose that $\mathbf{a}$ and $\mathbf{b}$ are $q \times p'$ rank q matrices. Then, provided that the inverses exist,

$$(\mathbf{A} + \mathbf{a}^{\mathrm{T}}\mathbf{b})^{-1} = \mathbf{A}^{-1} - \mathbf{A}^{-1}\mathbf{a}^{\mathrm{T}}(\mathbf{I} + \mathbf{b}\mathbf{A}^{-1}\mathbf{a}^{\mathrm{T}})^{-1}\mathbf{b}\mathbf{A}^{-1} \quad \text{(A.2.1)}$$

This remarkable formula shows how to modify the inverse of the corrected cross product matrix when one or more rows of a matrix are deleted or added. The most important special case is that of deleting a single row $\mathbf{x}_i^{\mathrm{T}}$ from $\mathbf{X}$. Setting $\mathbf{A} = \mathbf{X}^{\mathrm{T}}\mathbf{X}$, $\mathbf{a} = -\mathbf{x}_i^{\mathrm{T}}$, $\mathbf{b} = \mathbf{x}_i^{\mathrm{T}}$,

$$(\mathbf{X}_{(i)}^{\mathrm{T}}\mathbf{X}_{(i)})^{-1} = (\mathbf{X}^{\mathrm{T}}\mathbf{X} - \mathbf{x}_i\mathbf{x}_i^{\mathrm{T}})^{-1} = (\mathbf{X}^{\mathrm{T}}\mathbf{X})^{-1} + \frac{(\mathbf{X}^{\mathrm{T}}\mathbf{X})^{-1}\mathbf{x}_i\mathbf{x}_i^{\mathrm{T}}(\mathbf{X}^{\mathrm{T}}\mathbf{X})^{-1}}{1 - \mathbf{x}_i^{\mathrm{T}}(\mathbf{X}^{\mathrm{T}}\mathbf{X})^{-1}\mathbf{x}_i} \quad \text{(A.2.2)}$$

A version of this formula was given by Gauss (1821), and in several papers about 1950 (Bartlett, 1951; Plackett, 1950; Sherman and Morrison, 1949; Woodbury, 1950). Bingham (1977) used this basic formula in a wide variety of applications in regression. A discussion of the history of this type of updating, and generalizations of it, is given by Henderson and Searle (1981).

A closely related result concerns the determinant of a partitioned $q \times q$ matrix $\mathbf{Z}$, where

$$\mathbf{Z} = \begin{pmatrix} \mathbf{A} & \mathbf{B} \\ \mathbf{C} & \mathbf{D} \end{pmatrix}$$

and $\mathbf{A}$ and $\mathbf{D}$ are nonsingular. Then,

$$|\mathbf{Z}| = |\mathbf{A}|\,|\mathbf{D} - \mathbf{C}\mathbf{A}^{-1}\mathbf{B}| = |\mathbf{D}|\,|\mathbf{A} - \mathbf{B}\mathbf{D}^{-1}\mathbf{C}| \quad \text{(A.2.3)}$$

This result is attributed to J. Schur by Henderson and Searle (1981). It can be used to establish several useful updating and downdating formulae. For example, let $\mathbf{A} = \mathbf{X}^{\mathrm{T}}\mathbf{X}$, $\mathbf{B} = \mathbf{X}_{\mathrm{I}}^{\mathrm{T}}$, $\mathbf{C} = \mathbf{X}_{\mathrm{I}}$, and $\mathbf{D} = \mathbf{I}_m$, where the use of I as a subscript is as in Section 3.6. Then

$$\begin{aligned} |\mathbf{X}_{(\mathrm{I})}^{\mathrm{T}}\mathbf{X}_{(\mathrm{I})}| &= |\mathbf{X}^{\mathrm{T}}\mathbf{X} - \mathbf{X}_{\mathrm{I}}^{\mathrm{T}}\mathbf{X}_{\mathrm{I}}| \\ &= |\mathbf{X}^{\mathrm{T}}\mathbf{X}|\,|\mathbf{I} - \mathbf{X}_{\mathrm{I}}(\mathbf{X}^{\mathrm{T}}\mathbf{X})^{-1}\mathbf{X}_{\mathrm{I}}^{\mathrm{T}}| \\ &= |\mathbf{X}^{\mathrm{T}}\mathbf{X}|\,|\mathbf{I} - \mathbf{V}_{\mathrm{I}}| \end{aligned} \quad \text{(A.2.4)}$$

A.3 Residual correlations

Let $v_{ij} = \mathbf{x}_i^{\mathrm{T}}(\mathbf{X}^{\mathrm{T}}\mathbf{X})^{-1}\mathbf{x}_j$ be the (i, j)-th element of $\mathbf{V}$ and define ρ_{ij} to be the correlation between the i-th and the j-th residuals,

$$\rho_{ij} = \frac{-v_{ij}}{[(1 - v_{ii})(1 - v_{jj})]^{1/2}} \qquad i \neq j \tag{A.3.1}$$

If $\mathbf{x}_i = \mathbf{x}_j$, then

$$\rho_{ij} = -v_{ii}/(1 - v_{ii}) \tag{A.3.2}$$

The residual correlation for replicated rows of $\mathbf{X}$ is thus always negative and will be large only if the corresponding v_{ii} is large. However,

$$\rho_{ij} = -v_{ii}/(1 - v_{ii}) \geq -1/(c - 1)$$

where $c > 1$ is the number of replicates of $\mathbf{x}_i$. For replicated points, therefore, large negative correlations ($\rho_{ij} < -\frac{1}{2}$) can occur only if $\mathbf{x}_i$ is replicated twice.

To investigate the general causes of a large value for ρ_{ij}^2, we shall fix $\mathbf{x}_j$ and choose $\mathbf{x}_i$ to maximize ρ_{ij}^2 (Cook, 1979). The required calculations are facilitated by first writing ρ_{ij}^2 in terms of explicit quadratic forms in $\mathbf{x}_i$. Let

$$v_{kl(i)} = \mathbf{x}_k^{\mathrm{T}}(\mathbf{X}_{(i)}^{\mathrm{T}}\mathbf{X}_{(i)})^{-1}\mathbf{x}_l$$

Using (A.2.1),

$$v_{kl} = v_{kl(i)} - v_{ki(i)}v_{li(i)}/(1 + v_{ii(i)}). \tag{A.3.3}$$

and

$$v_{kl(i)} = v_{kl} + v_{ki}v_{li}/(1 - v_{ii}). \tag{A.3.4}$$

These expressions show how to update and downdate the elements of $\mathbf{V}$:

$$v_{ij} = v_{ij(i)}/(1 + v_{ii(i)}) \tag{A.3.5}$$

$$v_{jj} = v_{jj(i)} - v_{ij(i)}^2/(1 + v_{ii(i)}) \tag{A.3.6}$$

and

$$v_{ii} = v_{ii(i)}/(1 + v_{ii(i)}). \tag{A.3.7}$$

Finally, ρ_{ij}^2 may be expressed as,

$$\begin{aligned}\rho_{ij}^2 &= \frac{v_{ij(i)}^2}{(1 + v_{ii(i)})(1 - v_{jj(i)}) + v_{ij(i)}^2} \\ &= \left[(1 - v_{jj(i)})\left(\frac{1 + v_{ii(i)}}{v_{ij(i)}^2}\right) + 1\right]^{-1}\end{aligned} \tag{A.3.8}$$

This form is convenient since $v_{ii(i)}$ and $v_{ij(i)}^2$ are quadratic forms in $\mathbf{x}_i$ and the corresponding inner product matrix $(\mathbf{X}_{(i)}^{\mathrm{T}}\mathbf{X}_{(i)})^{-1}$ does not depend on $\mathbf{x}_i$ by construction. Since $\mathbf{x}_j$ is to be held fixed, $v_{jj(i)}$ is a constant. Thus, to maximize ρ_{ij}^2 by choice of $\mathbf{x}_i$ it is sufficient to maximize

$$f(\mathbf{x}_i) = \frac{v_{ij(i)}^2}{1 + v_{ii(i)}}. \tag{A.3.9}$$

If the model contains a constant term, as will usually be the case, the first term of $\mathbf{x}_i$ is constrained to be 1 and the maximum of $f(\mathbf{x}_i)$ must be taken with respect to the last p components of $\mathbf{x}_i$. Assume that the independent variables are measured around the sample averages in the reduced data set, and let $\mathbf{x}_i^{\mathrm{T}} = (1, \boldsymbol{x}_i^{\mathrm{T}})$ and

$$(\mathbf{X}_{(i)}^{\mathrm{T}}\mathbf{X}_{(i)})^{-1} = \begin{pmatrix} 1/(n-1) & \mathbf{0} \\ \mathbf{0} & (\mathscr{X}_{(i)}^{\mathrm{T}}\mathscr{X}_{(i)})^{-1} \end{pmatrix}$$

Then,

$$v_{ij(i)} = \frac{1}{n-1} + \boldsymbol{x}_i^{\mathrm{T}}(\mathscr{X}_{(i)}^{\mathrm{T}}\mathscr{X}_{(i)})^{-1}\boldsymbol{x}_j$$

and (A.3.9) can be usefully re-expressed as

$$f(\boldsymbol{x}_i) = \frac{\{[1/(n-1)] + \boldsymbol{x}_i^{\mathrm{T}}(\mathscr{X}_{(i)}^{\mathrm{T}}\mathscr{X}_{(i)})^{-1}\boldsymbol{x}_j\}^2}{(1 + [1/(n-1)] + \boldsymbol{x}_i^{\mathrm{T}}(\mathscr{X}_{(i)}^{\mathrm{T}}\mathscr{X}_{(i)})^{-1}\boldsymbol{x}_i)} \tag{A.3.10}$$

The largest possible value for ρ_{ij}^2 will obviously depend on the subset of R^p over which the maximum is taken. If the model contains functionally related terms (for example, x and x^2) the appropriate subset may be complex and will depend on the model. Here, we consider the unconstrained maximum over R^p by first considering subsets of the form $G(k) = \{\boldsymbol{x}_i | \boldsymbol{x}_i^{\mathrm{T}}(\mathscr{X}_{(i)}^{\mathrm{T}}\mathscr{X}_{(i)})^{-1}\boldsymbol{x}_i = k,\ k > 0\}$ and then maximizing over k. The effect of this is that for some models the derived maximum may not be attainable.

Using the Cauchy–Schwarz inequality, it can be verified that

$$\max_{G(k)} [f(\boldsymbol{x}_i)] = \frac{k}{[n/(n-1)] + k}\left[\frac{1}{(n-1)\sqrt{k}} + (\boldsymbol{x}_j^{\mathrm{T}}(\mathscr{X}_{(i)}^{\mathrm{T}}\mathscr{X}_{(i)})^{-1}\boldsymbol{x}_j)^{1/2}\right]^2 \tag{A.3.11}$$

which is attained at

$$\boldsymbol{x}_i = \boldsymbol{x}_j k^{\frac{1}{2}} / (\boldsymbol{x}_j^{\mathrm{T}}(\mathscr{X}_{(i)}^{\mathrm{T}}\mathscr{X}_{(i)})^{-1}\boldsymbol{x}_j)^{\frac{1}{2}}$$

The global maximum can now be obtained by finding the value of k which maximizes (A.3.11). This value is $k^* = n^2\boldsymbol{x}_j^{\mathrm{T}}(\mathscr{X}_{(i)}^{\mathrm{T}}\mathscr{X}_{(i)})^{-1}\boldsymbol{x}_j$.

Substituting k^* into (A.3.11) and the resulting expression into (A.3.8) and simplifying yields the final result,

$$\max_{x_i} (\rho_{ij}^2) = v_{jj(i)} \frac{n}{n-1} - \frac{1}{n-1} \tag{A.3.12}$$

which is attained at $\boldsymbol{x}_i = n\boldsymbol{x}_j$.

These results show that for ρ_{ij}^2 to be large either $\boldsymbol{x}_i$ or $\boldsymbol{x}_j$ must be a remote point. Otherwise, $v_{jj(i)}$ and thus (A.3.12) will be small. A second requirement for a high correlation is that one point must be (approximately) a positive scalar multiple of the other, $\boldsymbol{x}_i \cong d\boldsymbol{x}_j$ where $d > 0$. With $\boldsymbol{x}_j$ fixed, the value of $\boldsymbol{x}_i$ which maximizes ρ_{ij}^2 is $\boldsymbol{x}_i = n\boldsymbol{x}_j$. Moreover, since the right side of (A.3.11) is monotonically increasing in k for $k \leq k^*$, in any fixed data set the correlation between a remote pair of points which are (approximate) replicates will tend to be large. Finally, when n is large, high correlations will also occur when $\boldsymbol{x}_i \cong -d\boldsymbol{x}_j$. When $\boldsymbol{x}_i = -\boldsymbol{x}_j$ the cases lie on the opposite edges of the sampled region and $\rho_{ij} > 0$.

Bibliography

Abrahamse, A. P. J. and Koerts, J. (1969). A comparison between the power of the Durbin–Watson test and the power of the BLUS test. *J. Amer. Statist. Assoc.*, **64,** 938–49.

Abrahamse, A. P. J. and Koerts, J. (1971). New estimators of disturbances in regression analysis. *J. Amer. Statist. Assoc.*, **66,** 71–74.

Abrahamse, A. P. J. and Louter, A. S. (1971). One new test for autocorrelation in least squares regression. *Biometrika*, **58,** 53–60.

Aitchison, J. and Dunsmore, I. (1975). *Statistical Prediction Analysis.* Cambridge: Cambridge University Press.

Aitkin, M. and Wilson, G. T. (1980). Mixture models, outliers and the EM algorithm. *Technometrics*, **22,** 325–31.

Allen, D. M. (1974). The relationship between variable selection and data augumentation and a method for prediction. *Technometrics*, **16,** 125–27.

Anderson, R. L., Allen, D. M., and Cady, F. (1972). Selection of predictor variables in multiple linear regression, in Bancroft, T. A. (ed.)., *Statistical Papers in Honor of George W. Snedecor*. Ames: Iowa State Press.

Andrews, D. F. (1971a). A note on the selection of data transformations. *Biometrika*, **58,** 249–54.

Andrews, D. F. (1971b). Significance testing based on residuals. *Biometrika*, **58,** 139–48.

Andrews, D. F. (1972). Plots of high-dimensional data. *Biometrics*, **28,** 125–36.

Andrews, D. F. (1974). A robust method for multiple linear regression. *Technometrics*, **16,** 523–31.

Andrews, D. F., Bickel, P., Hampel, F., Huber, P., Rogers, W. H., and Tukey, J. W. (1972). *Robust Estimates of Location.* Princeton, N.J.: Princeton.

Andrews, D. F. and Pregibon, D. (1978). Finding outliers that matter. *J. Roy. Statist. Soc., Ser. B.*, **40,** 85–93.

Anscombe, F. J. (1961). Examination of residuals. *Proc. Fourth Berkeley Symp.*, **1,** 1–36.

Anscombe, F. J. (1967). Topics in the investigation of linear relations fitted by the method of least squares (with discussion). *J. Roy. Statist. Soc., Ser. B*, **29,** 1–52.

Anscombe, F. J. (1973). Graphs in statistical analysis. *Amer. Statistician*, **27,** 17–21.

Anscombe, F. J. and Tukey, J. W. (1963). The examination and analysis of residuals. *Technometrics*, **5**, 141–60.

Atkinson, A. C. (1973). Testing transformations to normality. *J. Roy. Statist. Soc., Ser. B.*, **35**, 473–79.

Atkinson, A. C. (1981). Robustness, transformations and two graphical displays for outlying and influential observations in regression. *Biometrika*, **68**, 13–20.

Atkinson, A. C. (1982). Regression diagnostics, transformations and constructed variables (with discussion). *J. Roy. Statist. Soc., Ser. B*, **44**, 1–36.

Bailey, B. (1977). Tables of the Bonferroni *t*-statistic. *J. Amer. Statist. Assoc.*, **72**, 469–78.

Barnett, V. and Lewis, T. (1978). *Outliers in Statistical Data.* Chichester: Wiley.

Bartlett, M. (1951). An inverse matrix adjustment arising in discriminant analysis. *Ann. Math. Statist.*, **22**, 107–111.

Bates, D. and Watts, D. (1980). Relative curvature measures of nonlinearity (with discussion). *J. Roy. Statist. Soc., Ser. B*, **22**, 41–88.

Bates, D. and Watts, D. (1981). Parameter transformations for improved confidence regions in nonlinear least squares. *Ann. Statist.*, **9**, 1152–67.

Beale, E. M. L. (1960). Confidence regions in nonlinear regression (with discussion). *J. Roy. Statist. Soc., Ser. B*, **22**, 41–88.

Beckman, R. and Trussell, H. (1974). The distribution of an arbitrary Studentized residual and the effects of updating in multiple regression. *J. Amer. Statist. Assoc.*, **69**, 199–201.

Behnken, D. W. and Draper, N. R. (1972). Residuals and their variance patterns. *Technometrics*, **14**, 102–11.

Belsley, D. A., Kuh, E., and Welsch, R. E. (1980). *Regression Diagnostics.* New York: Wiley.

Bickel, P. and Doksum, K. (1981). An analysis of transformations revisited. *J. Amer. Statist. Assoc.*, **76**, 296–311.

Bingham, C. (1977). Some identities useful in the analysis of residuals from linear regression. University of Minnesota, School of Statistics, Technical Report No. 300.

Bliss, C. I. (1934). The method of probits. *Science* (*Lancaster*), **79**, 38–39.

Blom, G. (1954). Transformations of the binomial, negative binomial, Poisson and chi-squared distributions. *Biometrika*, **41**, 302–16.

Blom, G. (1958). *Statistical Estimates and Transformed Beta Variates.* New York: Wiley.

Box, G. E. P. (1979). Strategy of scientific model building, in Launer, R. L. and Wilkinson, G. N., (eds.), *Robustness in Statistics.* New York: Academic Press.

Box, G. E. P. (1980). Sampling and Bayes inference in scientific modeling and robustness (with discussion). *J. Roy. Statist. Soc., Ser. A*, **143**, 383–430.

Box, G. E. P. and Cox, D. R. (1964). An analysis of transformations (with discussion). *J. Roy. Statist. Soc., Ser. B*, **26**, 211–46.

Box, G. E. P. and Cox, D. R. (1982). An analysis of transformations revisited, rebutted. *J. Amer. Statist. Assoc.*, **77**, 209–10.

Box, G. E. P. and Draper, N. R. (1975). Robust designs. *Biometrika*, **62**, 347–52.

Box, G. E. P. and Tiao, G. C. (1968). A Bayesian approach to some outlier problems. *Biometrika*, **55,** 119–29.
Box, G. E. P. and Tidwell, P. W. (1962). Transformations of the independent variables. *Technometrics*, **4,** 531–50.
Bradu, D. and Gabriel, R. (1978). The biplot as a diagnostic tool for models of two-way tables. *Technometrics*, **20,** 47–67.
Braham, R. (1979). Field experimentation in weather modification (with discussion). *J. Amer. Statist. Assoc.*, **74,** 57–67.
Brown, R. L., Durbin, J., and Evans, J. M. (1975). Techniques for testing the constancy of regression relationships (with discussion). *J. Roy. Statist. Soc., Ser. B.*, **37,** 149–63.
Campbell, N. A. (1978). The influence function as an aid in outlier detection in discriminant analysis. *Applied Statistics*, **27,** 251–58.
Carroll, R. (1980). A robust method for testing transformations to achieve approximate normality. *J. Roy. Statist. Soc., Ser. B*, **42,** 71–78.
Carroll, R. and Ruppert, D. (1981). On prediction and the power transformation family. *Biometrika*, **68,** 609–16.
Carter, O. R., Collier, B. L., and Davis, F. L. (1951). Blast furnace slags as agricultural liming materials. *Agronomy Journal*, **43,** 430–33.
Cleveland, W. S. (1981). LOWESS: A program for smoothing scatterplots by robust locally weighted regression. *Amer. Statistician*, **35,** 54.
Cleveland, W. S. and Kleiner, B. (1975). A graphical technique for enhancing scatterplots with moving statistics. *Technometrics*, **17,** 447–54.
Cook, R. D. (1977a). Detection of influential observations in linear regression. *Technometrics*, **19,** 15–18.
Cook, R. D. (1977b). Letter to the editor. *Technometrics*, **19,** 348.
Cook, R. D. (1979). Influential observations in linear regression. *J. Amer. Statist. Assoc.*, **74,** 169–74.
Cook, R. D., Holschuh, N., and Weisberg, S. (1982). A note on an alternative outlier model. *J. Roy. Statist. Soc., Ser. B*, **44,** (in press).
Cook, R. D. and Prescott, P. (1981). On the accuracy of Bonferroni significance levels for detecting outliers in linear models. *Technometrics*, **23,** 59–64.
Cook, R. D. and Weisberg, S. (1980). Characterizations of an empirical influence function for detecting influential cases in regression. *Technometrics*, **22,** 495–508.
Cook, R. D. and Weisberg, S. (1982). Criticism and influence in regression. In Leinhardt, S. (ed.), *Sociological Methodology 1982*, Chapter 8. San Francisco: Jossey-Bass.
Cox, D. R. and Hinkley, D. V. (1974). *Theoretical Statistics*. London: Chapman and Hall.
Cox, D. R. and Snell, E. J. (1968). A general definition of residuals. *J. Roy. Statist. Soc., Ser. B*, **30,** 248–75.
Daniel, C. (1959). Use of half normal plots in interpreting two-level experiments. *Technometrics*, **4,** 311–41.
Daniel, C. (1976). *Applications of Statistics to Industrial Experimentation*. New York: Wiley.
Daniel, C. (1978). Patterns in residuals in the two-way layout. *Technometrics*, **20,** 385–95.

Daniel, C. and Wood, F. (1980). *Fitting Equations to Data*, 2nd Edn. New York: Wiley.

David, H. A. (1981). *Order Statistics*, 2nd Edn. New York: Wiley.

Davies, R. B. and Hutton, B. (1975). The effects of errors in the independent variables in linear regression. *Biometrika*, **62**, 383–91.

Dempster, A. P. and Gasko-Green, M. (1981). New tools for residual analysis. *Ann. Statist.*, **9**, 945–59.

Dempster, A. P., Laird, N., and Rubin, D. (1977). Maximum likelihood from incomplete data via the EM algorithm (with discussion). *J. Roy. Statist. Soc., Ser. B*, **39**, 1–38.

Devlin, S. J., Gnanadesikan, R., and Kettenring, J. R. (1975). Robust estimation and outlier detection with correlation coefficients. *Biometrika*, **62**, 531–46.

Dixon, W. J. (1950). Analysis of extreme values. *Ann. Math. Statist.*, **21**, 488–506.

Dongarra, J., Bunch, J. R., Moler, C. B., and Stewart, G. W. (1979). *The LINPACK Users' Guide*. Philadelphia: SIAM.

Draper, N. R. and John, J. A. (1980). Testing for three or fewer outliers in two-way tables. *Technometrics*, **22**, 9–16.

Draper, N. R. and John, J. A. (1981). Influential observations and outliers in regression. *Technometrics*, **23**, 21–26.

Draper, N. R. and Smith, H. (1966). *Applied Regression Analysis*. New York: Wiley.

Draper, N. R. and Smith, H. (1981). *Applied Regression Analysis*, 2nd Edn. New York: Wiley.

Duncan, G. T. (1978). An empirical study of jackknife constructed confidence regions in nonlinear regression. *Technometrics*, **20**, 123–29.

Durbin, J. (1970). An alternative to the bounds test for testing serial correlation in least squares regression. *Econometrica*, **38**, 422–29.

Durbin, J. and Watson, G. S. (1950). Testing for serial correlation in least squares regression I. *Biometrika*, **37**, 409–28.

Durbin, J. and Watson, G. S. (1951). Testing for serial correlation in least squares regression II. *Biometrika*, **38**, 159–78.

Durbin, J. and Watson, G. S. (1971). Testing for serial correlation in least squares regression III. *Biometrika*, **58**, 1–19.

Ellenberg, J. H. (1973). The joint distribution of the standardized least squares residuals from a general linear regression. *J. Amer. Statist. Assoc.*, **68**, 941–43.

Ellenberg, J. H. (1976). Testing for a single outlier from a general linear regression model. *Biometrics*, **32**, 637–45.

Ezekiel, M. (1924). A method of handling curvilinear correlation for any number of variables. *J. Amer. Statist. Assoc.*, **19**, 431–53.

Ezekiel, M. (1930, 1941). *Methods of correlation analysis*. New York: Wiley.

Ezekiel, M. and Fox, K. (1958). *Methods of correlation and regression analysis*. New York: Wiley.

Farebrother, R. W. (1976a). BLUS residuals, Algorithm AS104. *Applied Statistics*, **25**, 317–19.

Farebrother, R. W. (1976b). Recursive residuals – a remark on algorithm A75: Basic procedures for large, sparse or weighted least squares problems. *Applied Statistics*, **25,** 323–24.

Feigl, P. and Zelen, M. (1965). Estimation of exponential probabilities with concomitant information. *Biometrics*, **21,** 826–38.

Filliben, J. J. (1975). The probability plot correlation coefficient test for normality. *Technometrics*, **17,** 111–17, correction, **17,** 520.

Fisk, P. R. (1975). Discussion of a paper by Brown, Durbin, and Evans. *J. Roy. Statist. Soc., Ser. B*, **37,** 164–66.

Fox, T., Hinkley, D. and Larntz, K., (1980). Jackknifing in nonlinear regression. *Technometrics*, **22,** 29–33.

Furnival, G. and Wilson, R. (1974). Regression by leaps and bounds. *Technometrics*, **16,** 499–511.

Galpin, J. and Hawkins, D. (1981). Rejection of a single outlier in two or three way tables. *Technometrics*, **23,** 65–70.

Gauss, C. F. (1821, collected works 1873). *Theoria Combinationis Observationum Erroribus Minimis Obnoxiae.* Werke 4, Section 35, Gottingen.

Geisser, S. (1965). Bayesian estimation in multivariate analysis. *Ann. Math. Statist.*, **36,** 150–59.

Geisser, S. (1971). The inferential use of predictive distributions, in Godambe, V. and Sprott, D. (eds.), *Foundations of Statistical Inference.* Toronto: Holt, Rinehart and Winston, pp. 456–66.

Geisser, S. and Eddy, W. F. (1979). A predictive approach to model selection. *J. Amer. Statist. Assoc.*, **74,** 153–60.

Gentleman, J. F. (1978). Moving statistics for enhanced scatterplots. *Applied Statistics*, **27,** 354–58.

Gentleman, J. (1980). Finding the *k* most likely outliers in two-way tables. *Technometrics*, **22,** 591–600.

Gentleman, J. F. and Wilk, M. B. (1975a). Detecting outliers in a two-way table: I. Statistical behavior of residuals. *Technometrics*, **17,** 1–14.

Gentleman, J. F. and Wilk, M. B. (1975b). Detecting outliers, II: Supplementing the direct analysis of residuals. *Biometrics*, **31,** 387–410.

Ghosh, M. and Sharma, D. (1963). Power of Tukey's test for nonadditivity. *J. Roy. Statist. Soc., Ser. B*, **25,** 213–19.

Gnanadesikan, R. (1977). *Methods for Statistical Analysis of Multivariate data.* New York: Wiley.

Grubbs, F. E. (1950). Sample criteria for testing outlying observations. *Ann. Math. Statist.*, **21,** 27–58.

Hamilton, D., Watts, D., and Bates, D. (1982). Accounting for intrinsic nonlinearity in nonlinear regression parameter inference regions. *Ann. Statist.*, **10,** 386–393.

Hampel, F. (1968). Contributions to the theory of robust estimation. Unpublished Ph.D. dissertation, Univ. of California, Berkeley.

Hampel, F. (1974). The influence curve and its role in robust estimation. *J. Amer. Statist. Assoc.*, **69,** 383–93.

Harvey, A. C. and Phillips, G. D. A. (1974). A comparison of the power of some tests for heteroscedasticity in the general linear model. *J. Econometrics*, **2,** 307–316.

Hawkins, D. M. (1980). *Identification of Outliers*. London: Chapman and Hall.

Hedayat, A., Raktoe, B., and Telwar, P. (1977). Examination and analysis of residuals: A test for detecting a monotonic relation between mean and variance in regression through the origin. *Communications in Statistics*, **A6,** 497–506.

Hedayat, A. and Robson, D. S. (1970). Independent stepwise residuals for testing homoscadasticity. *J. Amer. Statist. Assoc.*, **65,** 1573–81.

Hegemann, V. and Johnson, D. E. (1976a). On analyzing two-way AoV data with interaction. *Technometrics*, **18,** 273–82.

Hegemann, V. and Johnson, D. E. (1976b). The power of two tests for nonadditivity. *J. Amer. Statist. Assoc.*, **71,** 945–48.

Henderson, H. V. and Searle, S. R. (1981). On deriving the inverse of a sum of matrices. *SIAM Review*, **23,** 53–60.

Hernandez, F. and Johnson, R. A. (1980). The large-sample behavior of transformations to normality. *J. Amer. Statist. Assoc.*, **75,** 855–61.

Hinkley, D. V. (1975). On power transformations to symmetry. *Biometrika*, **62,** 101–12.

Hinkley, D. V. (1977). Jackknifing in unbalanced situations. *Technometrics*, **19,** 285–92.

Hinkley, D. V. and Runger, G. (1980). Analysis of transformed data. University of Minnesota, School of Statistics Technical Report No. 341.

Hoaglin, D. C. and Welsch, R. (1978). The hat matrix in regression and ANOVA. *Amer. Statistician*, **32,** 17–22.

Hocking, R. R. and Marx, D. L. (1979). Estimation with incomplete data: An improved computational method and the analysis of nested data. *Communications in Statistics*, **A8,** 1155–82.

Hogg, R. V. (1979). Statistical robustness: One view of its use in applications today. *Amer. Statistician*, **33,** 108–115.

Holland, P. and Welsch, R. (1977). Robust regression using iteratively reweighted least squares. *Communications in Statistics*, **A6,** 813–28.

Huang, C. J. and Bloch, B. W. (1974). On the testing of regression disturbances to normality. *J. Amer. Statist. Assoc.*, **69,** 330–35.

Huber, P. (1964). Robust estimation of a location parameter. *Ann. Math. Statist.*, **35,** 73–101.

Huber, P. (1973). Robust regression: Asymptotics, conjectures, and Monte Carlo. *Ann. Statist.*, **1,** 799–821.

Huber, P. (1975). Robustness and designs, in Srivastava, J. N. (ed.), *A Survey of Statistical Design and Linear Models*. Amsterdam: North-Holland.

Huber, P. (1977). *Robust Statistical Procedures*. No. 27, Regional conference series in applied mathematics. Philadelphia: SIAM.

Huber, P. (1981). *Robust Statistics*, New York: Wiley.

Jaeckel, L. A. (1972). The infinitesimal jackknife. Unpublished Bell Telephone Laboratories Report, Murray Hill, N.J., summarized in Miller (1974), *op. cit.*

John, J. A. and Draper, N. R. (1980). An alternative family of power transformations. *Appl. Statistics*, **29,** 190–97.

Johnson, D. E. and Graybill, F. A. (1972a). Estimating σ^2 in a two-way model with interaction. *J. Amer. Statist. Assoc.*, **67,** 388–94.

Johnson, D. E. and Graybill, F. A. (1972b). Analysis of two-way model with interaction and no replication. *J. Amer. Statist. Assoc.*, **67,** 862–68.

Johnson, W. and Geisser, S. (1979). Assessing the predictive influence of observations. University of Minnesota, School of Statistics Technical report No. 355.

Johnson, W. and Geisser, S. (1980). A predictive view of the detection and characterization of influential observations in regression analysis. University of Minnesota, School of Statistics Technical Report No. 365.

Kennedy, W. and Gentle, J. (1980). *Statistical Computing*. New York: Dekker.

Koerts, J. and Abrahamse, A. P. J. (1968). On the power of of the BLUS procedure. *J. Amer. Statist. Assoc.*, **63,** 1227–1236.

Koerts, J. and Abrahamse, A. P. J. (1969). *On the Theory and Applications of the General Linear Model*. Rotterdam: Rotterdam University Press.

Kruskal, W. (1968). When are Gauss–Markov and least squares estimators identical? A coordinate free approach. *Ann. Math. Statist.*, **39,** 70–75.

Lachenbruch, P. A. (1975). *Discriminant Analysis*. New York: Hafner.

Landwehr, J., Pregibon, D., and Shoemaker, A. (1980). Some graphical procedures for studying a logistic regression fit. Presented at American Statistical Association Annual Meeting, Houston, TX.

Larsen, W. A. and McCleary, S. A. (1972). The use of partial residual plots in regression analysis. *Technometrics*, **14,** 781–90.

Lee, J. C. and Geisser, S. (1972). Growth curve prediction. *Sankhyā A*, **34,** 393–412.

Lee, J. C. and Geisser, S. (1975). Applications of growth curve prediction. *Sankhyā A*, **37,** 239–56.

Little, R. J. A. (1979). Maximum liklihood inference for multiple regression with missing values: A simulation study. *J. Roy. Statist. Soc., Ser. B*, **41,** 76–87.

Lloyd, E. H. (1952). Least squares estimation of location and scale parameters using order statistics. *Biometrika*, **39,** 88–95.

Lund, R. E. (1975). Tables for an approximate test for outliers in linear models. *Technometrics*, **17,** 473–76.

Mallows, C. L. (1975). On some topics in robustness. Unpublished Bell Telephone Laboratories report, Murray Hill, N.J.

Mandel, J. (1961). Nonadditivity in two-way analysis of variance. *J. Amer. Statist. Assoc.*, **56,** 379–87.

Mandel, J. (1971). A new analysis of variance model for nonadditive data. *Technometrics*, **13,** 1–18.

Marasinghe, M. G. and Johnson, D. E. (1981a). Testing subhypotheses in the multiplicative interaction model. *Technometrics*, **23,** 385–93.

Marasinghe, M. G. and Johnson, D. E. (1981b). On the multiplicative interaction model. Unpublished.

Margolin, B. H. (1977). The distribution of internally Studentized statistics via Laplace transform inversion. *Biometrika*, **64,** 573–82.

Marks, R. G. and Rao, P. V. (1979). An estimation procedure for data containing outliers with a one-directional shift in the mean. *J. Amer. Statist. Assoc.*, **74,** 614–20.

Mickey, M. R., Dunn, O. J., and Clark, V. (1967). Note on the use of stepwise regression in detecting outliers. *Computers and Biomedical Research*, **1**, 105–9.

Miller, R. (1966). *Simultaneous Inference*. New York: McGraw Hill.

Miller, R. (1974). The jackknife – a review. *Biometrika*, **61**, 1–16.

Milliken, G. A. and Graybill, F. (1970). Extensions of the general linear hypothesis model. *J. Amer. Statist. Assoc.*, **65**, 797–807.

Moses, L. (1979). Charts for finding upper percentage points of Student's t in the range 0.01 to 0.00001. *Communications in Statistics*, **B7**, 479–90.

Mosteller, F. and Tukey, J. W. (1977). *Data Analysis and Linear Regression*. Reading, Mass. Addison-Wesley.

Nelder, J. and Wedderburn, R. (1972). Generalized linear models. *J. Roy. Statist. Soc., Ser. A*, **135**, 370–84.

Orchard, T. and Woodbury, M. A. (1972). A missing information principle: Theory and applications. *Proc. Sixth Berkeley Symp.*, **1**, 697–715.

Ostle, B. and Mensing, R. (1975). *Statistics in Research.*, third edition. Ames, Iowa State.

Pearson, E. S., D'Agostino, R. B., and Bowman, K. D. (1977). Tests for departure from normality: Comparison of powers. *Biometrika*, **64**, 231–46.

Pearson, E. S. and Sekar, C. C. (1936). The efficiency of statistical tools and a criterion of rejection of outlying observations. *Biometrika*, **28**, 308–19.

Phillips, G. D. A. and Harvey, A. C. (1974). A simple test for serial correlation in regression analysis. *J. Amer. Statist. Assoc.*, **69**, 935–39.

Picard, R. (1981). On the assessment of the predictive ability of linear regression models. Unpublished Ph.D. dissertation, University of Minnnesota, School of Statistics.

Pierce, D. A. and Kopecky, K. J. (1979). Testing goodness of fit for the distribution of errors in regression models. *Biometrika*, **66**, 1–5.

Pizetti, P. (1891). I fondamenti mathematici per la critica dei risultati sperimentali. Genoa. Reprinted in Atti della Univers di Genova, 1892.

Plackett, R. L. (1950). Some theorems in least squares. *Biometrika*, **37**, 149–57.

Pregibon, D. (1979). Data analytic methods for generalized linear models. Unpublished Ph.D. dissertation, University of Toronto.

Pregibon, D. (1981). Logistic regression diagnostics. *Ann. Statist.*, **9**, 705–24.

Prescott, P. (1975). An approximate test for outliers in linear models. *Technometrics*, **17**, 129–32.

Prescott, P. (1976). Comparison of tests for normality using stylized sensitivity surfaces. *Biometrika*, **63**, 285–89.

Prescott, P. (1977). An upper bound for any linear function of normed residuals. *Communications in Statistics*, **B6**, 83–88.

Reeds, J. A. (1976). On the definition of von Mises functionals. Unpublished Ph.D. dissertation, Harvard University, Dept. of Statistics.

Roy, S. N., Gnanadesikan, R., and Srivastava, J. N. (1971). *Analysis and Design of Certain Quantitative Multiresponse Experiments*. Oxford: Pergamon Press.

Rubin, D. (1976). Inference and missing data. *Biometrika*, **63**, 581–92.

Ryan, T., Joiner, B., and Ryan, B. (1976). *Minitab Student Handbook*. North Scituate, Mass. Duxbury Press.

Scheffé, H. (1959). *The Analysis of Variance*. New York: Wiley.

Schlesselman, J. J. (1971). Power families: A note on the Box and Cox transformation. *J. Roy. Statist. Soc, Ser. B*, **33,** 307–14.

Seber, G. A. F. (1977). *Linear Regression Analysis*. New York: Wiley.

Serfling, R. J. (1981). *Approximation Theorems of Mathematical Statistics*. New York: Wiley.

Shapiro, S. S. and Francia, R. S. (1972). An approximate analysis of variance test for normality. *J. Amer. Statist. Assoc.*, **67,** 215–16.

Shapiro, S. S. and Wilk, M. B. (1965). An analysis of variance test for normality (complete samples). *Biometrika*, **52,** 591–611.

Shapiro, S. S., Wilk, M. B., and Chen, A. J. (1968). A comparative study of various tests for normality. *J. Amer. Statist. Assoc.*, **63,** 1343–72.

Sherman, J. and Morrison, W. J. (1949). Adjustment of an inverse matrix corresponding to changes in the elements of a given column or a given row of the original matrix. (abstract) *Ann. Math. Stat.*, **20,** 621.

Shih, W. C. (1981). Case analysis for incomplete data problems. Unpublished Ph.D. dissertation, School of Statistics, University of Minnesota.

Sims, C. (1975). A note on exact tests for serial correlation. *J. Amer. Statist. Assoc.*, **70,** 162–65.

Smith, V. K. (1976). The estimated power of several tests for autocorrelation with non-first-order alternatives. *J. Amer. Statist. Assoc.*, **71**, 879–833.

Sparks, D. N. (1970). Half normal plotting. *Applied Statistics*, **19,** 192–9.

Srikantan, K. S. (1961). Testing for a single outlier in a regression model. *Sankhyā A*, **23,** 251–60.

Stanley, W. and Miller, M. (1979). Measuring technological change in jet fighter aircraft. Report No. R-2249-AF, Rand Corp., Santa Monica, CA.

Stefansky, W. (1972a). Rejecting outliers in factorial designs. *Technometrics*, **14,** 469–79.

Stefansky, W. (1972b). Rejecting outliers by maximum normed residual. *Ann. Math. Statist.*, **42,** 35–45.

Stewart, G. W. (1973). *Introduction to matrix computation*. New York: Academic Press.

Stone, M. (1974). Cross-validatory choice and assessment of statistical predictions (with discussion). *J. Roy. Statist. Soc., Ser. B*, **36,** 111–47.

Theil, H. (1965). The analysis of disturbances in regression analysis. *J. Amer. Statist. Assoc.*, **60,** 1067–79.

Theil, H. (1968). A simplification of the BLUS procedure for analyzing regression disturbances. *J. Amer. Statist. Assoc.*, **63,** 242–51.

Tietjen, G. Kahaner, D., and Beckman, R. (1977). Variances and covariances of the normal order statistics for sample sizes 2 to 50 in Owen, D. B. and Odeh, R. E. (eds.), *Selected Tables in Mathematical Statistics*, **5**. Providence, RI: Amer. Math. Soc.

Tietjen, G. L., Moore, R. H., and Beckman, R. J. (1973). Testing for a single outlier in simple linear regression. *Technometrics*, **15,** 717–21.

Tukey, J. W. (1949). One degree of freedom for nonadditivity. *Biometrics*, **5,** 232–42.

Tukey, J. W. (1957). On the comparative anatomy of transformations. *Ann. Math. Statist.*, **28,** 602–32.

Tukey, J. W. (1970). *Exploratory Data Analysis*, limited preliminary edition. Reading Mass. Addison-Wesley.
Velleman, P. and Welsch, R. (1981). Efficient computing of regression diagnostics. *Amer. Statisitician*, **35,** 234–42.
von Mises, R. (1947). On the asymptotic distribution of differentiable statistical functions. *Ann. Math. Statist.*, **18,** 309–48.
von Mises, R. (1964). *Mathematical Theory of Probability and Statistics*. New York: Academic Press.
Wedderburn, R. (1974). Quasi-likelihood functions, generalized linear models, and the Gauss–Newton method. *Biometrika*, **61,** 439–47.
Wedderburn, R. (1976). On the existence and uniqueness of maximum liklihood estimates for certain generalized linear models. *Biometrika*, **63,** 27–32.
Weisberg, S. (1974). An empirical comparison of the cumulative distributions of W and W'. *Biometrika*, **61,** 644–46.
Weisberg, S. (1980a). *Applied Linear Regression*. New York: Wiley.
Weisberg, S. (1980b). Comment on a paper by White and MacDonald. *J. Amer. Statist. Assoc.*, **75,** 28–31.
Weisberg, S. and Bingham, C. (1975). An analysis of variance test for normality suitable for machine calculation. *Technometrics*, **17,** 133.
Wilk, M. B. and Gnanadesikan, R. (1968). Probability plotting methods for the analysis of data. *Biometrika*, **55,** 1–17.
Wilks, S. (1963). Multivariate statistical outliers. *Sankhyā A*, **25,** 507–26.
Wood, F. S. (1973). The use of individual effects and residuals in fitting equations to data. *Technometrics*, **15,** 677–95.
Woodbury, M. (1950). Inverting modified matrices. Memorandum No. 42, Statistical Research Group, Princeton University.
Woodley, W. L., Simpson, J., Biondini, R., and Berkeley, J. (1977). Rainfall results 1970–75: Florida Area Cumulus Experiment. *Science*, **195,** (February 25), 735–42.
Wu, C. F. (1980). Characterizing the consistent directions of least squares estimates. *Ann. Statist.*, **8,** 789–801.
Yates, F. (1937). *The Design and Analysis of Factorial Experiments*. Bulletin No. 35, Imperial Bureal of Soil Science. Harpenden: Hafner (Macmillan).
Yates, F. (1972). A Monte Carlo trial on the behavior of the nonadditivity test with non-normal data. *Biometrika*, **59,** 253–61.
Zahn, D. (1975a). Modifications of and revised critical values for the half-normal plot. *Technometrics*, **17,** 189–200.
Zahn, D. (1975b). An empirical study of the half-normal plot. *Technometrics*, **17,** 201–11.

Author index

Subject index